Low Voltage Wiring

Low Voltage Wiring: Security/Fire Alarm Systems

Terry Kennedy
and
John E. Traister
(*deceased*)

McGraw-Hill

New York   Chicago   San Francisco   Lisbon   London
Madrid   Mexico City   Milan   New Delhi   San Juan   Seoul
Singapore   Sydney   Toronto

Cataloging-in-Publication Data is on file with the Library of Congress

McGraw-Hill

*A Division of The **McGraw·Hill** Companies*

3 4 5 6 7 8 9   DOC/DOC   0 9 8 7 6 5 4

ISBN 0-07-137674-7

The sponsoring editor for this book was Scott Grillo, the editing supervisor was Steven Melvin, and the production supervisor was Sherri Souffrance. This book was set in Century Schoolbook per the MHT design by Paul Scozzari of McGraw-Hill's Professional Book Group composition unit, Hightstown, N.J.

Printed and bound by R. R. Donnelley & Sons Company.

Previously published as *Security/Fire-Alarm Systems: Design, Installation, Maintenance*, copyright © 1996, 1990.

This book is printed on recycled, acid-free paper containing a minimum of 50% recycled, de-inked fiber.

Contents

Preface

In general, an alarm system is used to protect life and property. Therefore, any company or person who owns something of value has a valid interest in security and fire-alarm systems.

Security and fire-alarm systems are designed for every application imaginable—for structures as small as homes and roadside produce stands; large hotels and factories; security alarms for guarding automobiles; large national defense installations—the list is very long. And this book is designed to review the various security systems currently available to help the reader select the most appropriate system for the job.

This new edition adds a first part that surveys the business of security systems and computer information for the professional. It also includes a nuts-and-bolts survey of how to start or grow a business, and how to work in other areas of the low-voltage systems industry.

It then covers installation methods, techniques, and requirements to comply with the 1999 National Electrical Code. Part 3, A Deeper Understanding, reviews electrical circuits, the code, and print reading in depth.

The new edition also features a twenty-first-century Reference Section that guides the reader into a listing of references for information to be found on the Internet. To move fully into the new millenium, we have included a compact disk. The checklists in the reference section can be accessed simply and quickly from your own computer.

The abundance of solid information makes this book invaluable to anyone involved with security and fire-alarm systems.

John E. Traister and Terry Kennedy

The Security Systems Business

1

A Business Built on Contracts

In time, many field technicians want to go into the security and fire alarms business for themselves, small to midsize contractors want to grow their companies, and large firms need a top-down overhaul. This chapter reviews the basic mechanics of managing a security and fire-alarm contracting firm proactively, rather than haphazardly.

Overview of a Contract Service Business

The nature of most businesses related to the AEC (Architecture, Engineering, Contracting) industry is that profits are based on the development and performance of contracts. Many deals related to other types of business involve some variety of an agreement document, but construction industry contracts require complex participation from all parties: the principal customer, the design and engineering team, the general contractor, the subs, and the materials suppliers. As we all know, a site is chosen and a design team works with the principals to develop the earth and the built structures to skillfully achieve the goals of the owners. Naturally, the intention of the construction team is to produce the product that the architects and engineers have designed, to enjoy the sense of a job well done, and to make some money.

Many enterprises, like retail manufacturing or food services, allow the businessperson to select or design the product, then manufacture and sell it in a protected environment. For example, a bag of chips can be produced in a factory and the production process can be sheltered from the elements and a great deal of the work can be thoroughly boiler plated. Thus, the profits and future of the business can be predicted in some detail. Naturally, there will be swings: potatoes or corn crops might suffer from the weather, transportation costs might fluctuate, the customer base might tighten up, but basically an entrepreneur in these enterprises has a good deal of control over the process of providing for their clientele.

All of us associated with construction know that this is simply not the case with construction projects. Our structures are designed for the site and the building owner. Powerful design opinions can come into play, creating down time and change orders. Schedules can be built on wishful thinking in order to make the contract price more palatable to the owner during the bidding process. The project management team can be extremely lacking in diligence and professionalism. A good deal of the project can end up completely at the mercy of the elements. Because of these complexities, contracts are the backbone of subcontracting, serving as maps for all of the parties involved in the project.

When it comes to the art of writing solid contracts, a security systems business is similar to the rest of the trades in many ways. However, a security systems business does provide some types of protection that are not found in many of the other trades. The majority of the work involved with security systems occurs after the building is well underway. Typically, the majority of an installation is protected from the elements. The subcontract documents are fairly straightforward. It is relatively simple to define what is a change from the original plan and what is not. Typically, very little of the security system is subjected to the aesthetic whimsy of architects and owners.

This does not mean that the security systems contractor can be cavalier about the bidding and contract stage of the work. Many things can cause a bid to come in too low and lose money for a firm. The security system company's bidding process can cause a number of these twists: inaccurate takeoffs from the plans, missing overhead factors in the estimate, faulty predictions of the labor involved in the work. Many problems that can put a project in the red are out of the security systems contractor's control, for example: impedance of the trades caused by lack of professional construction management, incomplete plans, and changes during construction.

It is very easy for the security contractor to be so involved in the mechanics, the installation of the systems, that the importance of being good at business slips by the wayside regularly and all efforts are concentrated on completing the tasks in the field. This tendency is natural because so many subcontractors come from a background of fieldwork, rather than business training.

This chapter is designed to steer the security systems contractor away from the habitual tendency of letting business slide because of the demands of site work. Security and other light electrical systems are a continually growing part of our buildings. The goal here is to offer a bit of distance from the endless requirements of the field and some ideas for moving from the slant of the tradesperson into the mindset of a professional contractor.

With the Web, vast amounts of energy pouring into research and development for computers and other electronics arenas and the automation of buildings: security and all of the other low-voltage components of our built structures will continue to be a growth market. As well as the steady increase in contracting light electric systems, numerous other elements make electronic systems for buildings a good industry with plenty of potential for developing a long-term, strong business.

Because security systems subs typically gain experience as tradespeople, they might not have ever received any hands-on business training other than estimating, scheduling, and other one-dimensional office tasks. Because of this tendency, when mechanics start companies, they might have spent very little time dealing with the overview elements of business, such as long-range planning.

Thus, when a security systems person is starting or trying to grow a business, it can be a real stretch, trying to achieve the necessary overview of the company to integrate and streamline all parts of the whole.

This chapter is designed to help the security systems contractor develop an overview and arrive at techniques that can be used to gain distance from the demands of the daily maintenance of the business in an effort to attain a broader viewpoint of the operation. Developing the skills to achieve an overview can allow the contractor to pay attention to the balance needed to manage the daily work load while intelligently setting and realizing business goals.

The technical aspects of a security systems business are covered thoroughly in the later chapters. Obviously, the installation of systems is very important and the field work will always demand close attention. However, remember that the field work is more dramatic than office work and it tends to draw attention automatically. The countless tasks that occur in the office before a job even begins are not as visible as what happens at the job site. However, these office tasks keep a company on track after jobs are started. Before the first circuit is put together, dozens of tasks must be completed, such as bidding, landing the job, and preparing contracts. Good office systems should be in place before a job begins to promote professional and cost-effective performance.

To begin to develop overview of your security systems business, the main subject that needs to be addressed is a personal question: *What do I want from being a security systems contractor?*

To many people, this seemingly strange question doesn't really relate to taking care of a business. A person will often have negative thoughts about the whole idea: What does this have to do with getting jobs done? What difference does what I want make? We all know that business is just a lot of hard work.

It is very important to break out of this mindset and develop a proactive attitude to be really good at business. Forget the negative thoughts and begin to think about enjoying your own life. Relax and allow your thoughts to take a big stretch and explore what you really want to be doing, where you want your life to go. At first, just allow your mind to work with the thoughts, then take some time to go off by yourself: a fishing trip, a trip to a security systems convention, some camping, or whatever it takes to give yourself some distance.

The important thing, if you haven't already learned it, is to be honest with yourself. Keep working through the thoughts until you can recognize and begin to let go of negative thought patterns that hold you down. During this stage, you can keep what you are thinking to yourself. All that you are learning to do is to relax with yourself as a professional and see what you really want from life so that you can chart where you are going.

If you see that your life is just what you want, that the responsibilities you have taken on are fulfilling and your life is on track, then moving into a business plan is a fairly straightforward activity. But if you are strapped into commitments that don't really belong in your life, it's time to come to terms with yourself.

Perhaps you locked into security systems work and supporting your family at a very young age and wish that you had taken some time for yourself before making those long-term commitments. Allow yourself to admit this to yourself. You don't really need to tell others what you realize and there might be nothing you can do except make the situation as good for everyone as possible.

If you are in a situation like having dependent children, you must obviously be there for them. Responsibilities are a part of anyone realizing their long-term goals. You have to take care of your children and make the family life a healthy, fun part of your game plan. Tearing your life apart is not the idea here; living it more fully through living your work is what you will set out to achieve.

Use this part of business planning to be very straightforward with yourself—a straightforward approach will help you with all parts of your life, including drawing up security systems contracts.

Tell yourself the truth while you are thinking about your life: "I wish I had waited a bit and hadn't started a family when I was so young. Sometimes I wish that I could relax and do things that were more fun—be a professional fisherman or something besides just putting in these security systems. But, you know, there is also a very responsible person in me who knows how strong the love and commitment to family make me, how secure they make me, how simple life is when I care for others and am not too self centered."

The point is that in order to set long-term goals, it is best to know oneself closely because following through on the goals involves a lot of determination. If we are being ourselves, perseverance is much easier than if we are trying to make ourselves over. Once you can see what you really want to be doing, you can look at what you are doing now more clearly and see where you want to go.

Don't do anything at first, just look closely and get to know yourself. You don't have to talk about it with anyone right away, just note it all. You might be in for some real surprises. You might discover that all you really want to do is escape all the responsibility in life and just go fishing. But with time, if you accept yourself as you really are, you might see that some of your thoughts are simply fantasies about escaping all the work involved in a security systems business.

However, you might find that you simply stumbled into your business and that if you want to enjoy your life fully, some big changes are needed. Take your time and don't do anything rash; you must make any needed changes in a way that is healthy for yourself and your family. All of this careful assessment will help you develop get a real hold on your business, not just perform a business plan exercise that has no long-term meaning for you.

Perhaps you would rather be a grading contractor, working with heavy equipment, or have a restaurant because you have always enjoyed barbecuing. If this is really the case and you are not interested in the exciting developments in the world of building electronics, you will have time to think through your new plans and make the transition carefully. Then it will be time to discuss your goals with your family and develop a business plan that reflects how you want your life to progress. You can bring your loved ones in on the ideas and develop a long-term scenario that is right for everyone else who is involved.

This book focuses on the security systems business, but many consultants, such as the Small Business Administration, can help if you need to make more radical changes than growing or starting your security systems firm.

The SBA is an excellent place to look for help because the advisors have been through years of business and can help with the overview. However, don't use any consultant or coach unless you feel really good about working with that person. The development of the actual business plan is covered in Chapter 3, but now is a good time to start thinking about your goals in life.

Writing Solid Contracts

The typical subcontracting firm in the construction industry has developed a contract document as a boiler-plated form in their computer system. After a site visit, takeoffs, preliminary estimates, and a meeting with the principals to hammer out questions, the contract phase begins. Even if the general contractor insists on using their boiler-plated contract, every question should be ironed out in the conditions of the job. Every condition that is not worked through before the contract is complete can develop into a legal battle in this age of lawsuits. Well-designed contracts can put a damper on legal action by working through questions that could lead to disagreement as snafus arise during the on-site work.

Not only is the advance survey of potential problems important at this stage, but probably more important, is the opportunity to begin a sense of working together on the project rather than approaching it adversarially. This is the most opportune period to develop an idea of what all parties expect from the security systems installation. Go through every paragraph and all of the line items in the conditions section of the contract may very well. This is not a time to rush through; developing solid contracts may very well be the most important part of the work. Therefore, you must set aside the time to pay close attention. It is advantageous to complete the contract, set it aside, and review it again in a few days when your mind is fresh.

The first round of precontract meetings is used to hammer out the initial questions. After the first draft of the documents is presented to the client for review, a second sit-down conference should be called. Everyone who is a decisionmaker in relation to the security systems should be called to this meeting with their drafts of the preliminary contract. This meeting should be thought of as a very important event—the entire group now has an opportunity to air their concerns, explore potential conflicts, and work through any potential problems. Figure 1.1 is included here for the security systems contractor to use as a beginning set of guidelines for what must be covered during the contract phase. The CD contains an active file for this and all of the checklists in the book. Our information is simply a beginning for the security systems contractor.

The American Institute of Architects (AIA) has developed the most widely used of all boiler-plated contracts. Also, CSI (the Construction Specifications Institute) has worked out a numbering system for all of the categories of construction and has devised specifications for the various types of work.

With the widespread use of computers, developers, general contractors, and the other parties for whom security systems subs might work have taken off from the AIA format and worked with their lawyers and insurance vendors to create boiler-plate formats. These formats are so common that security system contractors can be tempted to sign off before doing a careful review of the contract. A typical, boiler-plate agreement right off the computer of the developer or general contractor can look safe enough at first glance.

But as the sub looks more closely at what the terms state, commitments that are not so obvious might jump out from the fine print. The clauses in the contract agreement itself might stipulate the attachment of other documents: recitals, conditions of the contract, supplements, drawings, specifications, requirements to supply shop drawings and cut sheets from vendors, addenda, scope of work, and any number of other defining matter that could affect the subcontractor's profits.

In the end, what appears to be a simple contract document might be loaded with stipulations that need to be covered in the estimate. Looking through the agreement carefully, the security systems contractor is apt to unearth clauses that bind the company on important matters like risk, liability, or warranties, which complicate the overall and supplementary conditions of the required performance. These matters can add greatly to what is required of the subcontractor and introduce a great deal of liability for the security systems firm.

A very basic statement that requires that the security systems subcontractor must review the work of the design team and confirm that it is correct for the job will take the responsibility for design and place it directly on the shoulders of the security contractor. One typical clause is for the principal to demand that the security systems sub must take responsibility for the performance of the various vendors' products for a given amount of time. The wording of this type of clause might very well go way beyond all of the various specs written into the manufacturer's product warranties. This might put the sub in a situation in which the contractor is required to stand behind conditions from which the vendor is exempt. Throughout the entire bidding process, your team must be familiarizing themselves with all risks that are involved with accepting the project. A number of steps can give you the overview required to make sound, business-like discussions when taking on jobs: carefully read through

Checklist for Contract Review		
	Item	Notes
✓		
	This job has been thought through step by step	
	Each question and potential problem has been discussed with the other parties.	
	Potential additional work has been discussed	
	Change orders have been surveyed and discussed—must be written, signed, and dated. Requests must be issued for any changes to proceed.	
	The contract contains clauses that cover delays not caused by our firm. Completion dates altered by events beyond our control: weather, labor shortages at other firms, accidents, acts of God, and impedance of the trades, will not cause penalties for our company.	
	The payment-release schedule, including change orders, is addressed thoroughly and precisely	
	All insurance issues (including bonds, coverage, and liabilities of the various parties) have been reviewed and addressed in the contract documents.	
	Plan specifications, product cutsheets with their availability and prices and their acceptance by the principals and availability have been reviewed in detail.	
	The provision of utilities, accessibility of work areas, and impedance of the trades have been reviewed	
	Any penalties or interest that could arise have been considered and factored into estimating	
	Construction defects and consequential damages to our work product have been covered with possible future construction defect lawsuits in mind	

Figure 1.1 Checklist for contract review.

all documents, watching for phrases that will force you to shoulder responsibilities beyond the ordinary requirements of a job well done, and use a veteran construction attorney for an analysis of any clauses about which you have doubt.

When choosing attorneys, it is important to use one with a strong construction background. Attorneys are like any other vendor of a service, if they are not highly skilled in the construction industry as their practice, they will not be cost effective. If your attorney is highly skilled with construction documents, their knowledge should keep your legal fees lower than those from counsel with a general practice. Also remember, that your staff can lower the time required by the attorney and their staff by doing plenty of homework before meeting with the attorney; simply supplying them with all of the needed documents and even highlighting the items that concern your needs will expedite the lawyer's work.

The idea is that a security systems business is based on contracts; paying close attention to all contracts will make the company strong. Walking away from a contract might be more important than signing—your money is made when the contract is signed. Naturally, performance in the field is extremely important, but a well-executed contract paves the way for making the fieldwork successful.

A very important part of a business based on contracts is preconstruction, round table meetings. Careful review of construction documents, careful take-offs and a real world estimate are very vital parts of the project. However, after performing all of those tasks with care and review with a competent construction attorney when needed, the precontract and an additional prework conference can pave the way for successful execution of the contract documents. At these meetings, all changes and clarifications can be stepped through and initials put to documents. If the installation moves into a worst-case scenario, this puts you in a strong position for defending your position in court.

However, always remember that this is a team effort and the more you enlist all of the other players on your team: the contractor, the vendors, the architect, the door and window installers, the more smoothly the fieldwork can be executed. This will also lessen the chances of being wrapped up in all of the down time and expenses inherent with legal battles.

Set up a system that steps your firm through all of the phases, from the time the request to bid comes in, and provides a thorough list of questions for working through at all preconstruction meetings. The following list is similar to the contract completion checklist that you reviewed earlier in this chapter. However, it is designed to be used after the initial contract work is tied up and final matters need to be hammered out at round-table meetings. As with all of the checklists, a live version of the list as a Microsoft Word table is included on the CD—use the blank spaces and modify what exists to a format which is tailored specifically for your firm (Fig. 1.2).

A good way to remember that building is a team effort is to think about scheduling. Preconstruction meetings are invaluable for scheduling; they are

Checklist for Preconstruction Conferences		
Item		**Notes**
✓		
	All questions about the drawings, details, and specifications have been listed for discussion	
	All questions, restrictive clauses, and contingencies related to the requirements of the security systems have been listed	
	All prep work that needs to be done has been listed	
	All equipment requirements have been listed for the meeting	
	All questions related to specified products have been listed	
	All past and potential design questions for the structures have been listed	
	All potential delays including access, weather, materials acquisition, overlapping of trades, site access, framing repair, etc., that could tie up your work have been addressed	
	Questions related to unexpected labor costs have been listed	
	All cleanup issues, including environmental and recycling, have been addressed	
	All opportunities to sell extras and change orders have been addressed	
	A benefit and cost savings list of reasons for using our firm is ready for meeting	

Figure 1.2 Checklist for (pre)construction conferences.

the best tool to iron out questions related to the integration of the trades. As all low-voltage systems teams know, if one phase of construction has been delayed because of poor project management, it might cause a complete shutdown of your own installation.

No subcontractor wants their mechanics on the job if they are not able to make progress. If your crews must stop work because the framers are still trying to finish partition walls, your firm could lose money and run into penalties from completion clauses in the contract. Small delays can cause an exponential cost increase while the project is being built out. They can also cause construction defects because everyone begins to rush through their work. This is a particularly burdensome situation because construction defect litigation does not typically occur immediately. And losses will not be seen for years to come.

Scheduling an installation should begin as the estimator takes the job apart, calculating labor for the bid. Many types of scheduling software are available, from simple timeline packages to complex Critical Path software. Large security companies with a number of huge installations on the board (commercial buildings, industrial projects, subdivisions, large condo clusters) might require one of the sophisticated Critical Path programs.

It is wise to start rough scheduling during estimating. This practice enables the estimator to notice quirks that could add additional costs in the field. It can also aid in listing questions for the contract phase and the related round-table meetings. Starting scheduling when estimating will also flag the estimator that the main work of scheduling is often rescheduling because slow downs occur at the job site. It is important for a scheduling package to make updating the schedule very simple—punch in the changes and the software updates the entire timeline automatically.

An excellent book on schedules is McGraw-Hill's *CPM in Construction Management* by James J. O'Brien and Fredric L. Plotnick. To review scheduling software, simply go to the appendix on the CD and request that the companies offer you a look at their products. And set up a policy for taking advantage of the round-table meetings for going through potential production snafus with the other subs that will be involved in the work.

Marketing

We have all heard that word of mouth is the most powerful method to market our work. Many security subs have experienced this first hand. Someone calls and says that an old client has recommended your firm. You give them a price and they don't even shop for other bids. In this age of rushing through things, trust is a priceless commodity. Word of mouth is by far the most reliable and inexpensive form of marketing. As you work with the ideas in this book and start looking closely at what you want out of life, examine your company exactly as it works at the present time. Look carefully at your current base of customers and sort out the people with whom you already have a solid working relationship. This should include business owners of course, but remember that architects and general contractors can often bring in jobs more frequently than any other source. When you start to rework your marketing and sales department, start with what is right under your nose.

With a beginning of understanding your goals, you can take a closer look at marketing. The game plan is described at length in Chapter 3. It is also wise to develop a solid forecasting and leveling system. With forecasting and leveling, you can begin to see some patterns in your cash flow.

A number of questions need to be answered. For example, you will want to know if you can increase sales with your business working as it does right now. You will also have to think about whether you could take care of increased sales with your business as it stands. And, you might want to give a try to raising sales right away, just to see how it would work.

Figure 1.3 shows checklists that you can use to examine important parts of your company. We have offered some suggestions on the first checklist. Utilize the ideas if they work for your company. If not, they will work as an example of what to check into and you can use the second list, adapting it to fit your firm's needs. The CD contains them as Word for Windows files and you can put them right to use on the computer or print them out and work on them by hand, editing on the computer as you progress.

A great many books cover all parts of business and marketing is no exception. In order to expand your insight into the subject, look at some of them. Remember that the only worthwhile ideas and theories are those that apply immediately to your business. A great many marketing theories, any number of ideas, are only valuable if they are simply implemented in a direct relationship to your specific needs.

It is wise to remember that sales and marketing are two different activities, but they often overlap. *Sales* is getting signed contracts on the desk, but the way you handle yourself during the bidding and contract work are an important part of *marketing*. If you are a true professional, the people you are working with will take notice and they will remember your presentation and behavior when another job comes along. If your firm satisfies the various parties and they tell friends, you have brought in qualified leads, which is a big part of marketing.

After you have taken a close look at your current customer base, start a policy of doing mailers with tasteful cards that announce changes at your firm. When you introduce a new product, a new service, or anything that can be of service to your clients, let them know with a tasteful card. This will help them remember to give you a call.

Next, start contacting them for a list of friends and customers with whom you can make contact. Don't expect overnight results—it is the long haul, the steady implementing of this simple process that brings results. Include all of these people in your announcements and call them. Keep increasing the size of the database and sending out regular announcements about your firm.

While you are implementing announcements, review all of your local periodical advertising. There are many professional periodicals, the Yellow Pages,

Checklist for Market Development		
Goals	**Time Frame**	**Notes for Expediting**
Expand marketing contacts to raise profit from existing business by 20%	9 months	Contact existing clients, architects, and general contractors ➤ Flyer announcing new plan to old clients ➤ New brochure to new prospects ➤ Follow-up phone work
Continue to expand marketing contacts to raise profit from existing business by 20%	12 months	Get contracts from the friends of existing clients, plus new architects and general contractors ➤ Flyer introducing the company ➤ Send out a security systems newsletter ➤ Put an information site for clients on the Net
Review other low-voltage systems work	1 year	What businesses can we launch easily? ➤ Check out computers ➤ Check out photovoltaics
Explore markets for consulting business to raise total cash flow by 10%	1 year	Bar Association and insurance companies and find out where they locate their security systems consultants and experts ➤ Develop a brochure ➤ Do a mailing ➤ Try ads

Figure 1.3   Checklist for market development.

	Checklist for Market Development	
✓		

Figure 1.3 *(Continued)*

small newspapers, and the Internet. Do test runs with these media and keep tabs of the results. If they are effective, keep using them. Give them about three months, remember that marketing is not an overnight effort, but requires diligence and consistency.

Word of mouth will always be responsible for the most completely successful segment of any firm's marketing. Again, it will not usually happen overnight, but there is no more powerful marketing.

Many other simple, cost-efficient forms of marketing can be easily implemented in your security systems business: speaking to organizations about security; providing workshops and seminars about building security; and advertising in small publications, such as local, free newspapers. The ideas are numerous and they are only limited by your imagination.

Sales

Most parts of the construction industry rely on solid estimating to close deals. Every experienced security systems contractor is fully aware that a comparison of the costs presented by the bidders is how most decisions are made. This is often the sole factor involved with your firm getting the contract. At the same time, every security professional knows that low bids are not always the best way for the principals to choose their subcontractor—the low bidder might not do quality work. Still, selecting the low bidder is how the choice of the security contractor is typically made.

Thus, having a smooth, accurate estimating procedure in place is very important to the business. Chapter 2 is devoted to estimating and it looks closely at computers. The estimate is an essential part of sales. With computers, the estimator can use a digitizer, simply point at circuits, and the software will enter the facts, costs, markups, and all the data. If the computing system is well designed, the information about the client will only need to be entered once and the software will carry it on through contracts, schedules, change orders, billing, your announcements for marketing, and even on to the next bid.

In today's marketplace, it is likely that your competition will all be armed with accurate estimates when they meet with the principals to close their proposed deals. However, one tool will give the quality security systems contractor an edge: widespread, local knowledge that your firm does quality work. Here, we return to word-of-mouth marketing and sales. When you have a strong history of good performance, those past customers can close the deal simply by talking with new customers briefly. One of the most important tools for raising sales volume is to prepare methods for giving the new client a brief history of those clients and an easy way for them to make contact with your solid customers from the past.

When the estimates are tight in the bidding process, a chat with a satisfied past customer can give you the contract. It makes no difference whether you receive most of your requests to bid from typical sources: architects, general contractors, product manufacturers, word of mouth, advertising, or a combination thereof, being able to present a group of satisfied customers can close the deal. In fact, it can be the sales tool that makes the difference.

The checklist on the next page illustrates an example of items that should be covered during the sales process. You can adjust it to fit with your company specifically or use the blank checklist (Fig. 1.4) to adapt all of the steps to your security firm. The checklists are on the CD and you can simply boot up the files and go to work.

Insurance

One of the most important parts of your business is all of the various types of insurance that are needed to safeguard it from the many unforeseens that are inherent with security contracting. The following pages present a list that can help you step through all of the insurance that you are currently involved with and can be used as a flag for a review while carefully exploring all of your insurance needs.

This list provides a very simple look at the typical types of insurance that a subcontractor uses. Review it carefully and think about any additional types of insurance that you might need. It is advisable to find a good insurance attorney and go through your current understanding of your policies, carriers, coverage, and required coverage, and allow them to suggest any other forms of insurance that you might need.

It is also advisable to take a look at areas of the law, related to insurance in which you might not be well versed. It is important to understand coverage, construction defect litigation, the execution of insurance claims, and how liability could affect you as a businessman. An excellent source of information is McGraw-Hill's *Construction Claims Manual for Residential Contractors* by Jonathan F. Hutchings.

It is wise for the security subcontractor to select a member of the staff to watch your insurance portfolio on an on-going basis. This person should make use of your insurance attorney's knowledge from time to time for a thorough understanding of policies and insurance coverage issues. Using a lawyer might sound like an unnecessary expense, but remember that you are looking at a long-range game plan to set up your business so that you can make healthy profits and enjoy it as well. Knowing where you actually stand with insurance is some of the most valuable knowledge you will ever gain. The costs of the insurance, the lawyer, and staff time should be factored into your bidding process to cover the expense. The net assets that contractors accumulate through years of work can be devastated in later years by one construction lawsuit, an accident, or any unforeseen incident that has not been analyzed in relationship to insurance coverage.

Unemployment insurance

Every area of the country requires that companies keep current unemployment insurance. The premium is in the form of taxation and must be taken care of on a quarterly basis. The levy is derived from the company's track record of claims filed by staff. A strong security contracting firm takes good

Sales Expansion Plan		
Activity	**Time**	**How to Expedite**
Install fast computer estimating system to boost existing business by 15%	9 months	Contact existing clients, architects, and general contractors ➤ Follow up on leads within 1 day ➤ Develop a tickler system for follow up ➤ Review all sales-presentation documents ➤ Redo those that are not high quality
Continue to improve on % of closes to presentations by 20%	12 months	Try new methods for improving sales ➤ Offer a maintenance incentive ➤ Hand out a copy of the maintenance newsletter ➤ Try using a dedicated salesperson
Check outside businesses that install low-voltage systems	1 year	Try selling and installing on existing jobs ➤ Computer circuits ➤ Photovoltaics ➤ Send staff to trade shows to contact vendors
Go to consulting seminars and meetings to learn more syntax for closing slaes 10%	1 year	Visit past customers: ➤ Send a brief newsletter about security ➤ Do some of your work for free ➤ Keep your hours low

Figure 1.4 Sales expansion plan.

Sales Expansion Plan		
Activity	**Time**	**How to Expedite**

Figure 1.4   *(Continued)*

care of employees and businesses and often has a strong employee retention rate. The cost of the tax, which is from 1 to 4 percent of the payroll, should be factored into the estimating software.

FUTA

The federal government's tax for unemployment is FUTA, which varies, but, on average, factors at 0.8 percent of payroll expense. Consult with your accountant to make certain that the percentage is correct in your bidding process each year.

Social Security and Medicare (FICA)

FICA is another expense set by the federal government. Be certain that you have a very good accountant so that this figure is current and updated annually in your estimating procedure.

Worker's compensation insurance

Security systems contracting is one of the more safe areas of the construction industry, but accidents can always happen. Your state mandates that you carry worker's comp and it is your responsibility to search for good policy carriers who offer a fair price and have a healthy track record of covering their claims. Worker's comp insurance pricing is computed on a percentage basis. The wages of a staff member are factored with risks involved with the person's particular job. In other words, a field employee is more often exposed to the danger of accidents than the office staff and the cost of the policy is figured on this basis.

Liability insurance

Most of the contracts that you sign will require that you provide a certificate of liability coverage before you begin to execute the job. The costs for contractor liability insurance are rated by factoring in payroll, local of the company and business, the nature of the business, the company's history of claims, and the dollar amount of coverage required by the security contracting firm. It is especially important that the designated insurance person review the insurance carrier's position with the insurance attorney and that the liability insurance costs are factored into the estimating process.

It is also wise for your staff person to shop for this insurance, but not to make decisions solely on the basis of cost. The fine print in the insurance forms should be carefully evaluated, along with all of the clauses in any contract documents.

Insurance costs are very important. The premiums should simply be regarded as a cost of doing business as matter of factly as the cost of gasoline for vehicles is approached. Shopping and understanding insurance are an on-going process. This cost should be factored in with insurance costs. Lawyer fees, accountant fees, updating software, settling claims, employee time, and your own time are all real-world expenses related to each and every job. The cost of this part of your business is very important and is a part of the labor burden that must be reflected and updated on a regular basis as a part of your estimating process. Neglect of this work can lead to many problems that will always show up as a loss of profit and make running the business chaotic, rather than smooth.

Subcontractors

One of the main purposes of this book is to aid the security systems professional in having a strong sense of where they intend to go with their business life and how to get there. One of the most important parts of a rewarding career is learning how to work smart, rather than hard. It is an old cliché in business, but often the old adages are full of truth—that's why they become clichés.

A great many of the readers will go into the business of security systems contracting during their professional history. This chapter is designed to emphasize the importance of knowing that this is a business built around contract documents. Contracts begin with estimates. An important part of estimating is realizing what your firm is set up to do well. For example, the average security systems firm would never consider taking on the entire electrical contract for a building project—it's much better to stick to the security work and get in and out of the job.

Along this same line of thinking, in some parts of your business, it would be much wiser to subcontract with another company, just as the general contractor does with your firm. Many of the people who own security systems businesses have come up from the field, starting their own companies. The daily work load demands a huge amount of attention. If the contractors did not come from a business background, they might never find time to quit being field technicians and get to running their own company.

It is not always easy to learn to work smart rather than hard, but the more a business owner does delegates tasks to others, the easier it becomes; and after a person senses how much more smoothly things can run, it often becomes a natural thing to do.

Naturally, a good deal of delegating responsibility to others occurs in-house with your own staff. However, remember that assigning tasks to other firms could make your business run much more smoothly. The simplest way to grasp this concept is to look at tasks that are ordinarily subcontracted with others. Accounting is an excellent example. A contractor can track payables, including payroll and receivables in house, but the end-of-year accounting is often taken care of by the accountant.

There are many advantages to jobbing out work. One of the most obvious is that the payroll burden doesn't grow, which is a very big expense with any company. No matter what tasks they perform: consultants, lawyers, installers of special products, maintenance of installed systems, running cable, patching walls, the performance of staff can be increased because they can spend time improving their own performance, morale can be boosted, and the company can pay close attention to its main work—organizing and overseeing the installation of security systems.

What to do in-house and what to job out is always an important decision. However, it can be tested by experimenting with firms on a one-time basis and continuing if the attempt is successful. The way to test the waters is to continue to return to thinking about the ways in which you can implement a new overview of your own life and working smart, rather than hard. Typically, the fewer employees required to run a smooth business, the easier it will be for you to be an effective businessperson.

Accounting

Accounting is a very important part of any business. It is similar to the central nervous system of a person's body with all of its nerves that are continually monitoring what is going on with all of the body parts. The accounting system, coupled with scheduling, if run effectively, will be able to recognize failures in the business system on a daily basis.

With the advent of inexpensive computers, accounting has changed radically. The daily tasks of tracking payables and receivables still demands a good deal of attention when data is being entered and analyzed. But now, a number of the more grueling tasks, such as extending columns of figures, are done by the system in a fraction of the time that they can be done with adding machines. Computers don't suffer from fatigue and they don't make mistakes if the information is input correctly.

In a small firm, one person might be able to take care of basic accounting input with the assistance of an outside accounting firm and the owner reviewing the reports from the system. However, one knowledgeable person should have time to ensure that all of the accounting is working properly. It is important that this person has knowledge of accounting and is not burdened with such excessive, extraneous tasks as answering telephones or dealing with people who walk in the door. One of the worst patterns that can develop from this type of management is that receivables fall behind for long periods of time and are never brought into a timely routine of collection.

The important decision is to realize that business is a part of your life, not your entire life. Business should only be one of the parts that carve out a good life for you. Apply this to your accounting discussions. Having balanced books and understanding where your cash is at are as important in your firm as having a well-run field. Hire competent, self-motivated people, take good care of them and make certain that their tools are excellent and well maintained. Back your accounting staff up with a strong outside accounting firm and supply them with powerful, properly installed software.

Your security systems business is based on contracts and accounting is one of the best tools for evaluating how skillful you are at estimating. The many reports now available at the push of a button from today's contracting software can diagram the financial position of your company in moments. Chapter 3 takes a close look at how the construction industry has been completely changed by computer systems.

Cash Management Forecasting and Leveling

These sound like terms from an MBA and, of course, they are used by MBAs. In reality they are simple ideas. The simpler they are kept, the more effective they are. *Leveling is the process of smoothing out the ups and downs in the cash your company earns from performing on contracts.* Forecasting is looking at your past jobs, finding a pattern of ups and downs in cash flow, and being able to predict, with some accuracy, what is coming in the future. Naturally, many unforseens can crop up in the future and affect your earnings: the economy, competition for bids, and weather cycles, to name a few. However, with good forecasting and leveling tools in place, the businessperson can recognize typical cycles in their business. For example, especially wet winters tend to come in cycles of a few years of very wet seasons then a series of dryer years.

In the end, well-managed sales can be a big tool for leveling your cash. Chapter 4 looks at how security systems businesses can take on other profit centers that are closely related to their current work load. If a company has more than one source of income, one of the other operations can tide the company through when there is a downturn in installing security systems. For example, upgrading and maintaining security systems that have been previously contracted can be increased when there is a downturn in the building market.

As a part of working smart, utilize a slowdown for marketing the maintenance and upgrading business while at the same time looking for more customers during the next upturn in construction.

Setting the goal of a good life for yourself and carrying through on the commitment requires getting your business by the reins. When the owner takes proactive control of the business, the business can get much simpler to manage. One of the main tools for taking charge is the ability to collect data about the business, analyze it, and take the necessary actions to control the cash flow.

Without fast access to accurate data and the ability to analyze it, and put the needed actions into place to respond to the results of the analysis, managing your business will always be more difficult than if you have these systems in place. Security systems contracting is a business based on contracts and one of the key elements to writing solid, profitable contracts is fast, accurate estimating. One of the first areas where the ability to gather and analyze data quickly is very effective is on the estimate sheet.

As you continue to look at your life and your business, you might decide to expand. Cash crunches are one of the main obstacles that companies run into during growth. You have probably experienced them already, during slowdowns in the construction cycle. Your business will undoubtedly run into cash crunches. Forecasting and leveling are very efficient tools for dealing with cash crunches. One of the worst parts of cash crunches is the effect on morale, which can slow down efficiency and cause even more loss than the absence of work.

Such questions as next year's cash flow or even the ability to meet a new payroll can be addressed with a good deal of accuracy if reliable forecasting and leveling tools are in place. Your related business can be pumped up, lines of credit can be activated, and your business can keep running effectively.

The newest versions of quality job cost-accounting software provide straightforward, rapid methods for getting reports from the accounting and estimating data. With the use of the reports provided by a system, the peaks and valleys in your cashflow will become obvious.

Understanding the causes of the peaks and valleys requires a bit more scrutiny. Looking closely at estimates and scheduling might reveal that a number of errors are occurring. For example, labor overhead figures are not set high enough and estimates of time to perform several different field tasks are consistently too low. Also various trends can create ups and downs. For instance, when there has been a slowdown in construction in the winter, you might have continually had a valley and no other source of revenue to level cash during the bad weather.

With this knowledge at hand, you can become proactive and spend the down time building up and upgrading maintenance service that levels your cash during the winter. Forecasting and leveling are common business ideas and there is a great deal of information is available about them. However, it is good to remember that simplicity makes work much easier than complexity. Don't get sidetracked into becoming a business student unless you decide that you want to do so. The Small Business Administration can be a great help with this type of information and your accountant can also be a valuable resource.

This book is not meant as an extensive business book. Our goal is to cover the elements of security systems contracting. It is important for the businessperson to have forecasting tools as part of your being able to realize the goals you set. Figure 1.5 introduces a simple review of your cashflow and how to analyze it for forecasting and leveling. An empty list follows for quick copying if you want to start from scratch. Both lists are available on the attached CD.

Forecasting and Leveling		
	Change	**Activity**
✓		
	Bring job cost accounting software up to date. Be sure reports are accurate. Begin to understand trends for forecasts.	Review with accounting. Buy new software if needed. Have accounting produce reports.
	Isolate where we make money and when we lose it.	Get staff to recognize these patterns.
	Find training.	Get accounting staff to forecasting and leveling training and go with them.
	Grow the other sources of income like a maintenance company.	Find out how this part of the business is doing.

Figure 1.5 Forecasting and leveling.

Forecasting and Leveling		
	Change	Activity
✓		

Figure 1.5   *(Continued)*

2

Estimating the Installation of Security Systems

Sound cost estimating consists of a complete takeoff of the materials and equipment for the job and then an estimate of the labor required for installation. After materials and labor are calculated, direct expenses, variable job factors, taxes, overhead, profit, and any additional costs are added to determine a "selling price" for the project. Chapter 2 shows how this all comes together.

How to Achieve Accurate Estimates

The steps necessary to prepare a cost estimate for a given security-system installation will normally be:

1. *Takeoff:* The count of all security devices, smoke detectors, magnetic contact devices, and similar items, and the measurement of all perimeter wiring, feeders to security control panels, anchors and fasteners, etc.
2. *Listing the Material:* All items accounted for in the takeoff should be listed in an orderly sequence on a standard pricing sheet, as will be described later.
3. *Applying Labor Units:* Determining the proper labor unit from proven labor-unit tables and applying them to the various materials or labor operation under the labor-unit column on the pricing sheet.
4. *Finalizing:* The summation of material dollars and labor hours, the assignment of job factors and dollar values to labor hours, and the determination of overhead and profit.

Material takeoffs

A quantity survey or material takeoff consists of counting each of the components of the installation: door contacts, bells, lockswitches, alarm panels, telephone dialers, etc.). These quantities are entered in their appropriate space on a material takeoff form. An example of a sheet is shown in Fig. 2.1. Naturally, the sample is not intended as a cast-in-concrete example. Each company has their own unique approach to estimating and as long as materials, labor, and overhead are all covered and the system is fast and accurate, the terms and the style of the form are immaterial.

Estimating today is mostly done on the computer and the formats for the take-off lists and presentation bid sheets from the various types of software tend to have their own unique appearance. For simple takeoffs, the form in Fig. 2.1 can be carried into the field and a bid can be presented to the customer on the spot for a rapid closing of the sale. However, it is not wise to do this type of instant estimating for large, complicated jobs.

More complex and detailed estimates are typically done with a computer. Many types of computers can now be taken into the field with ease. They can be connected to up-to-date, accurate prices from suppliers by modem and print a professional-looking estimate, right in the field.

Except with the most bare-boned operations, computers are now necessary for estimating. To make your systems fully effective, a great deal of knowledge is needed. It is rarely cost effective for the security systems contractor or the staff to attempt to research, purchase, hook up hardware, and install the software so that it works effectively.

Some computer consultants specialize in contractor systems: estimating, links to suppliers, takeoff devices, scheduling devices, job cost accounting, time billing, change orders, contracts, and linking the field to the office. The problem is finding specialists who have a broad knowledge of hardware and software and know how to select it and install it effectively. All parts of the business should be interactive. It is best to have the person lay out their idea of how the entire operation will work in detail, then compare the plan with that of several other consultants. Check the person's track record with care and then implement one part of the system (such as estimating) and observe the smoothness of the consultant's work. If all goes well, you will never lose money from having a highly automated business.

Laying out a security company's use of computers in a highly effective manner and implementing the hardware and software requires a tremendous amount of knowledge. Many hours can be lost and the wrong systems installed if a contractor tries to use in-house staff or consultants who have not been carefully researched to automate the firm. The best approach by far is to put the company's time into researching hardware, software, and consultants. Full automation can have a staggering effect on a security systems contractor's ability to produce.

Computers follow the same procedures that contractors have done by hand, with the use of calculators, in the past. Looking closely at Fig. 2.1, you will see that each room or area of the building is listed in the columns marked "Location," and then the number of items per Location is entered in the squares below ("Number of Items Per Location"). These items are totaled and priced. The total net price of material, labor costs, sales tax, etc. are then extended to obtain a bid price.

ALARM SYSTEM PROPOSAL

FOR: Name _____   Date _____

Street _____ City _____ Phone _____

LOCATION

PREPARED BY:
NAME _____
ADDRESS _____
CITY _____ STATE_____ZIP_____
PHONE _____

NUMBER OF ITEMS PER LOCATION	ITEM	TOTAL QTY	MODEL NO.	UNIT PRICE	TOTAL PRICE
	CONTROL PANEL				
	POWER SUPPLY				
	EXIT/ENTRY DELAY				
	INDOOR REMOTE STATION				
	OUTDOOR REMOTE STATION				
	PREALARM STATION				
	OUTDOOR SOUNDING DEVICE				
	INDOOR SOUNDING DEVICE				
	DIALER				
	DIALER POWER SUPPLY				
	DIALER SWITCHING MODULE				
	DIALER CONNECTOR CORD				
	DIALER TAPE CARTRIDGE				
	ULTRASONIC DETECTOR				
	PHOTOELECTRIC SYSTEM				
	SMOKE DETECTORS				
	HEAT SENSORS				
	MAGNETIC CONTACTS				
	GLASS BREAK DETECTORS				
	SWITCH MAT				
	MISC. WIRING SUPPLIES				

TOTAL NET PRICE	
INSTALLATION	
SALES TAX	
TOTAL PRICE	
DEPOSIT	
BALANCE	

Figure 2.1 Alarm system proposal.

Some estimators do a very detailed material takeoff, listing all circuits separately and including such small items as wire connectors and fasteners, all of the smallest details. Others take off the major items of material for an entire building, listing only the different types of materials separately and giving a lump-sum dollar value for small incidentals. Computers can do this very quickly, using either method. It really doesn't matter which procedure is used, as long as the estimator has sufficient information from which to make a complete list of all materials required to complete the installation, whether they are listed separately or as assemblies. With this take-off list, the estimator can apply labor units to the installation, price the materials and order them. If your firm intends to purchase handheld computers that can estimate in a building, at the architect's office, or on a tailgate, discuss this form with the software company and the installer of the software to see if it can be duplicated closely by the software.

It is common for the material takeoff to begin by the counting of all the parts (e.g., magnetic contacts, glass-break detectors, smoke detectors, heat sensors, and sounding devices). The estimator continues by counting all alarm panels, telephone dialers, and all other major equipment. With this material out of the way, the measuring of all wiring, window foil, etc., occurs using a digitizer, a rotameter, or an architect's scale to determine the distances shown on the blueprints.

The actual mechanics of a material take-off procedure are comparatively simple and will become routine in a very short time. The sooner the estimating procedures become routine, the sooner the estimator will be able to make rapid and accurate take-offs. However, it is important that care is taken with counts and measurements, no matter how often estimates are created.

The estimator should remember that the circuit lines on floor plans represent only the horizontal portion of the various runs. To accumulate the vertical runs in the system, a scaled section of the building, showing the elevations of the various floors and ceilings, should be used. Then, the mounting heights of alarm stations, sounding devices, etc., can be marked to scale on the drawing of the building section. During the measuring process, whenever a point is reached where there is a vertical section of the run, the digitizer, rotameter, or any measuring device is traced over the proper vertical distance on the scaled cross section. This will continually accumulate the vertical distances and the horizontal distances on the floor plans. When the measurements are taken with an electronic estimating device, such as a digitizer, they are automatically dumped into the estimating software for automatic extensions.

Several other aids will help the estimator produce an accurate take-off. One is to use different colored pencils when checking off runs of cable as they are measured—a different color for each wire size, cable type, or section of the runs of cable. For example, black could be used to indicate the detection circuits, blue for control circuits, and red for annunciation circuits.

Listing the materials

While taking off the various alarm components from drawings, the estimator must list the items on pricing sheet forms so that costs can be obtained for the various items and labor units as they are added and extended. To help make this operation easier for both the estimator and the purchasing agent, the listings should be made in an orderly sequence on the pricing sheets. With the use of computers for estimating, this task is greatly simplified. The estimator simply selects a task from a list and punches in a quantity. The computer does the rest, typing in the line item with an accurate cost and extending the unit cost by the count that the estimator has provided. Needless to say, this type of automation, coupled with the reduction in errors when using computers, can quickly pay for a portable system.

When the take-off has been done properly, the estimator will immediately have two items of valuable information: a brief description of each outlet, component circuit run, equipment, etc., and the quantity of each item listed. From these descriptions, the estimator can determine the exact quantity of materials and the necessary labor hours to completely install the system, provided that he or she has a good knowledge of actual security/fire-alarm installations and building construction.

It is wise to visit the job site to do the take-offs or require a meeting there with the architect and the general contractor to actually visualize the project. At this meeting, a less-obvious activity at this point in the bidding process is that the estimator is looking at how the job will be built out. This is actually when scheduling first begins—each circuit and installation and potential snafu must be noted by the estimator so that it can be translated into the estimating and become a part of scheduling.

Individual parts, such as fasteners, hangers, wire connectors, etc., are rarely noted on drawings. Therefore, the estimator will be required to make an educated guess, based upon an understanding of the project's requirements, past experience with other projects, and the use of good judgment.

With the use of computers, assemblies can be built that add the parts for various types of linear and square foot runs, right into the extension of a measurement. For example, the estimating team can build a template in the estimating software that includes every cost involved with installing each linear foot of all the types of cables that your company uses. These prices can then be broken down further to installations in various areas: above dropped ceilings, through sheetrocked walls, on masonry walls—the lists can be very thorough and they can be easily adjusted until they are very accurate. When electronic measuring and take-off devices are used, estimating can be highly simplified, cost effective, and free of errors.

There is really no easy way to accomplish an exact material take-off, but when a systematic method is used, experienced estimators are able to produce very accurate estimates, rarely omitting important items. The extent of the detail with which items of material are listed can vary to fit the contractor's particular needs, but always remember that the more exacting and detailed estimating is, the more often deals will be competitive and closed with contracts that are real moneymakers.

Estimating installation costs

Determining the amount of labor that will be expended on a particular project is much more involved than the mere application of labor units. The pricing of materials also requires an intelligent analysis of situations at the site that might require more products than a typical installation and problems that require materials unique to the particular project.

The estimator must also be fully aware of the many nuances involved with interpreting quotations by suppliers or price services. The majority of estimators obtain firm quotations from one or more security equipment suppliers, including the manufacturers of security/fire-alarm equipment. When computers are used, the pricing of more than one supplier can be checked, sometimes by e-mail.

Requests for quotations on special materials should be made as early as possible, although suppliers might tend to wait until the last few minutes prior to bid openings before giving the quotation. This, of course, is done to prevent underbidding by another supplier. Again, the computer can be a big help in keeping abreast of the latest prices.

When a quotation is received, the contractor or installer should check over the list of items on the quotation carefully. Suppliers do not normally guarantee that the items will meet with the project's specifications, nor will they take any responsibility for errors. Substitutions are common these days, and it is the contractor's responsibility to be certain that all items quoted will meet with the architect's or engineer's specifications.

If there are questions of this nature when the bidding is within tight time constraints, contacting the architect and discussing the situations can sometimes cement an on-going relationship through trust developed by your firm's concern for a job well done. Be sure to explain that you can bid competitively with the other subs and will provide a cut-sheet for the architect—if it is not the correct product, the price might change. This creates an equitable situation for all parties and any reputable, intelligent architect welcomes a can-do attitude.

Whenever possible, the contractor should obtain a guarantee of the quoted price for a definite period of time. Most suppliers will stand by their quotations for approximately 30 days. But what happens if it takes six weeks to award the contract? There is a good chance that the materials quote will go up and the contractor will have to pay more for the material than the price used in estimating. Therefore, the contractor should try to determine exactly when a particular job will be awarded (this is not necessarily the date of the bid opening) and then obtain a guarantee that the quote will be good until at least that time.

While waiting for a quotation from suppliers, labor units should be applied to the take-off. A *labor unit* is a time figure that indicates the time required to install, connect, or otherwise make usable a given item of material or a given labor operation. These units are used by the majority of contractors who must quote a firm lump-sum price to obtain security or fire-alarm systems work. The units are normally based upon worker-hours or a percentage of a worker-hour. For example, 1.50 worker-hours indicates $1\frac{1}{2}$ worker-hours; that is, the labor required for a particular operation will take one worker $1\frac{1}{2}$ hours to accomplish.

Labor units are applied to each item of material and then extended and totaled to give the total worker-hours required to complete the project. The value of the labor in dollars and cents is then determined by multiplying the total computed worker-hours by the installer's average hourly rate of pay. These factors are all included in software databases and can easily be changed as the costs of doing business change.

A separate unit of labor should be provided for the installation of each item of material or labor operation performed. This unit should be broken down further to apply to varying working conditions. For example, if a labor unit is given for installing 100 feet of two-conductor #22 cable at ground level, it stands to reason that installing the same amount of cable 20 feet above the floor will require more worker-hours. Scaffolding would have to be set up and moved into the area, and the workers would be required to spend more time carrying reels of wire to the scaffold platform (see Fig. 2.2). Even if the work could be done with a ladder, some additional time would be required to move the ladder and climb up and down as the movement occurs.

Figure 2.3 offers several conditions that affect labor operations. Any number of conditions, such as weather, are specific to your firm and must be given consideration in preparing any bids. This table is on the CD and can readily be adjusted for the estimating team to use when assembling estimates.

During the bidding of labor for any security or fire-alarm system installation, the estimator must take into consideration several factors often overlooked by the inexperienced person. For example, the labor unit must include layout instructions, material handling, the actual installation of the material, coffee breaks, visits to the rest room, etc. If the labor units used do not include all of these items, the contract must make allowances to cover them. From these statements, you can see that the amount of time required for a worker to install a given item might not be an adequate basis for determining an accurate labor unit.

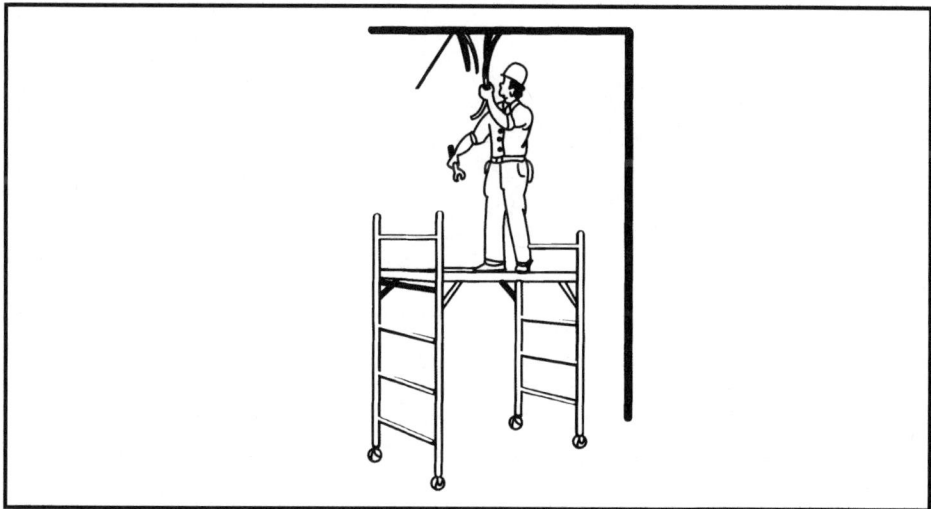

Figure 2.2 Working at heights requires more worker-hours than working at ground or floor level.

Conditions That Affect Labor Costs		
Condition	**Description**	**Notes**
The type of building construction.		
Height of the installation above normal working areas.		
The weight of the material or equipment.		
Performance of the general contractor (if known).		
The availability and proficiency of the workers.		
Is the wiring to be concealed or exposed?		
Is the installation in new or existing buildings?		

Figure 2.3 Conditions that affect labor costs.

The contractor or estimator is continually faced with having to use good judgment, which can only come from experience when dealing with labor units. At first, the selection of the proper labor units might seem like a difficult task, but after some experience in the field, the estimator will be able to choose the labor units most applicable to the particular project or portion of a job. When working with computers, these situations can be input into the database as time related to specific situations.

The first step in arriving at the most accurate total estimated labor for a given job is to take off and list the material items on the pricing sheets, segregated in accordance with the installation and building conditions. On larger and more complex jobs, the different categories applicable to each type of material can be expanded in line with the different specific installation conditions.

The second step is to apply the labor unit specifically related to that particular installation condition for the size and type of material involved, depending on the extent of segregated listing of the material and the extent of segregation of the available labor data. However, there is no point in listing the materials on a segregated basis if segregated labor data is not available or if the estimator does not adjust the existing data to account for the specific conditions. Anything less than a segregated listing of the materials, in accordance with the varying installation conditions and the application of related labor data reduces the accuracy of the total estimated labor.

Once the choice of labor unit has been made, the mechanics of labor-unit entry consist of merely copying the appropriate labor units from whatever source is available and entering the units in the labor-unit column on the pricing sheet opposite the proper item of material or labor operation. Computers make it very easy to update this information by comparing estimates with time cards for the actual time required to perform various tasks and how special conditions at the site have affected the labor bid. Not only do the computers make building these templates of exact labor costs simple, they make it easy to select them from the company's database and apply them to the current estimate.

After all of the labor units have been applied on the pricing sheet, they should be extended and totaled. This operation involves little more than elementary mathematics, but many errors can be made. The estimator, therefore, should be extremely careful at this point. One decimal point in the wrong place can mean the difference between a profit and a loss on a project. Unless projects are very simple, it is a severe loss of estimating time to avoid the use of computers for this difficult, error-prone task.

No bid should ever be turned in without checking the figures. Preferably, the person making the initial take-off should check through the figures, then someone else should review them. One method of checking column totals is to add them first from top to bottom and then from bottom to top. Computers are very accurate, but even when they are used, it is wise to print the estimate and have another person review it for data-entry errors.

In any case, sufficient time should be allowed for checking the figures because errors often result from hasty last-minute efforts to complete an estimate to meet a specific bid time.

Summarizing the estimate

Summarizing the estimate is the final accumulation of all estimated costs: labor, materials, job factors, direct cost, overhead, and profit. Determination of the final quotation is one of the most important steps in preparing the estimate, because one mistake in the final summarizing can affect all of the accuracy with which the previous steps have been handled. Computers take care of all of these tasks quickly and accurately, but the final estimate should be printed and reviewed for input and conceptual errors and omissions.

It is wise to put computers at the top of your list of things to do if yours are not already providing excellent service for your work. The summary information in you computer software will probably be presented differently than in Fig. 2.4. But we have included this example of a hand-produced summary in order to clarify what to demand from your software. If you need to perform a hand-produced estimate because of computer down time, you can modify this form to meet the needs of your company. It can serve as a sound guide to accurately summarize the estimate for practically any security/fire-alarm system installation.

Direct job expense and overhead

A thorough understanding of both direct job expense and overhead is necessary so that they can be included in the final estimate to defray unforeseen costs. In general, direct job expenses are those costs (in addition to labor and materials) that have to be paid for as a direct result of performing the job. In other words, if the job were not performed, these costs would not occur. Overhead expense, on the other hand, is all costs that have to be paid whether the particular job is being done or not.

An estimate is not complete until all direct and overhead expenses have been added to the other items entering into the cost of the project. Direct job expenses are relatively simple to calculate if you are paying close attention to the bidding and all items of expense are included that relate directly to the job at hand. Calculating overhead, however, is a different picture altogether.

Many contractors take their previous overhead figures and apply them to work new estimates. This might result in an accurate estimate, but in most cases, the overhead will change during the performance of the work being bid. Therefore, the contractor should analyze the anticipated future overhead for all jobs being bid at the present. Analysis is one of the areas where computers are unsurpassed. They can retrieve information quickly and accurately, and process it into forecasts in a flash.

Another factor to be considered during estimating is the size of the job. In most cases, a small job will cause a higher percentage of overhead than a large

BID SUMMARY SHEET

JOB _____   PAGE _____
_____   OF _____
_____
_____

ESTIMATED BY _____ CHECKED BY _____ DATE _____

DESCRIPTION	MATERIAL				LABOR		
	Quantity	Unit Price	Per	Amount	Unit	Per	Amount
	TOTAL						

MISCELLANEOUS		RECAPITULATION	
		Material Cost	
		Hours Labor @	
		Hours Labor @	
		Direct Job Expense	
		Total Prime Cost	
		Overhead Expense	
		TOTAL COST	
		Profit	
		Selling Price	
		BID SUBMITTED	

Figure 2.4 Typical bid summary sheet.

job. However, the contractor cannot assume that this will always be true—especially in the case of specialized projects.

When the estimate has been completed to the point of adding the overhead, the known data should include cost of materials, cost of labor, and direct job expense. The overhead is often determined by one of the following methods:

1. The overhead for the year may be divided by the gross sales volume for the year to find the overhead as a percentage of the gross sales volume. This percentage is then applied to the prime cost of the job.
2. The overhead expense for the year can be divided by the total cost of labor, material, and job expense for the year to find the overhead as a percentage of the prime cost.

In some cases, the estimated annual volume differs from the past annual volume for which overhead based upon accounting records is obtainable. Whenever situations arise where projected overhead and past overhead vary, the estimator must determine by simple proportion, the applicable average overhead percentage of prime cost and apply this percentage using the job-size scale.

This is often done by estimating the overhead percentage on the basis of past recorded data, adjusted to future volume and size of work. If the job being estimated represents a change in general work pattern or is a special type of job, the estimator must make an intelligent analysis of all the conditions and further adjust the estimated overhead percentage to be applied as accurately as possible.

Overhead is often treated as a percentage that is typical of bidding practice in the locale where the contractor works. And the overhead of competition must be considered in order to produce effective estimates. However, if competing contractors are operating on a "shoe string" basis and you are not capable of winning bids against them without suffering losses, you must rethink what type of installations you want to perform. There are times when "low-ball bidding" can create havoc in a local market, however, this is usually a temporary condition that rectifies itself as the competing contractors discover that they are not seeing a profit from their work.

The most accurate way to attach overhead to estimates is to have good records of profits and losses. Computers simplify this task, but it is still important to watch how each factor contributes to the price of your bid.

Completing the summary

Completing the summary involves only the inclusion of such miscellaneous items as wire connectors, tape, and fasteners (Fig. 2.5). Most contractors would agree that listing these items serves no purpose. Therefore, on most projects, an allowance for these items is made rather than listing with the price of each individual item. This allowance is usually in the form of a lump-sum figure, a percentage gained from experience, or an educated guess. As a rule, ½ of 1 percent is sufficient for all projects, except highly specialized ones. Once this figure has been determined, the dollar value should be entered in the appropriate space on the summarizing form. With the use of computers, this can be done automatically.

The contractor will also be required to calculate miscellaneous labor costs on many projects. Such conditions as overtime, labor disputes, and special installations will make the inclusion of extra labor necessary. Although past job per-

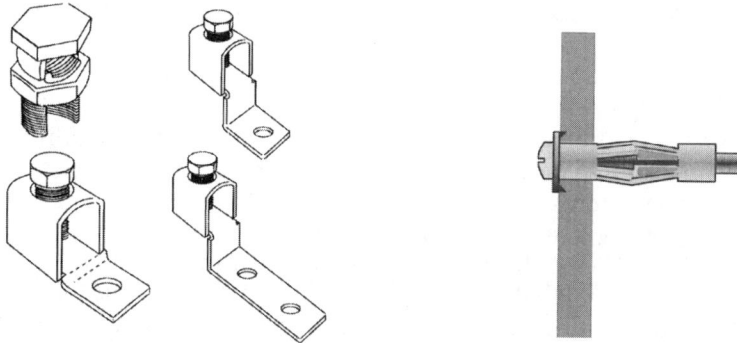

Figure 2.5 With the exception of very specialized projects, wire connectors, fasteners, and similar items are lumped together and figured as a certain percentage of the total installation cost.

formance can be considered, no set rule is used to calculate this figure exactly. It is a matter of experience and good judgment.

The subtotals of the dollar value of the labor, material, subcontractors (if any), and direct job expenses are totaled to give the total prime cost. The percentage of applicable estimated overhead determined (as previously discussed) is applied, and the dollar value of the overhead expense is calculated. This, added to the prime cost, provides the total gross dollar cost for performing the job.

The percentage of profit to be included in the estimate is either determined by the contractor alone or after consultation between the contractor and the estimator, taking into consideration the type and size of the job, the character of the competition on the job, and the desirability of obtaining the job.

Some contractors prefer to apply a fixed percentage of profit to all estimates. Others vary the percentage in accordance with the factors indicated above. Some do not use a percentage-based add-on, but determine the dollar value of the profit desired on the basis of a certain amount for each worker-day required by the job or by allowing a flat sum.

Certain items of cost that in a true sense are direct job expenses, but against which it might not be desirable to assess a profit (for example, sales taxes, excise taxes, and payment and performance bonds). If these items have not been included previously in the estimate, they must be added into the final price.

The total estimated price is calculated by totaling the gross cost, profit, and other items. Normally, the total estimated price or the nearest even figure is determined to be the amount of the bid. In too many instances, when the contractor or estimator becomes uneasy over the competition on the job, the amount of the price they actually bid bears little resemblance to the total estimated price. This type of desperate bidding should be eliminated from the company as it is overhauled—many times, the contractor actually makes more money by passing on a job than taking it and losing a large sum of cash, rather than making a profit.

Too much emphasis cannot be put on the necessity of including in the summary the proper allowances for direct job expense, job factor, nonproductive labor, labor productivity factor, overhead expenses, and profit. Estimates that are properly summarized and carefully reviewed will typically provide enough income from that job to pay for all costs, both direct and indirect, and to maintain a margin of profit. Bidding on a hit-or-miss basis, rather than with accuracy is the road to ruin, especially in the 21st century when bidding is so tight because of the use of electronic estimating. When each job proves to be reasonably profitable, the entire business operation is successful and the owner is closer to realizing his goals in life.

Fig. 2.6 is designed to be used by the estimator to review what has happened on the job site and helps prevent an oversight of materials or labor line items while preparing the bid.

Labor units

The labor units in Fig. 2.7 are the result of averaging the figures of several proven estimating manuals. These units are based on workers experienced in the security/fire-alarm field and, furthermore, that modern handheld power tools, wire strippers, and other time-saving devices are assumed to be used in the installation.

As all people who bid on contracts know, labor units are the most complex factor in estimating, and no one list of units will accurately forecast the work of all contractors. Worker-hour values are given in Fig. 2.7 for the most commonly used items of security/fire-alarm systems. Interested contractors can adopt these units and modify them appropriately to suit their own operations; the table is available on the CD.

Remember that these labor units reflect the work of experienced contractors. If it is necessary to increase any of them drastically, there must be a reason (like time-consuming work habits or use of devices or materials that slow down installation). It is imperative that the reasons for time delays are searched out and methods for improving operating efficiencies are worked through with your staff. Labor units are grouped by levels of difficulty because different types of working conditions are often encountered.

Many installation variables can affect the labor cost for any given type of outlet, contact installation, wire-pulling operation, etc. For all practical estimating purposes, a large number of these conditions can be divided into four installation situation groups. On the chart, these groups are designated as 1, 2, 3, and 4. Group 1 represents the least amount of work and group 4 represents situations where the greatest number of worker-hours are required for the same labor operation listed in all of the groups.

Group 1: All working areas are open and readily accessible to workers. Work above grade levels requires no scaffolding, only stepladders.

Group 2: Includes the installation of security/fire-alarm equipment, contacts, and wiring in areas that are partially accessible, but require minor fishing of

Checklist for Estimating the Site Visit		
✓	Item	Notes
	Double check all crucial measurements, especially those above grade.	
	Check the condition of the framing and make notes.	
	Search carefully for any prep work that needs to be done.	
	Check the condition of windows, concrete, and any special areas that are involved in placement.	
	Check for any parts of the old systems.	
	Check for potential water or chemical effect on the installation.	
	Review all possible causes for work slowdowns.	
	Review for all possible additional needs for labor.	
	Double check the feasibility of the desired location for the central control box.	
	Review all finishes for installation of the finish hardware.	

Figure 2.6 Checklist for estimating the site visit.

Items	Group			
	1	2	3	4
Alarm bells	0.50	0.70	0.90	1.10
Alarm panels	1.75	2.00	2.25	2.50
Door cords	0.70	1.00	1.30	1.65
Exit/entry delay module	0.45	0.50	0.70	0.85
Fire-alarm station	0.50	0.70	0.90	1.10
Foil blocks	0.40	0.75	1.00	1.25
Heat sensors	0.80	1.00	1.25	1.50
Horns	0.70	1.00	1.25	1.50
Indoor remote stations	2.50	3.30	5.50	7.70
Indoor mounting devices	0.70	1.00	1.30	1.65
Lockswitches	1.00	1.50	2.00	2.50
Low-voltage cable, 2/c (per 1000 ft)	8.50	12.00	14.00	16.00
Low-voltage cable, 3/c (per 1000 ft)	9.50	14.00	16.00	18.00
Low-voltage cable, 4/c (per 1000 ft)	10.50	15.50	17.50	19.50
Magnetic contacts, pair	0.40	0.75	1.00	1.25
Mechanical contacts	0.30	0.50	1.75	0.90
Motion detectors	1.30	1.80	2.10	2.30
Outdoor alarm bells	1.25	1.75	2.00	2.30
Outdoor remote stations	3.00	3.50	4.00	4.50
Photoelectric cells, recessed	1.20	1.75	2.00	2.30
Photoelectric cells, wall-mounted	1.00	1.30	1.75	2.00
Power supplies	0.30	1.50	2.00	2.50
Relays	1.20	1.70	2.10	2.50
Sirens	1.00	1.50	2.00	2.30
Smoke detectors	0.80	1.00	1.25	1.50
Telephone dialers	2.00	2.50	3.00	3.50
Transformers	0.45	0.50	0.70	0.85
Windows foil (per 100 ft)	4.50	5.00	5.50	6.00
Window sensors	0.70	1.00	1.30	1.65

Figure 2.7   Security/fire-alarm system labor units (worker-hours).

cables in concealed partitions. Installation of surface molding to conceal wiring also will fall into this group.

Group 3: These wiring situations usually involve the installation of concealed wiring in partially inaccessible areas, such as crawl spaces, where the working room is partially limited. Other situations include notching of firestops or diagonal bracing to get cables in finished wall spaces, installing wiring on masonry walls where furring strips have been applied, and installing wiring in attics or basements where both horizontal and vertical surfaces have been closed.

Group 4: The most difficult situations include cutting through masonry walls, removing finished floor boards to route wiring, removing baseboards and door/window trim to permit routing of new wiring, or cutting and patching of finished surfaces to conceal new wiring.

To illustrate how a typical estimate is performed, take the commercial building shown in Fig. 2.8. The design criteria are:

1. Security and fire alarm system is to be provided.

2. System is to be local type.

3. Distance to street is 12 feet.

4. Distance to tenants on each side is approximately 1 foot.

5. There is a dimly lighted alley in the rear of the building.

6. Front and rear doors are used for egress and entry.

7. All control stations are to be inside.

8. There are no watchdogs in the building.

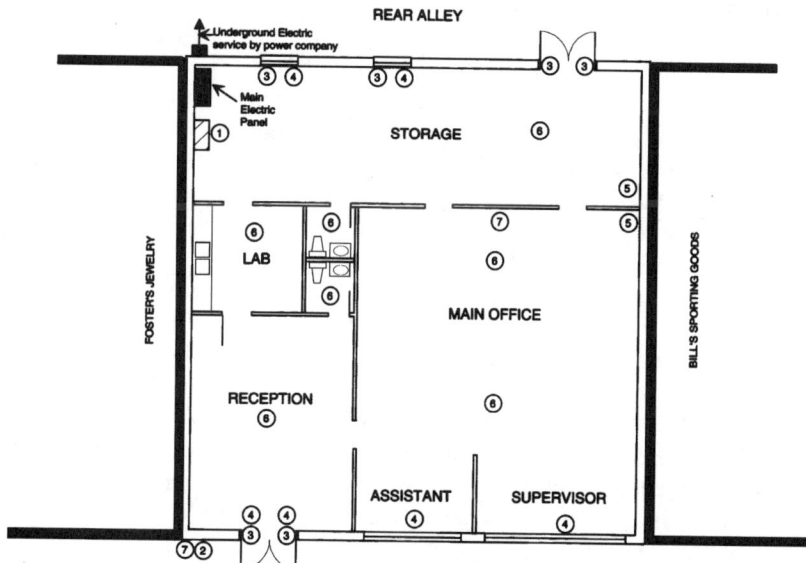

Figure 2.8 Floor plan of a small commercial building.

9. Security system is to be armed only with windows closed.

10. Rear alley is to be lighted during a security alarm.

11. Window types:
    a. Front: fixed plate glass
    b. Rear: double-hung

12. Door types:
    a. Front: double plate glass with steel casing
    b. Rear: double steel

The security/fire-alarm system was laid out as shown in Figure 2.8. The large numbers correspond to the following equipment:

1. Control panel

2. Remote station (siren speakers above acoustical ceiling)

3. Magnetic contacts

4. Glass-break detectors

5. Photoelectric detector

6. Smoke detector

7. Fire horn

The materials take-off begins with listing all major components by catalog number, quantity, and price of each item (Fig. 2.9). Use a rotameter, scale, or a digitizer, and calculate the total footage of wire. Then use a lump-sum figure for miscellaneous items (connectors, fasteners, etc.). When completed, extend the figures as shown in Fig. 2.9.

This estimate sheet is not to be taken as a real-world estimate. It is simply an example of how the process works. Remember that prices vary throughout the country—they tend to be lower in rural areas and grow as the locale moves toward urban areas. However, there is no set rule for pricing. Only knowledge of your area and the performance of your crews can be relied upon.

Also, if you are using a digitizer and computer estimating software, the formats of the estimate sheets vary in their data entry and final presentation formats.

Refer to the labor units in Fig. 2.7 and enter proper units in the appropriate column. Extend the labor units, then total both the material and labor columns. Enter the material dollar value and labor worker-hours under the recapitulation columns. Stock estimating software will have built-in methods for applying these formulas. Ordinary spreadsheet software can also be used to build templates like these. However, even stock formulas for the amount of labor required to perform a task should be adjusted to the real-world time required by your firm for performance of the work.

Multiply the worker-hours by the average hourly rate, calculate the labor add-ons and other direct job expenses (if any), and enter them in the proper spaces.

BID SUMMARY SHEET

JOB _Page Health Dept._
611 Mossy Oak Drive
Overall, Virginia 22658

PAGE 1
OF 1

ESTIMATED BY ___JET___ CHECKED BY ___RRU___ DATE ___10/7/95___

DESCRIPTION	Quantity	MATERIAL Unit Price	Per	Amount	LABOR Unit	Per	Amount
051 Control Panel	1	360.00	ea.	360.00	1.75	ea.	1.75
413 Battery	1	36.60		36.60	0.30		0.30
216R Indoor Remote Station	3	39.00		117.00	2.50		9.50
290 Prealarm Station	3	27.00		81.00	0.50		0.50
581 Outdoor Sounding Device	2	40.00		80.00	1.25		2.50
205 Indoor Sounding Device	1	20.00		20.00	0.70		0.70
550 Siren Driver	2	60.00		120.00	1.00		2.00
304 Relay	1	15.00		15.00	1.20		1.20
2025 Photoelectric System	2	300.00		600.00	1.20		2.40
295A Smoke Detectors	7	60.00		420.00	0.80		5.60
441 Magnetic Contacts	6	4.00		24.00	0.40		2.40
170 Glass Break Detectors	6	45.00		270.00	0.70		4.20
171 Adapter (delay)	1	27.00		27.00	1.00		1.00
217 Emergency Switch	2	3.00		6.00	1.00		2.00
221 Transformer	1	20.00		20.00	0.45		0.45
18/2 Wire	700 ft	0.20		140.00	8.60	M	6.02
Misc. (fasteners, connectors)	Lot	27.00		27.00	4.00		4.00
TOTAL				**$2363.60**			**42.36**

MISCELLANEOUS		RECAPITULATION	
		Material Cost	$2363.60
		42.36 Hours Labor @ $20.00	847.20
		3 Hours Labor @ $25.00	75.00
		Direct Job Expense	250.00
		Total Prime Cost	$3535.80
		Overhead Expense 15%	530.37
		TOTAL COST	$4066.17
		Profit 15%	609.22
		Selling Price	$4676.09
		BID SUBMITTED	$4676.09

Figure 2.9 Completed estimate sheet.

Total the dollar values of material, labor, and direct job expense to obtain a total prime cost. Finish the estimate by adding overhead and profit to obtain a selling price. Figure 2.9 shows that the total selling price is $4056.42.

Estimating with Computers

In the quest to shorten the lengthy process of estimating the installation costs of security fire-alarm systems, two major developments have occurred: the assembly concept and computerized estimating. Both developments have the potential to provide great benefits to installation contractors, although reaping the full benefits of these improvements takes a bit of effort, particularly because it requires a good deal of concentration to tailor computer systems to meet the specific needs of each contractor. And, if the contractor expects to be fully effective, a lot of research might be required.

The assembly method was first popularized by the Estimatic Corporation in the 1950s (Estimatic was also a pioneer in computerized estimating long before anyone else was doing it). The company used the concept that virtually every type of electrical/electronic symbol used in construction drawings could be summarized as a specific list of materials.

For example, the assembly for a common duplex receptacle would include the receptacle, the finish plate, a box, a plaster ring, screws for fastening the box to the framing, a grounding pigtail, a couple feet of #12 wire in the box, an average of two wire nuts, and two $\frac{1}{2}$-in. EMT connectors. Thus, the assembly includes everything indicated by the duplex receptacle symbol on the plans.

This is typical of all assemblies, where there could be many thousands of combinations of different types of receptacles with different types of finish plates, different types of plaster rings, and other devices. Even raceways or cable assemblies can be broken down into assemblies (for example, three #12 THHNs in $\frac{3}{4}$-in. EMT, three #6s, and one #10 in 1-in. EMT, etc.).

Thus, the contractor develops a full set of assemblies and prices them all in relationship to the company's locale and labor pool. When the estimator takes on a job, all of the symbols and raceways on the plans can be counted, and then priced as assemblies (so many type XYZ assemblies, so many type ABC assemblies, etc.).

Computers are amazingly efficient for this type of task. Assemblies can be built with ease and changed (either by component or labor value) with very little trouble. With the use of competent computer people, the pricing can even be linked directly to the databases of several vendors.

Because all the assemblies are pre-priced and pre-labled, the estimator no longer needs to count or price most individual parts. When it is time to order material, the computer can develop shopping lists automatically by sorting through the assembly database.

Then, the estimator can use e-mail to deliver the list as orders for suppliers. The sophistication of computer equipment has now made it so this can be done directly from the jobsite and the estimate can be directly downloaded to accounting to open a new job file. Certainly, computerized estimating (Fig.

2.10) has made a noticeable impact on the trade, with the computer able to perform all the tedious mathematic functions. It can save an estimator a lot of work and, at the same time, it eliminates the likelihood of errors in mathematics. These are the main benefits of computerized estimating. Other benefits can be reaped, but none are nearly as important.

The take-off requires a good bit of attention, but it can be highly streamlined and made very accurate with the use of digitizers and counters that dump figures directly into software for extension and addition to the bid sheets.

Writing up the estimate (entering it into the computer) has also been highly improved. At first glance, this would seem to be automatic, compared to the old method of writing down each item and assigning it a cost and labor rate, but other factors come into play with the computer that require time and careful attention.

First of all, are the computer codes. Estimating systems use code numbering for each particular type and size of material. For example, [3/4]″ EMT might be code number 27.7; 1-in. EMT might be code number 27.11; #12 THHN solid could be code number 246.94; and so on. So, now the estimator must enter not only a description of the item, but a code number as well. This is not always necessary, but almost all estimators enter a description so that they can double-check their work. And because there are so many different items in electrical construction that no one remembers all the code numbers, it is necessary for the estimator to look up code numbers for most of the items.

This process speeds up as the staff turns out estimates. The most common codes are soon committed to memory and large wall charts, short-cut screens on the computer, and digitizer tablets that rest on the table make it simple to obtain all of the code numbers at a glance.

Figure 2.10 Computer estimating has had a noticeable impact on the trade.

Thus far, the best method developed for itemized estimating is to use a combination of assemblies and the computer. In the best of these systems, the estimator does the standard type take-off with a digitizer that enters the total quantities into the computer. Then, the computer takes all of these assemblies, breaks them down into their individual parts, and prints out one complete bill of materials.

You get the best of the assembly system's benefits and you don't have to update the prices of your assemblies continually because the computer only needs the quantities of materials that make up each assembly, not the prices. After the whole job is broken down to a single material list, the computer will provide a material price and labor-hour figure to each item (not to entire assemblies) if you desire that it breaks things out in this manner.

Of course, the task of updating the material prices in the computer remains, but given the make up of the industry at this time, there is no worthwhile alternative to spending some time doing this. However, things are speeded up by e-mailing the list to the vendor for an up-to-date quote or setting this up on a regular basis by computer.

Although computerized estimating systems have been around for a number of years now, only in the past decade have the prices of both the computer equipment and the programs decreased to where the average contractor can afford them. Actually, for a number of years now, computers have been making the transition from a "nice tool" to "standard equipment" in security systems contractors' offices.

Remember when purchasing any computer system that flexibility is crucial, as is proper training in the use of the system. Any computer system that you use for estimating should be capable of being fully integrated with computer programs that your company is using for accounting and management. So much is going on with computers that keeping abreast of hardware and software can be a full-time job. Are you in the security systems business or the computer business? We all know the answer to that question. Put your time into what you do well, make it easy on yourself and your staff, and put the time you would spend researching computer systems into researching high-quality computer consultants who specialize in fully integrated, office and field installations for the electrical industry.

3

Managing a Security Systems Business

Management is an art. It is the activity that causes a company to make it or to fail. If it comes from natural interest, rather than just making do, it can propel a company through hard times and promote success.

The art of management is the major factor that determines whether a business will be successful or fail. Other parts of a business can influence its strength: location, products and staff—but if management is strong, those parts will be well chosen and implemented.

Management is very important in security systems contracting for many reasons. For example, performance is always on a fast-track timeline as a building project nears completion, the work of other trades can affect the work of the security systems contractor in very crucial ways, and change orders and unforeseen snafus can require rapid adaptation as the job progresses.

Managing a security systems operation is complex: collecting past-due invoices, keeping abreast of the latest information about components, watching the progress of installations, coordinating cash on hand with payables, negotiating contracts, selling new jobs, taking care of the company's equipment, and working through staffing problems. The list goes on and on. It paints an awesome picture of responsibility.

Why Are You in Business?

Typically, people are in business for two main reasons: they are doing something in which they have a great deal of interest and they want to make money. Too often in our society, people are simply there because they need the money and don't have any other goals. If you are in business solely for the purpose of making money, you will probably be more successful in a franchise or distribution business like owning a 7-11 or a beer distributorship. This type of company is much less complex and vulnerable to the marketplace than a business based on contracts. The profits are straightforward, management is not so complex, and the profits can be invested wisely for much quicker gains.

As we explore the subject of a game plan in more detail, you will see that it is a much stronger position to enjoy one's business than to simply be showing up to squeak by and make some money. A life of drudgery, not at all a requirement in life, is a rough way to get by.

If you do, in fact, enjoy contracting with all of its difficulties, the first step in overhauling your business is to pay close attention to how effective you are as a manager. Effective management can make your position as a security systems contractor much more enjoyable.

Dozens of books are available about management and classes and seminars are everywhere. However, a lot of the information is designed for people who are students, theorists, or are in middle management. If a person is involved in directing a security contracting business, they are usually strapped for time; any management ideas that they intend to incorporate into their work need to be succinct.

One important thing should happen as a part of overhauling your business—it is the backbone of sound management. Look at the business carefully and divide it into its various operations. Don't hurry through this task. After you understand what the various operations are: estimating, job cost accounting, purchasing, installations, break them down into parts. Next, take each of the major parts (even the most mundane sounding, like safety) apart carefully. Then, analyze where they work smoothly and where they need smoothing out. The highly successful manager filters each of the ingredients out, looks at that task carefully, and then establishes company protocol by integrating each of them in a simple, efficient manner.

Establishing procedures simplifies and speeds up production, but it is important to remember to stay flexible. Don't lock into new procedures so tightly that you lose the ability to shift approaches. It is very important to maintain the skillfulness to change the way tasks are handled whenever improved techniques can be found, to grow and change with the business' needs on a continual basis.

As you smooth out each part of the business, remember the next important part of management. Assign tasks to people with the least amount of responsibility and use subcontractors and outsource work where it makes sense. When you assign tasks to your staff, don't ever think of them as being low on the totem pole. Remember that a "go-for" is very important to the execution of your jobs. And, a person pushing a broom can often observe what is going on at a job site more clearly than the foreman, the superintendent, the lead man, or the owner of the company.

People with light loads of responsibility are naturally standing back with a limited amount of information in their minds. They are able to watch. Powerful managers take advantage of this condition by having discussions with people who have this unique overview, listening to them, and bringing all of the staff into the loop of success.

One of the biggest mistakes in the construction industry is for the principal to think that she or he can perform all of the tasks better than anyone else, then to hang on to the chores for dear life, refusing to hand them down to other personnel.

This causes many problems, two of which are a breakdown in trust and communication between the boss and other employees, and a chaotic mindset for those in charge. Trying too hard causes a muddled perspective throughout the company. If the boss is always in a hurry and can't pay attention, many employees will imitate this behavior, avoid the bosses, or rebel by slouching through their work.

If the owner refuses to let go and continues to perform tasks that others could be doing in their own way (which might even be better than the owner's) management will never reach its optimum. If this common management error is not contained, the principal is continually losing sight of the big picture while wasting time correcting minor problems.

When you begin to look at your business closely, first observe any habits that are a part of your firm's management, such as not being able to pass on authority. One of the best tools for being able to see your business clearly and making your life smoother is to make time to get away from it all, to get some distance. Also, this time can be very valuable for meeting new people and marketing the company. Naturally, this does not mean that the boss should be off having fun all the time and not paying attention to the details of the business. On the contrary, handing down tasks and taking time off opens an opportunity to listen carefully to that person pushing the broom. Is the broom the correct size? Are the dumpsters being dropped in a labor-saving location?

By letting go of tasks and listening with attention on a regular basis, a savvy security systems contractor can not only fine tune the business, but can gain and keep the trust and respect of employees. Allowing people to have control and work at their best motivates them to feel pride in a job well done and creates a more enjoyable, efficient, and profitable work place.

Growing a Security Systems Firm

As you begin to put management systems in place and adjust them, you will most probably think about your business dreams. Most everyone has fantasies about having more cash. A contractor often thinks that the more jobs there are, the more money there will be. This is true to a certain point, but it can also be the road to hell paved with good intentions, as the old cliché goes. Growth is a crucial part of a business; it causes a great many firms to belly up. The first impulse is to simply beat the bushes, hunting for more jobs. And naturally, more jobs are required for growth. But a lot of details must be worked out and a great many questions asked as one begins to venture down the path of growing a business.

Two of the primary questions are: where do I look for business and can I take care of it if I get it? Another old saying is—be careful what you wish for. Nothing is more frustrating than having so many jobs that your performance has dropped because you, your staff, and your capital are spread too thin.

The first step for growth is to pay close attention to overhauling your management systems until they are fine tuned. It is imperative that you are a skillful manager during a growth curve. While you are revising your management approach, begin to think about how you really want to grow. What interests you and what might be fun for you?

In our look at forecasting and leveling, we reviewed the idea of having separate profit centers that are peripheral to contracting security systems. Many of them are covered in more detail in Chapter 4. The whole idea is to develop sources of revenue that are closely related to your main business, but can be effective when it is suffering from a slowdown.

An excellent example of a peripheral business is moving into the installation of other low-voltage systems. Photovoltaics are an example; these low-voltage systems use silicon wafers to produce a flow of electrons from the sun. Also, there is a good deal of information in the appendix on the CD that will take you directly to the Web sites of companies.

Developing New Services

As mentioned, Chapter 4 takes a closer look at some of the many side businesses for a security systems contractor. These endeavors are often a part of the contractor's current business.

For example, we mentioned photovoltaic systems and your firm has most probably used light-driven sensors in installations. In fact, you might have used PV panels as an energy source for past jobs. They are straightforward and fairly simple to understand. With the Internet available, research on any new subject is much simpler than it was when a contractor had to depend on suppliers, books, and organizations for information. The company that was trying to move into another field had to depend on a handful of people for further knowledge. But with the Net, spend a bit of time searching and the information provided by one source can be cross-referenced with knowledgeable groups from around the globe. Equipment can be priced and checked for performance quickly and with a good deal of accuracy.

If we use photovoltaics as an example of a low-voltage system that can be pumped up as a growing source of business, the first step is to increase your company's current base of knowledge: global research, sources of supply, benefits of photovoltaics, what the marketplace looks like, who is currently using them, what the demographics look like for short-term growth, and how the growth curve for their use looks for the future.

As we have stated, one of the most important criteria, aside from ease of implementation and similarity to what your firm is currently doing (large industrial, commercial, or residential installations) is that the work is of general interest to yourself and part of your staff. Having a natural interest in a product or service (for example, photovoltaics) will make learning curves and other difficulties much smoother for you and your staff to overcome.

After choosing a particular service and product and doing the basic research, you will need to find some beginning projects. One of the simplest ways to find work is through the suppliers and manufacturers. They are often eager to meet reliable installers who show an interest in their product line.

There are several other major incentives for using the vendor as a market source. First, the technical staff of the manufacturer can be invaluable when you are wading through the mechanical snafus in your first few contracts. Another advantage to this approach is that vendors offer a potential for on-going business that requires little or no marketing expense. Working and finding jobs with them is much less time consuming than other forms of marketing. And, they might be in need of suppliers of their products, which can turn into an additional source of income that is very closely related to the other parts of your company.

During your research, contact several of the manufacturers, get further information, and talk with their marketing and sales groups. Again, using photovoltaics as an example, you should feel good about developing a relationship with one of the vendors. You explore their track record and talk with other installers in various localities and with satisfied customers. And you crunch the numbers for realistic, potential earnings.

The manufacturer gives you a first client and you are very careful about the installation, possibly not doing better than a break-even position financially on this first deal. Consider the loss a part of research and development, and marketing. Then do more projects with the suppliers that are moneymakers.

After several installations, you will have acquired the knowledge and the buzzwords of the field and have a bit of experience for estimating and performance after getting contracts.

From there, the same techniques that are used to expand your security systems business fall into play. One of the first, easiest, and least-expensive techniques is to send out an announcement to your current client base.

The business owners for whom you have worked might be looking for ways to reduce their utility bills. The manufacturers you are working with will often have a good understanding of ways that photovoltaics can pencil in as cost-savings products for your customers. Their marketing departments will be looking at government and utility incentives for using efficient energy sources. Your firm can explore the offers for rebates on energy efficient systems with the Department of Energy, state and local governments, and utility companies. Any available incentives can be a part of your announcement. These ideas are explored further in Chapter 4 and they are much simpler to research, now that you have the Internet as a resource.

Announcements are the first and simplest approach to marketing. Your past customers might be interested in the idea of saving on utilities. Next, explore how to integrate photovoltaic power sources into your current jobs as reliable, quiet circuits. Along with these efforts, your announcement should go out to the architects, systems designers, and engineers with whom you have a track record for quality installations of security systems.

We have used photovoltaics as an example because of their straightforward relationship to the family of low-voltage electronics and equipment found in buildings. No matter what systems (automation, computer wiring, or photovoltaics) you choose to work with, the techniques for launching into the market are very similar.

After you have used the preliminary techniques to find the first series of contracts, pursue marketing: hire tradespeople who already know the market and have contacts, speak to local groups, get to know people who are involved in the fields you choose, and continue to make announcements. Don't expect the floodgates to open up instantly with a huge cash flow. In fact, rapid growth can often be a real hindrance to steady, well-managed growth. Plan to be consistent and dogged in your pursuit of a new market. And always adjust your company to be able to handle the new business in a diligent and professional manner.

Setting Goals

Before exploring all the steps involved in developing a final business plan, always remember that a really successful businessperson is one who enjoys their life, not someone who is simply a slave to a business. To achieve this position as you are overhauling your firm, you must get some overview and see what you are doing with your life in general. A business plan is a tool; it is most often used to go to lenders in the pursuit of financing. The business plan is most often used during the beginning phase of a business, when the owner is first getting started. But they are also used to expand a current business operation. However, before you even begin to put the business plan document together, take all the time you need to look closely at what you are doing now. What do you really want to be doing with your life? How does your current business work? Is it something you find satisfaction with on a day-by-day basis? Are you sick of what you are doing? Would you rather be making money from one of your hobbies, like camping or computers? Take some time to look into these questions with care. Tightening up a business and growing it are demanding undertakings, and after you have done all of that work, the effort of running your business will still be required.

Setting lifetime goals could be approached by closing down to the other parts of life, like family, and simply plowing along like a tractor. However, the idea of looking at your life and thinking through a game plan that makes sense for you, is to be able to adjust to the various phases of setting the goals, working toward their maturity, enjoying your family, and being able to go with the flow when changes in life occur.

With a calm, healthy approach, you can look at what you are doing now and think about where you would like to be in 10 years. Review these thoughts carefully; don't set abstract, unrealistic goals. Keep them simple and be very honest with yourself.

Taking a Look at the Future

If being a systems contractor is in fact what you want to be doing, then you are in a real-world position to think about the future—you won't just be hobbling along from day-to-day, resenting what you are doing. From this strong position, you can begin to put a plan in place and pursue it with real enthusiasm. This type of genuine interest is what you need to get through the many challenges you will face as the plan progresses.

After you are certain that you want to pursue a career in security systems, imagine how you would like to be spending your days twenty years from now. Would you like to be financially independent, retired, and traveling across the country with your loved ones? Do you want to be semiretired and managing the big decisions of the company you have built? Do you want to be the CEO or the chairperson of a large, public corporation? It is very important that you understand yourself fully. The goals can, and probably will change as you move through life, but you must know what you currently see as your future.

When you have worked through these questions with yourself and your family in an honest way, you can set the methods to achieve your goals. The most common way is to get a clear picture of the ideas you see at the twenty-year period, and then work back to the present. Understanding what you want in the future frees you up to examine what you are doing now and what your intentions are for this year, the next five years, and your entire career. It gives you a method for seeing just how important your life is to you and those around you. This understanding can give you a sense of purpose and make you much more powerful at focusing on what you need to take care of daily.

For most people, once they start to look at their lives with some perspective and see that the decisions are really up the them, their everyday lives begin to have a different tone to them—they have a more positive approach. While developing your goals be sure to keep in mind that your life is what is taking place today. Don't get bogged down with the past and don't forget about enjoying your life today and getting lost in your goals. This is a delicate balancing act, but once you get the feel for it, smoothing it out can be part of the challenge.

Pursuing your goals isn't an aim that means passing up what is taking place in your life right now. Be sure to start setting aside time for things that you love to do. Spend more time with loved ones and pursue things that you enjoy doing. Schedule these activities just like work and it will give you more sense of control, refresh you, and make the steps in achieving your goals much easier to attain. This can help to move you over to a proactive attitude and away from a sense of being stuck with all the tasks that are inherent with running a security-contracting firm. You can move from being someone who is staggering through life into the position of taking the wheel.

Remember that life never really gets better or worse—some people win the Lotto and don't know what to do with the sudden wealth and end up broke and miserable. Some people with a modest living retire fulfilled and with plenty of money to enjoy the type of life they have chosen for themselves. Everything has to do with our perception of what we are doing. And as has been said many times, a positive attitude will give us power and help us find a sense of fulfillment.

Figures 3.1 and 3.2 are designed to help you start filling in your goals. They are preliminary to business plans and the various steps involved in fine-tuning your business or the business you intend to start. Use the charts to fill in your long-term goals. The CD contains versions of the lists that you can print or copy as files and work on them at your leisure. Allow your ideas to change as you get a handle on your goal. You will use the goals to lay out the plan for the next year and all of the planning you need to do. The list is just a beginning with some ideas of how you might want to look at planning. A blank page follows for those who already have lots of ideas to put in place. Remember that you can print this list and take it with you through the day and over the weekend and holidays.

Setting Short-Term Goals That Guide You Toward What You Want

Take your time working through your goals. You don't have to make this a painful process—keep your business running like it is while you analyze it and take a close look at your life. As the long-term goals fall into place, you start to wonder how you are going to put the ideas into action.

It might turn out that you want to leave the security systems business altogether. Or, you might decide that you would prefer to give up your own business and draw a paycheck with a reliable company so that you can put more time into your online electronics supply business. There are many ways that you can pursue changing. You might adapt some of the ideas in the book to the new field, you can go to consultants, the Net, or the Small Business Administration.

If you intend to stick with or start a security systems business, consider a few things as you implement your long-term goals. Obviously, the primary thing to zero in on is exactly what business you want to strengthen. For example, this work might be supplying and installing residential security systems in subdivisions. You might have been stumbling along with this business for some time, dealing with competition and the tight margins that the developers squeeze subs into accepting.

For a great many subcontractors, their business has been so tight and difficult to maintain that they feel as if there is no hope beyond simply getting through the day. The first thing that is needed if you are working hard like this and feel like you are on a treadmill is to move away from the desperate position of taking any work that comes along.

The following ideas will get you started with supplementary cashflows. For more ideas, see Chapter 4. Add scenarios of your own if they come to mind:

Long-Term Goals Worksheet Zeroing in on What I Enjoy		
Real World	**Things I Really Like to Do**	**How Can I Make Money at It?**
	Security Systems	Want the business and really do enjoy the challenge of securing a building. Will grow the contracting and work with my son on our online electronics business, too.
	Softball	Too personal and can't make enough money at it.
	Electronics Supply on the Net	Find more products, get a Web tech to send it out to more browsers.
	Security Systems Consultant	Will talk to a couple of the lawyers at the Downtown Business Association and check it out on the Net.

Figure 3.1 Sample worksheet for outlining long-term goals.

What I Really Like to Do		
Real World	**Things I Really Like to Do**	**How Can I Make Money at It?**

Figure 3.2   Sample worksheet for determining what you like to do.

- Transitioning from a small roofing business to a larger one
- Transitioning to a major, incorporated roofing business and going public
- Maintaining position as a small roofer, the owner of an excellent company
- Starting new businesses that run simultaneously with the roofing business

Naturally, these goals are unique to each person, but setting goals is what will chart the road map for your success, no matter what path you choose. The main attitude that you can change in your business life is a position of pro-active, rather than re-active management. You are going somewhere with your life and you now have a reason to tighten up your business. This approach will be set against a backdrop of long-term goals and their relationship to every day tasks.

Now that you have a sense of resolve based on long-term plans that really mean something to you, it is time to look at how to put them into action. Short-term goals that guide you toward your game plan must be thought through and implemented. Figure 3.3 is an example of how a person might fill out the blank copy that you will find on the following page and on the CD.

Developing a Business Plan

Having roughed out your long-term goals and begun the short-term system for implementing them, we come back to business plans. Let's say that you decide on the security systems consulting business shown on the sample chart, a security systems maintenance business, and to grow your contracting business. You might see a need to bring some extra capital into the picture.

A business plan is a very common tool that all types of businesses take to lenders for raising funds. Whether for the transition of a small enterprise to a larger one or the beginning of a new project, the business plan is a must when you go out to look for funds. If you are going to a lender, you will most probably be required to have one. If not, it might be a good idea to do one for yourself in order to evaluate the potential for the business as a moneymaker. Don't make it complicated. Keep it simple. Figure 3.4 shows an example of a simple plan for the security systems contractor. It could be used to grow your business or to add an ancillary business like a parts business or the maintenance arm of your company. The following table is an example of the outline of a very simple plan.

Many types of information are available about business plans. There are books that will offer other formats for a simple plan. Plenty of software is available to guide you through a business plan and even print out clean presentation copies. The Net has a tremendous amount of information for free.

It would be best to get help with a first business plan. The input from others will flag you on concepts that you might be missing and help you avoid costly mistakes. Take your time when you seek advice. Listen to consultants, attorneys, advisors, and advice of any type carefully, but don't swallow any-

Short-Term Game Plan Starting Long-Term Goals		
Goal	**Time Frame**	**How Will I Implement the Goal**
Raise profit from existing business by 20%	9 months	Contact existing clients, architects, and general contractors.
Raise profit from existing business by 20%	12 months	Get contracts from the friends of existing clients, plus new architects and general contractors.
Raise cashflow from online electronic equipment business by 50%	1 year	Look for more products and hire a consultant to help me get on search engines.
Raise total cashflow by 10%	1 year	Check with the bar and online for advertising my expertise.

Figure 3.3 Sample worksheet for determining short-term game plan and starting long-term goals.

	Short-Term Game Plan Starting Long-Term Goals	
Goal	Time Frame	How Will I Implement the Goal?

Figure 3.3 *(Continued)*

Implementing the Business Plan	
Item	**Notes and Information We Need**
Owner's financial statement This is found in your accounting software or with your accountant	
Marketing plan How we will get knowledge of the service out there	
Goals for the new enterprise The plans for growth and income	
Management Describe who will manage the business and their credentials	
Products and services Exact description	
Market forecast How much business is out there?	
What are the financial projections How much of the market will we capture?	

Figure 3.4 Sample worksheet for implementing the business plan.

thing as the gospel truth. Ask lots of questions and always try to get more than one viewpoint.

Most of the time, business plans are created to go to a lender. If you are going to move right into an ancillary business, you might not need additional funds. But you might want to be prepared in case you do end up needing an infusion of cash as you launch the work. A close look at your current business and your plans for growth will always be useful for understanding your position.

Employees Are the Heart of the Business

Once you begin to understand why you are in business, what your business means to you, and where you want to go with it, you can begin to set long-term goals and put short-term goals into place.

This proactive approach can make work much more challenging, exciting, and a whole lot more fun. If you are pursuing your natural interests, all of the challenges can blend into the process, rather than being an on-going, endless cycle of emergencies.

With the goals in place, you can take the company apart and look at every operation carefully—the key step to strong, effective management. However, no matter how diligent and effective you are, you will still be completely dependent on the people who work for you.

This is always the case; there are no exceptions—people are the heart of any business. Although the logistics are extremely important, they will not help without an effective group of people implementing your plans.

The people who work with you are entirely dependent upon your ability to be powerful as a manager. A number of management principles make a work force effective. The first is to hand the work over to the person with the least amount of duties who is capable of performing the task. The more that you are capable of delegating responsibility, the more effective your firm will run.

An often-overlooked management premise is the simpler the performance of tasks, the more smoothly the company will run. This is part of the reason that successful managers take an operation apart to see how effectively different tasks are approached. Keep all tasks as simple as possible and make it easy to perform them.

A clear example that will strike home for everyone is tools. We have all been on a job site and seen a subcontractor's crew with a broken-down truck, old tools that are out of repair, dull drill bits, and lacking extra fasteners or parts. This type of a crew will invariably take an exponentially longer time to complete their work than a well-tooled, efficiently supplied group.

A very important part of this cycle is the mindset of the well-supplied crew; they are much more confident than the piece-meal crew. They don't waste time worrying, jerry rigging, arguing, and at other pointless, ineffective activities. They mobilize at the job site and get right to work. By the end of the day, they will have achieved obvious, tangible progress—and the installation will be high quality, not in need of alteration.

In this day and age, the owner of a company can make an immense impact on staff performance by simply keeping the work as easy as possible through the use of cutting-edge technology. Tools and supplies are very important and should always be current, easy to access, and in good repair.

And now computers and other electronic devices need to be considered tools and treated like drills, saws, and any common tool. The computers and electronics continue to change the AEC (Architecture, Engineering, Contracting) industry at a mind-boggling rate. For example, workers stand in the middle of a room and lay out the position of a run of raceway on the walls in a building— the laser level makes the run straight and true.

Computers Are Now a Necessity

As suggested, the days of guesstimating a job's cost and client's bill are long past. It is no longer a humorous situation for a subcontractor to keep information on envelopes and scrap paper under the seat of the pickup or in the glove box. Bidding and competition are just too sophisticated and tight now.

In the electronic age, the security systems subcontractor needs to be completely abreast of the times in order to be effective at the art of management. A one-time investment of several thousand dollars for computers, software, a laser printer, an internet connection, and the other items needed for a system, can pay for itself quickly in estimating alone.

With this new millennium, top-notch management requires computers. Estimating is only one part of the workload that computers will handle, and buying computers requires one major outlay of capital, with a relatively inexpensive on-going upkeep.

After the purchase, they can save on your business' largest expense, labor. A well-designed and implemented computer system can save on hiring in all parts of your business, both in the field and the office. If the computer system utilizes peripheral electronic devices like digitizers for taking off plans, and dumping the results into the estimating software automatically, the estimating process can be speeded up dramatically. This use of electronics can make it possible to raise the volume of estimating while growing the size of your company. If the growth is managed with care, the volume of business can be raised to a level that will easily support more estimating staff without cramping the cash flow of the business.

A great variety of activities involved with a security systems contracting business can be handled quickly, accurately, and inexpensively with the use of computers. The main criterion for choosing systems will be the tasks that you require the computers to perform and the size of your firm.

The first step in upgrading your existing systems or installing new computers is to evaluate what you want the computers to do. When you begin thinking about the process, be extravagant—if you suspect that electronic devices could streamline a task, chances are it can be done. Think all of these things

What Our Computers and Electronic Devices Should Do		
Item	**✓**	**Notes**
Take-offs with a digitizer that dump the information right into the estimating software.		
Extend estimates automatically.		
Connect with customers and vendors.		
Have a database of labor and overhead costs.		
Shop materials electronically.		
Open an account for each job that carries through to change orders and billing.		
Track change orders.		
Handle aged billing.		
Communicate between office and field.		
Print profit and loss reports for every job as it is going on.		
Update schedules as they change.		
Communicate from field to architect's and general contractor's offices.		

Figure 3.5 Sample worksheet for itemizing computer needs.

What Our Computers and Electronic Devices Should Do		
Item	✓	Notes

Figure 3.5   *(Continued)*

through and make a list of what you would really like to have computers do for your firm.

Figure 3.5 roughs out some ideas to get you started. Use the version on the CD to type in what you would like to have the software do for you. Make copies and hand them out to all of your staff. The effective manager communicates closely with staff at all times—especially when vast changes will affect their lives at work.

Select a group of people from each part of the operation: field, office, and estimating, and have them work together on a list of all of the tasks for the computers. Next, start interviewing accounting and estimating software vendors about their products and asking them for the names of consultants in your area.

Do not rush this process—this is the most important step in automating your office. Let your staff participate, but don't let them take charge of automation; it is too big of a drain on productivity. Let them talk through what needs to be performed by computers and electronic devices and get it on the lists, but get a consultant to select and install the systems.

Selecting the correct consultant will make all the difference in the world. Talk with a number of them, taking your time to discuss what they all have to say. Check their references and, preferably, visit their clients and check out what the consultant has already done. The consultant can make a computer installation a pleasant experience, but they must be highly skilled and easy to work with throughout the process.

With some time and energy invested, you should be able to develop a good idea of what your staff sees computers and electronic equipment doing for you. Don't hedge your bets at this point in the game—go all the way with ideas. Getting wild with ideas will not hurt a thing; in fact, the whole point is to streamline the workload. What is really important (we can't emphasize it enough) is the choice of a consultant that can make things happen quickly.

Now, we will think about some of the basic ideas related to what automation does for a security contracting company. If you look at what takes place at your firm on a regular basis, you will probably realize that a good deal of your work is repeated continually. Someone calls with a potential new project. Checks come in from jobs. Problems arise in the field. Invoices go out. Payroll must be out by Friday. Invoices come in from vendors. The list of tasks that you and your staff take care of in a week's time is long, but you will notice the repetition.

Automation can boilerplate a great many of the tasks, simplify them, and perform them with speed and accuracy that no human being can provide. At this point, remember that the equipment, software, and consultants aren't what really costs when it comes to electronic automation—the staff's time at the keyboard is what will really cost money. Time working at the computers will be a part of payroll and all business people know how large the payroll burden can be.

The point is that the installation of electronic equipment, like any purchase of tools is not a place to reduce costs. The more keystrokes and tweaking of the equipment that your staff is involved with after the installation, the higher the payroll burden related to getting tasks done with computers and the less valuable your automation efforts.

The primary tool for taking care of your list of tasks will be the software. It is the first thing to shop for and choose—the hardware should be chosen to fit the software. Your choice of software should allow your staff to work as closely to the way they already work as possible. Figure 3.6 can be used to interview software providers and consultants. Boot it up from the CD and adjust it or the blank one to suit your company. Again, the wise manager brings all of the staff in on an important project like automating the company.

Make Change a Friend, Not an Enemy

Management involves a great deal of finesse in order for it to be powerful. Any number of tasks must be delegated and they must be performed in a diligent and professional manner. Attempting to think so many business demands through at the same time can be overwhelming. It is important to break the operation down into easily comprehended components and establish routines for the expeditious execution of all of the tasks.

When one thinks of the company in components, it is much easier to develop techniques for taking care of business. However, after systems are put into place and running smoothly, there can be a tendency to fall back on one's laurels and trust that everything will work out alright. When this type of a mindset takes over, management often becomes rigid, unable to change, and attrition sets in—the company becomes dated, and loses market share, and the dynamics of a well-run company.

Management can count on change without reservation. Strong managers learn to thrive on change. They even instigate it. Naturally, systems are needed, simple, smooth methods for handling tasks, but always keep abreast of change in the marketplace, staff, and ways that current systems can be improved.

The security systems marketplace will continue to change rapidly in this new century. Low-voltage systems will become more and more centralized in buildings as a part of the building automation package. Welcome the changes and use the Net and modern tools to stay ahead. Bring your clients into the loop and offer your knowledge as a part of your service—make change an important part of your business, a challenge that keeps it fun and vital.

✓	How Our Company Works	
	Item	**Notes**
	Receive a request to bid.	Want to enter it into the system once and not have to reenter when the accounting trail starts.
	Estimate the project.	The estimator wants take-offs to be automatic, but with a way to adjust them, a shopping list to be prepared automatically.
	Estimate presentation.	The software can e-mail a professional estimate directly to the customer.
	Customer accepts proposal.	The job becomes an account easily.
	Job progress is monitored.	The software reports on progress as the job continues.
	Accounting trail for job and all jobs provided.	Management reports are readily available from software. Redoing the original schedule is easy.
	Accounting facilitates bringing receivables forward.	Accounting is made much easier so that aged accounts are flagged and invoices and envelopes printed automatically. Payables are also flagged and easily printed.
	Project becomes a part of the marketing database.	Marketing becomes an on-going process. Announcements are addressed right out of the database.
	The status of all jobs and cash are easily accessed at any time.	Reports and financials can be easily accessed for effective management.

Figure 3.6   Sample worksheet for showing how your company works.

4

Profit Centers: Service, Maintenance, and Ancillary Cash Flow

When it comes time to grow your company or get a handle on the peaks and valleys of your income, having different profit centers as a part of the contracting business can be a real benefit. The use of low-voltage systems in buildings will continue to grow as we move into the new century, and the security systems contractor is in an excellent position to enjoy healthy profits from this trend.

For those of you who are serious about taking charge of your business, setting real-world goals for yourself, and moving on into a day-by-day approach to furthering those goals will find that small, simple, ancillary cash flows can be invaluable. The hard work of forecasting and leveling steady and solid growth can be moved along and simplified with parallel businesses to really enhance your current assets. This book is about security systems contracting and we are staying with that information and limiting the number of ancillary businesses that we address, but there are many more. Chances are strong that you will find more profit centers related to your existing business than you would ever want to pursue.

Profit Centers as a Way of Business

Because you are reading this book, you most probably have your own security systems business or are thinking about starting one. A great many contractors who move from the field into the world of having a business tend to work hard on the field end of their company and forget to pay close attention to business. Some businesses are successful with this approach and earn a comfortable living for the owner—even though the owner manages both the field and the office. If this is what you have chosen, if you like that personal touch of a small business, and if the income works for your family, it can be a satisfying way to make a living.

However, if growth or having more time for yourself are among your goals, start looking at the various parts of your business as separate profit centers. For example, if your firm provides both burglar and fire-alarm systems, as well as complete systems that provide all types of security for a building, you have three obvious profit centers: burglar alarms, fire alarms, and complete systems. During the overhaul of the business, you can break the operation down into how each of the tasks for these separate centers is handled.

Naturally, simplifying each part of a business makes it much easier to examine with clarity. Breaking your business down into profit centers can also offer a simple method for adding thrust to the parts of the business—profits and losses in each field can be examined closely and the reasons for success and failure can be pinpointed. For example, it might be better for your firm to avoid large, all-around contracts and stick with household burglar-alarm systems.

The information can be extremely useful for marketing plans and improve the close-to-lead ratio of the sales performance of your group. Interestingly, the boundaries of your profit centers will begin to emerge as you look closely at your operation. One profit center that is frequently overlooked by contractors is the change order. Some companies see change orders as a nightmare, a situation that they hope to avoid at all costs.

In theory, design drawings and specifications are finished documents that illustrate exactly what is to be built. The design and engineering team executes the prints and specs for the building owner. These plans and specifications are the blueprint for what is to be built; each party submits a cost for their component and the building project is started.

However, anyone with building experience knows that the original plans and specifications are almost always changed during a construction project. Additions to the structure; alterations in the building design by architects, engineers, and owners; and on-site requirements for changes in the design because of unforeseen construction problems can all lead to major changes.

In a perfect world, the building construction would be orchestrated by an on-site project manager who is extremely knowledgeable and diligent, the weather would be dry and comfortable, all of the subs would be team players, and there would be no surprises for the security systems subcontractor when he or she arrives.

But we all know that this ideal situation is not what we find during construction. Changes can be simple, like having to run extra wire, or extreme, like having to upgrade the entire system for the client after the project is drywalled. The change order work might even cause the security contractor to bring in extra subcontractors: carpenters, drywallers, and touch-up painters.

The important shift in thinking is to learn to welcome the changes—think of them as a profit center. Think of ways to expand your ability to execute timely changes for the benefit of your clientele. Advertise this effective part of your service. And while you are on a job, look for and suggest changes that would improve the security of the building for the owner.

The main thing is to learn to think of COs (change orders) as a positive way to add to the bottom line of a job. Think of them as a separate profit center. Be eager to help your customers out with suggestions and any changes they desire. Be sure and sit down with the general contractor's project manager, the owner, any involved subs for important changes and iron out everyone's part in the work. The round-table, teamwork approach will avoid many potential problems.

Overhaul your CO system so that changes can be implemented quickly and executed in a manner that slows the job as little as possible. E-mail and portable printers for laptop computers can make this a very quick operation. The project manager for the general contractor talks with the supervisor for the security contractor. The security supervisor pops up a change order form on the computer, fills it in, sends any needed drawings or cutsheets back and forth to the architect by e-mail, and the cost is agreed upon. A document is printed and signed right there on the tailgate of a pickup, then e-mailed to the security sub's office for integration into the accounting for the job.

Make change orders quick, profitable, and a parallel profit center as a part of your on-going bidding process. Reviewing bid specifications carefully and then applying for requests or addendums through submittals to the principal and design team for alteration or replacement of faulty design criteria by way of change orders is just one example of how COs can be sold. A pile of change orders can add a good deal of money to the bottom line of any estimate; and this money is often earned at terrific profit margins.

Pay close attention to developing the sale of change orders as an ordinary part of doing business and always get a valid signature to back up the request.

Selling Extras

The simpler you make your business in a professional, diligent manner, the more easily and smoothly it will run. To realize the long-term goals you are setting, forecasting the peaks and valleys in cash flow and being able to level out the valleys will probably do more to help you with realizing your future than any other factor involved with your business.

Extras are immediate, parallel business that is already in place and can be boosted with a minimum of effort. The selling of extras can affect your firm's bottom line handsomely at the end of the quarter and they can add to profit in other ways. For example, keeping abreast of system upgrades is an excellent reason to reach out to old clients and make new contact—it then serves both as a marketing tool and can actually lead to some healthy sales.

At first glance, you might think that change orders and extras are the same thing. Naturally, one can look at them both as COs and approach the issue in that manner. We have separated them in an effort to break your business down into its simplest parts and because COs and extras originate from two different sources—even if they do overlap at times.

A change order most often originates from the job itself. In other words, errors in concrete, framing, mechanical layout early in the job are not noticed until the security sub is running wire or even doing finish work. Or perhaps the building owner discovers that some design element is not going to function properly for their company when the building is completed. The project manager talks with the various subcontractors who will be involved and requests change orders.

At this point, you might offer some extras that are beyond the scope of the change order and the building owner might decide to incorporate them, but the impetus for the COs came from the building project and the proposal for extras originated with you. The purpose here is not to pick nits, but to break your business down into components and streamline them for improved revenues.

Change orders require a rapid response and skilled execution by the technicians, but they might only require research and staying up with the market on a limited a basis. Also, change orders are often limited by the nature of the change and the desire to save money, avoiding cost overruns.

But selling extras is a wide-open platform for business. Your jobs are like temporary marketplaces. If you come to the bidding table with such an important part of the building as the security systems are for the typical owner, you already have an excellent playing field. One tendency of subs is to simply apply what the system designer calls for on the plans and in the specification documents to the bid, shop costs, and not pay any more attention to the system.

This can be a healthy approach if you use it for the first part of the bid on common jobs—it can weed out competition. However, once you see that you are talking real world with the owner, you can suggest extras. This approach can save money on the expensive problem of bidding and losing the jobs.

But on larger jobs and ordinary jobs that you have locked in, it is wise to have a thorough knowledge of all of the latest security system techniques and products. With the Internet at your fingertips, research is very simple and it can be conducted on a global basis—price, the most cutting edge of products, and new capabilities of systems can all be a powerful sales tool for your firm. The well-designed extras program can add a significant amount of pure profit to your firm's margins every year.

Related Businesses

Warranties and maintenance contracts as additional revenue

As all security systems professionals know, the development of electronics is moving at a fantastic rate. Though our lives are supposed to have been simplified by progress in these complex times, security systems professionals must often cover a lot of bases. For example, when relating to construction defects and workers comp, the businessperson must be a security systems professional, and at the same time, have a good deal of legal savvy. When parts of a system fail and claims on warranties arise, the security professional can be caught between all of the parties: the general contractor, the system designer, the manufacturer, and the customer. Manufacturers' product warranties give the customer rights and protection against failure of the product for a designated time period and limit the warranty to a set of prescribed conditions. But clients most often go to the security system subcontractor to pursue warranty problems.

When the conditions in a warranty describe actions of maintenance and care of the product that must be carried out by the owner, failure to do so typically sets up escapes out of the warranty for the manufacturer who has recourse for the denial of any responsibility and can void the warranty entirely. Manufacturer's conditions very greatly; they can include release clauses related to improper use of the product, lack of maintenance and care by the owner, and even damage by third parties—the conditions can even list grounds for dismissal of the entire warranty.

The contract documents also have conditions and the subcontractors can easily find themselves the recipients of a cross complaint and embroiled in the middle of a legal battle. When a system failure occurs, the owner pursues the developer or the general contractor to honor the warranty. The developer or the general contractor pulls all parties, including the manufacturer, into the case. In the litigious climate of the current times, it is imperative that the security systems contractor reviews warranty conditions carefully. It is wise to use a construction lawyer to review important warranties (those on expensive products) the first time they are used. The lawyer can help the sub understand how to treat the product at each job and to provide clauses in contracts that relate to the equipment. The lawyer will be able to see unique conditions that can cause responsibility for vendor's products to be transferred partially or entirely to the shoulders of the systems sub.

Commercial law is complex, but it's best to understand some basic rules. For example, every piece of equipment that is installed in your projects must fit under the umbrella of implied warranty as a part of the vendor offering the described performance of the equipment to the public. In other words, it should do basically what the manufacturer says it will do. If the item is a switch, it must be able to interrupt the flow of electricity in the manner in which that type of switch is supposed to provide switching. Any use of the switch for tasks other than those suggested by the manufacturer can relieve the manufacturer of responsibility. Typically, stock phrases in the warranty documents address this situation. Disclaimers are legal and can stand up in legal actions. This is an important reason for having a lawyer review warranties on expensive equipment that the systems sub uses frequently.

When you bid a job, be sure that you know what is warranted and what is not. Do not make any representations to the owner that items or situations are covered by the warranty that are not under its umbrella. Subcontractors should make it standard practice to have the lawyer prepare a document that all customers sign, which makes it clear that the client understands the warranty situation. Any defining document like a warranty understanding can be very helpful if lawsuits are filed at a later date. Many clauses need attention. For instance, the manufacturer often limits repair costs (the totaled sum of any repairs over the lifetime of the warranty) to a sum that does not amount to more than the cost of the product.

All of the laws, contract clauses, and warranties are very important. As you revamp your business, it is important to put all of these legal matters and their fine-tuning on track. Litigation can devastate a contractor's net assets many years after the money from the contract that triggers the legal action has been completed. This requirement can seem like an unfair burden for the subcontractor, but rather than treat it as a problem, it is much better to assume a can-do attitude and turn the situation around: make warranties a positive cash flow, rather than a pain in the neck.

As all security subcontractors know, systems are very often neglected: maintenance is not regular and well executed. Lack of diligent maintenance can cause components to degrade, systems to fail, and the damage from faulty maintenance can be overlooked and difficult to prove in construction lawsuits. However, lack of maintenance and warranty complexities can be addressed proactively and turned around to add to the company's cash flow.

Maintenance Contracts as a Profit Center

The second part of this book moves out to the field and covers the many aspects of job-site operations. Chapter 9 reviews the subject of maintaining security systems in some detail.

Maintenance is an area of construction that requires attention. Anyone familiar with the way buildings are built, the way they work, and the effects of time, can tour the average structure and find many areas where the components of a building need attention.

Though security systems are not as obvious as stucco cracks or the buildup of earth at foundations, they are certainly no exception to the need for maintenance. Whether the task is as simple as bringing up the electrolyte level or cleaning the terminals on a battery, or as major as finding that the resistance is continually going down in wire insulation and the wiring must be replaced, security systems are in continual need of maintenance.

On-going inspection and service on a quarterly basis are required for the long life of systems. A diligent service routine can result in a big savings for building owners over the life of the system. And that is just the immediate cost of a security system. If the security of the building, including life-threatening events, such as armed intrusion or fire dangers, are factored in, the need for professional, diligent maintenance is a given, not an extra.

An important factor of the maintenance equation is that the costs of installations will continue to rise. If the owner is not engaged in a regular maintenance program by a diligent and professional firm, it is most probably because they are not aware of all of the factors that are addressed in Chapter 9. Often, the responsible parties are simply not aware of the need for quality maintenance.

Many assume that the warranty has them covered, and therefore assume that there is no need to provide maintenance.

The security systems contractor needs to make addressing the building owner's need for knowledge of the importance of maintenance an on-going part of the installation and upgrading of systems. After the system is installed, there must be a meeting with the principals for instruction in the use of the equipment. As a part of that meeting, all warranties should be explained with care and the security systems contractor should have a written contract prepared for on-going maintenance.

All of the trades involved with the construction of our building stock should implement this practice: mechanical, roofs, siding, doors and windows, all of the parts of the building should be inspected regularly. Warranties should be contingent on these measures. Maintenance required by the owner's team should be clearly explained and required for a standing warranty agreement.

Selling warranty and maintenance contracts

At first glance, this might sound like an extra burden for the subcontractor, but that is not a proactive approach to the situation. The need for maintenance presents a win-win situation for your company and the building owner's asset. As seen, all of the trades have a limited responsibility to guarantee that they have installed what they said they would install. At the same time, the security systems subcontractor has a chance to offer a valuable service for their clients. This can be an excellent source of business parallel to the main business of contracting to install new or retrofit systems.

You will have to think this through and determine what your firm can offer and what your particular clientele needs. The idea is that you must provide a limited warranty, however, your clients can purchase a more extensive warranty if they prefer.

This might seem a bit risky at first, but the extended warranty package should only be available to those customers who choose to sign up for your company's maintenance service contract. This service product will also have to be designed specifically to fit your company's customers and the types of systems you install and work with. It would be wise to develop the warranty and maintenance business from what you know, although it can also be marketed to clients for whom you did not install their systems.

The basic idea, however, is that your firm will visit the building and perform a visual survey and use test equipment where required on a quarterly basis. The person from your firm who performs the inspection will provide the client with a signed inspection sheet and an estimate for service and repairs.

This proactive approach benefits all parties. Let's face it, the average building owner is simply not aware of the urgency for inspection and maintenance. If a security system is to be kept functioning at its maximum efficiency, they must be informed by their security systems contractor.

Develop a marketing brochure that includes what needs to be done, what will happen if it is not done, and photos that illustrate severe situations that arise from the lack of on-going maintenance. Security systems are an excellent part of construction for providing this service; they are very important to the owner and complex enough that the owner will be concerned that the technicians who provide service are knowledgeable. Be sure to make the owner fully aware of how difficult it is for professional building maintenance firms to hire people who are both knowledgeable and diligent about the intricacies of security systems.

The basic, proactive approach is that your customers might purchase an extended warranty from your company, but only if they sign up for the maintenance service. This closes two sales and the sale of any future work the security systems might need before they require replacement.

The arrangement is excellent; you have developed an ancillary cash flow and you can make a real-world commitment to the promises in warranties because you know that your own technicians have provided a sound installation and are watching it closely.

Upgrading systems can be a strong profit center

When going through the process of bidding on the security contract for a new structure, closing the contract very often depends on a low estimate. However, in retrofit work, this is not always the case. The owners are concerned about security and safety and they want to be able to depend on the provider. The business owner wants to be able to enjoy the benefits of a trouble-free, long-life installation. This can be a real boost for your company as you overhaul the way it works.

When putting in security systems in existing businesses, expanding what is in place or updating an old system, many business people will go to their friends for suggestions when they choose their contractor. As a result, it is essential for the security systems contractor to establish strong customer relations in their locale. Naturally, this need fits perfectly with all of the ideas related to word-of-mouth marketing.

Security systems have become a basic commodity in the business world and the systems installed today will soon be in need of upgrading. Systems wear out and technology brings forth new capabilities on a regular basis. Upgrading is a real opportunity for the security professional to develop an on-going income that is a valuable source for leveling the business. The most important aspect of this continual cash flow is keeping a solid group of technicians serving the clients in the field and a simple, well-managed sales effort that is enhanced by word-of-mouth marketing.

Figure 4.1 is designed to help the principals of your firm begin to design a retrofitting profit center. The table is on the CD and can be printed out for all of your staff to edit and add their contributions. It can also be copied to a file on the hard disk for reworking.

✓	Item	Notes
	Retrofitting Checklist	
	Which security products are best suited to our typical customer's needs?	
	Which products offer the best performance?	
	We have discussed the needs of our various types of clients carefully.	
	What types of structures can we most efficiently service?	
	What extra low voltage services do our retrofit customers most often need?	
	How can our customers become an integral part of marketing?	
	What types of extras can we sell with our systems?	
	Sell an extended warranty for mutual protection.	
	Are we prepared to provide a maintenance plan for retrofits?	

Figure 4.1   Sample retrofitting checklist.

Security Systems Consulting

Within the last 20 years, construction defect lawsuits have surfaced as a tidal wave, affecting all parts of the AEC (Architecture, Engineering, Contracting) industry. Many security systems subcontractors have been forced to deal with this immense web of complex and drawn-out civil actions. They are worth millions upon millions of dollars each year. A major part of that money is not for settlement of the legal actions, but for the discovery that leads to the settlements. Much of the money is paid out in legal fees for the law firms themselves. But another hefty amount is paid out for the experts who work for the defense and the plaintiff attorneys and the insurance carriers to determine what is really defective with the work product at the site.

If your firm has been sued directly or in a cross complaint, you are well aware of what a long, drawn-out process CD litigation can be. The business chapters have been designed to suggest that contractors develop a proactive approach for running a security systems business; and it is important to apply the idea to potential lawsuits. The technique discussed in the maintenance contract section of this chapter, for example, basing warranties on a maintenance contract, can be a real help in moving your contracting business away from the nightmare of drawn-out litigation and the revenue that can be lost in the process.

However, in order for the ideas to work well, they must be applied diligently and consistently. They must become as much a part of the daily activity of your firm as estimating or shopping vendors for good prices on equipment.

A great many liability issues are related to security: the typical worker's compensation liability; simple construction defect situations, like water intrusion at windows from faulty drilling for cable runs; much more intense consequential damage to property; and even loss of life. The responsibility is hefty but the intelligent contractor can harness the construction litigation industry by becoming a security systems expert. This can provide a healthy ancillary source of cash flow.

Typically in construction defect lawsuits, the plaintiff has run into problems with their structure. The complaints most often begin with the building owners finding water leaking into the structures. The owner or a group of owners retain a lawyer. The plaintiff lawyer hires an expert team and if security systems are a part of the structures, a party who is well versed in security systems is hired as the expert. The plaintiff's attorneys typically file suit against the developer or general contractor, who hire lawyers, who hire more experts. The developer sues the subcontractors in cross complaint and their lawyers hire more experts and file notice with the insurance carriers who might hire more lawyers and experts.

This is a very brief rundown of this situation. Naturally, the circumstances can vary greatly, but it is obvious that CD lawsuits require lots of hours from the experts. The time billing for this work is typically set at a substantial hourly rate for the security systems expert and there can be a good deal of it available.

CD work is only one part of working as a security systems consultant. Many other activities can be performed by your firm: inspections of systems for people who are buying buildings with the security systems already in place; review and design of security and other low-voltage systems for the architects and engineers; expert witnesses in discovery and at trial for construction defect litigation, as well as fire and other damage discovery for insurance companies.

The consulting revenues will become a part of your general revenue. At the same time, you will reap an excellent side benefit from this work—you will understand what your security systems contracting company should avoid during installations. Another advantage to consultancy work is that you will be exposed to a great many people who are related to construction projects and buildings: realtors, architects, building owners, and a host of others. Being a trusted professional is one of the prominent components of word-of-mouth marketing. It doesn't happen overnight, but through the years, being known as an expert in your field can create many long-term relationships that are financially and socially rewarding.

How consulting works

Typically, you will be hired to investigate the quality and effectiveness of security systems that are already installed. Most often, a couple of approaches will be required: a visual walkthrough of the structure for a survey of the existing system with some minor use of test equipment—or a thorough survey of all of the equipment and the wiring. The thorough study of the system might require intrusive investigation; you might have to cut areas of the walls open to understand what is actually going on behind the surface.

The two types of inspections are typically known as nondestructive (nonintrusive) and destructive (intrusive) site inspections or investigations. As with all parts of the AEC industry, terminology varies with different people. The big difference is that walking the structure and looking at the system and doing some minor work with test equipment is less time consuming than a thorough inspection that requires cutting into walls. Naturally, the cost for intrusive investigations is a good bit higher than for nonintrusive work.

Unless you are hired by a building owner to look at a system, a good deal of your work might be related to lawsuits. Most often the plaintiff team does a simple investigation of the building after the owners first retain their lawyer. The plaintiff's group will do a cursory investigation of the site and create a first list of some of the construction defects that are spotted. They will make this the Preliminary Defect List that is often filed with the lawsuit. After all of the parties that have been sued have had time to review the preliminary defect list, they will typically perform their own walkthrough of the site. By this time, the plaintiff's list of defects might have grown. After the defense team surveys the complaints of the plaintiff experts, they will most often attempt to work through an understanding with the plaintiff experts, laying out what parts of the building need to be cut open in order to understand the real nature of the alleged defects. The defense team may then walk the site for a round of nondestructive testing and both parties will meet to discuss their findings. At that time, they often agree on what parts of the buildings need to be cut open in order to understand what has actually caused the alleged defects.

After some sort of an agreement is reached relating to what is needed in the way of intrusive work, the parties typically schedule the first round of intrusive investigations. When the intrusive work is underway, the work product of the security systems contractor who did the original installation will be examined very carefully: for example, any parts of window jambs (where it is alleged that the security systems contractor has drilled holes that have allowed water to leak into the building) will be cut open to see exactly what work product is allowing water to intrude. The security system investigation will involve whatever is needed to understand the defects. For example, the experts might also use test equipment and even open walls to examine the condition of cable runs.

A simple rundown of the consulting field illustrates how much work each lawsuit involves. And, as mentioned, lawsuits are only one part of consulting. A great deal of business is available, but it might not be the type of work you enjoy. You might prefer the business of installing systems, rather than analyzing them. Naturally, each firm has to make their own personal decisions about what work they want to pursue. But if your firm is interested, a good bit of consulting work is available and the industry will continue to grow as a profit center.

Simple Testing of Systems

Fire alarms, burglar alarms, and computer networks: low-voltage wiring has become the central nervous system of our buildings. As we move further into the 21st century, our structures will become more and more dependent on complex low-voltage wiring. With the energy crunch and the need to protect our environment becoming more obvious, pressing us to take action for financial reasons, buildings will become more automated. Light switches will be controlled for conservation, heating zones will be regulated with specific care, windows and shingles will provide photovoltaic power for quiet circuits: the list is long and it will grow.

The spread of all of these automated processes will produce a tremendous growth in the need for technicians capable of fast, accurate troubleshooting. All of the many devices that will enter the marketplace—equipment as diverse as air-filtration systems, electrical-use monitoring, and zone air-handling products will all demand attention to keep them in good working order.

There will be a widespread need for a new type of low-voltage specialist, capable of offering the building owner a one-stop service. Rather than searching the Yellow Pages and trying to find problem solvers for all of the different types of equipment (a burglar alarm troubleshooter, a photovoltaics repair group, and someone to fix an automation panel), the owner will want to call a group that is equipped to run down any of the many problems that could arrive during a busy work day and fix it.

This new breed of in-and-out specialist will not be a jack-of-all-trades, but a true low-voltage problem solver. This is not the same service as the consultants mentioned previously. This work will typically involve the use of the new digital test equipment for nondestructive testing. The field crew will have to have an understanding of a number of the systems in a building. They will take care of rapid fixes, but they will not typically be the same team that does retrofitting or other major work. They will be dispatched and arrive in the building quickly, capable of running fast tests and making simple repairs. They could be a part of your firm's maintenance team.

This is an excellent time to begin such a service. The advertising is fairly straightforward (the Net and the Yellow Pages, for example). These simple advertising methods can easily be augmented with speaking at business organizations, updating groups on what's out there in the marketplace, how it works, and can be implemented by the building or business owner.

With attention and a diligent approach to the services offered the customer, this type of a service can be an excellent source of receivables that have no aging attached to them—the customer can pay the technician on the spot. This type of revenue is an excellent source of cash used for leveling the business—it is not a seasonal operation and not at the mercy of the ups and downs of the construction industry.

Selling Security and Other Types of Accessories

Here, we are referring to the many items that can be added to systems and enhance the performance of them. Many extras are simply never brought to the attention of people in other businesses. That the addition of motion detectors to a system, which the security systems sub is bidding on or servicing, could add some badly needed protection to a retailer, might have completely slipped by the system designer and never have been known by the business owner.

Obviously selling accessories can be a real asset for the customer. Sales of security accessories can be included in the work at hand or at any time during the performance of the contract. The enhancement of the system can add to the security and safety of the owner, making their business more secure from shrinkage, damages, and the dangers of fire. Obviously, this is a much-needed service.

As municipalities, governments, and organizations rewrite and change requirements, and codes, new equipment is mandated for our built structures. With the Internet in place, the wise contractor will use staff members who enjoy research and learning to keep abreast of the latest in laws, requirements and products. Being at the cutting edge of information will become increasingly important in determining who gets the jobs and who has bid similarly, but can't offer the benefit of knowledge.

Staying abreast of the industry need not be considered a burden; in fact, it should be acknowledged as a part of doing business. Sending employees to trade shows and learning seminars is one of the most simple ways to encourage their natural need to feel involved in the business. It can be an invaluable aid in word-of-mouth marketing. When building owners, architects, and general contractors discuss a project with technicians who are abreast of the latest information related to low voltage, they will be impressed. If the person is simply putting in hours and doesn't have any real overview or natural enthusiasm for the work, the security systems company immediately loses some valuable word-of-mouth marketing.

Accessory sales can add hundreds, even thousands, of dollars to the net profit of a job. Typically, the labor-intensive work is a part of the bid price and the accessories will involve very little extra labor once a job has been mobilized. The nature of the accessories can vary greatly; they can be simple mechanical devices like fusible-link smoke hatches or complex electronic devices like enhanced computer-operated central controllers.

For the savvy security systems contractor, there are numerous benefits to investigating and promoting accessories sales. Obviously, the first benefit is the ability to increase the revenue from jobs. A less-obvious possibility for a company that is diligent and capable of tight organization is the ability to meet quantity discounts from vendors, warehouse items that are needed continually by the contractor, and to use the computer effectively so that only those items that are really needed are warehoused. The others are purchased through vendors who can provide the latest on-demand service.

Decisions should be made with the help of all of the staff who are involved with the particular operation. For example, it might be advantageous to stock trucks with certain accessories so that the client can actually see the item in order to close the sale—or it might be better to keep the stock sample in a warehouse to take out to the job when needed. The more care that is put into these details, the more successful the operation will be.

As your firm grows and you begin more parallel businesses, the more accessories you will be exposed to, and the more you can enjoy profits by selling to your clientele. For example, if you decide to learn more about photovoltaics and other energy-efficient products, you will be given many opportunities to introduce them to clients along with your security business. Many times, the photovoltaic system can offer an excellent, additional standby electrical supply for such delicate equipment as computer-driven control boxes, automation boxes, and actual computer workstations.

As we move into the 21st century, all of the low-voltage products that are now available will be enhanced and many new items will appear. It will be an excellent market, ripe with many opportunities for the savvy security systems contractor.

Computer systems

As you expand and explore other low-voltage systems as potential profit centers, any number of ideas will undoubtedly present themselves to you and your staff. It is important that you take your time jumping into them. Size each possibility up for its potential net profits. There must be interest and enthusiasm among the field technicians. If they have a natural interest in the subject, the learning curve will be much simpler.

Start with endeavors that are relatively simple, that parallel your work on security systems. In this day and age, working with CPUs (Central Processing Units)—the name for main computer chips—is a growing part of the security industry. Some type of a CPU runs the vast majority of installations. Running cable to tie CPUs together is also a part of large security installations.

Because networking of CPUs for computer workstations and file servers is very similar to tying any small computing devices together, it is not a far stretch to incorporate them into your field work.

There might be a bit of adjustment with suppliers and the need to purchase minor tools, but basically the operation involves running wire through a building. The architects might use a consulting engineer or the building owner might use a computer contractor to take care of the installation. There might not be a schematic of where all of the workstations and the CPUs will be located. But the basic process is very similar to running wire for the security system.

Your firm will bid and then the contract will state that you will provide the cables and connectors and run them from the location of the file server to all of the workstations and the locations for peripheral devices like scanners, printers, plotters, and any other devices that will be a part of the computer network.

You can work through these simple installations with a lead technician on your staff that has a strong interest in learning the process. Do the first couple of installations with that person and then give him or her full charge. The same person or team can also take charge of all computer runs for security systems. If you continue to grow this business and do retrofits at sites where no security work is going to occur, you might decide to develop a team that specializes in small-computer networking.

Computer rooms

Expansion and exploration of the marketplace for parallel businesses might lead you to investigate more complex undertakings that have potential for higher revenues as profit centers. Again, it is crucial that you pay close attention to what you intend to do. Remember the old saying: "be careful what you wish for." One of the most painful things a contractor ever experiences is landing a contract that is full of potential, laden with a good deal of possible profits and then, because the contractor is not financially strong enough or does not have the staff to execute it properly, starts losing money.

The reference here to computer rooms is to large, complex computer installations. Obviously, it is very important to size up this situation with caution and take your time jumping into the field. As with any part of your business, employee interest and enthusiasm will be a big help in keeping your firm vital and abreast of the competition.

You are probably not yet familiar with many parts of the heavy-duty computer room. However, a number of the components might already be a part of your firm's day-to-day work product. Running data and voice cables is one of the biggest parts of major computer room installation and connection with the rest of the building. Taking care when fishing cables through a structure is already a skill that is well developed for the security systems technician. The scope of this work varies greatly—from simple drop add-ons to total, full service installations with many drops. You might decide to hire a person who is skilled at these installations to handle the field.

Your company can work on many types of installations: fiber optics, video, and cable TV, all combined with security work. The computer work can start with small drops or runs that your firm has not done before. From there, you might want to explore other components of major computer work (such as power conditioning).

Cleaning up power from the vendor is a part of security work and is simply applied on a larger scale with major computer work. The basic product is the same: clean, reliable electricity. A thorough knowledge of products used to cleanup and guarantee power can never hurt a security systems contractor. And you might wish to become a distributor for these are accessories, which will make supplying them even more profitable. Uninterruptible power supplies (UPS), power distribution systems (PDS), motor generators, frequency converters: your firm is probably already informed about some of these products. The services might be of a great deal of interest to your group and you might decide to add more to your list: power-quality analysis, electrical installation, and surge protection, for example. You might even want to learn to work with a sub and supply raised floor systems. At times, they can be very handy for security work. Installations range from huge to small computer and clean rooms. As building technology advances, systems will continually be integrated.

Low-voltage systems

The NEC doesn't really put a definition on low voltage and limited energy, and the coverage in later chapters is limited. Fiber optics use light for transmissions, rather than bouncing electrons, so they don't fit in the code's voltage equation. These situations have arisen because of the rapid changes in technology—and the compounding of changes has only begun. This century will see changes that we haven't yet anticipated.

For now, the contractor will have to learn on his or her own and work with the code as well as it can be done. Sound systems, window lighting, auxiliary lights, audio equipment, security, automation, low-voltage lighting: the names of the various types of circuits refer to the tasks they handle and the list will grow as technology continues to advance.

When thinking about moving into other parts of low-voltage work, there are so many directions: sound studios, sound broadcast, voice activated. The list is long and will continue to get longer. And the types of clients are growing as technology expands in their fields (hospitals, restaurants, factories, hotels, movie and TV buildings, temporary setups, office buildings, etc.).

Security contractors already work with the parts of fire-alarm systems: elevators, sprinklers, voice communications, etc. and moving on over to fiberoptic cabling is a simple leap. Without a voltage load, they are not typically a problem when run in the vicinity of power cables.

In article 780, the NEC reviews the flat cables that the vendors have devised for automating homes. The product is clever, allowing the installer to run coax and electrical wiring in the same flat cable. Even though this technology is currently slowed by financial constraints, it is very likely a pioneering wiring system. Knowing how to work with automation of buildings at the cutting edge will become more and more important.

The security contractor is already aware that parts of the communications work like phone wiring; moving further into this field is fairly straightforward. Chapter 8 in the NEC works with the beginnings of this industry, which will grow by leaps and bounds.

All of the various components of buildings will become more integrated, as with the flat cable attempts in home automation. What standards will take over the market will evolve from the current applications, such as CEBus using the conductors to send signals to automation devices.

With multimedia and all of the things that are transmitted going digital, like huge sound files, there will be more need for broadband signal transfer. For example, a building can run its video and audio through cables. The code has already begun to approach this in Chapter 8.

Obviously, there is plenty to keep abreast of, and the information seems to expand overnight. This can seem overwelming, but it is important to avoid an attitude or overload and view new low-voltage concepts as potential profit centers.

Building automation

The term *building automation* is typically used to describe the control of various systems in commercial and industrial structures. The control of lights, security systems, and HVAC are typical automated systems, but all components of a building could be automated to one level or another with existing technology.

Typically, a computer runs the control devices all over the structure and does very thorough database tracking of when all activities transpire. It can often compare these mechanical activities with accounting facts and on-screen, color graphic displays of the data, as well as printed graphs and spreadsheets.

The information can be tracked over long cycles and include information related to changing settings and devices in the system. Needless to say, this is invaluable information for facilities managers, product designers, engineers, architects, and most anyone who works in a field related to buildings.

The number of uses for a controlling CPU is virtually endless. A common use is monitoring indoor air quality and upgrading the air quality automatically when it is off. This service has a notable effect on curtailing employee absence from illness and the cost savings can be estimated by relating the cost of absentee employees to well employees.

The systems can store maintenance schedules and even signal the maintenance team at a distant site. This is an excellent situation for the maintenance group at your company: sales are closed by computer. The detail can be extra-ordinary—even flagging when items should be replaced.

The commercial and industrial building automation systems have been around for several decades. They provide management from the central controller: air conditioning energy savings, other green building monitoring, light control, security, fire, telecom, office automation, prevention-directed maintenance and operation of machinery—the direction that all buildings will move toward.

The daily management of a building is the work of the CPU, cutting staff and giving the various operations time and knowledge for effective management. Data is stored and sorted, operations are smoothed and automated, adjustments are simplified—reports are automated. The labor, fuel, and repair savings are substantial: 20 to 50% of costs generated by older methods.

The use in commercial structures is older than in homes, but the automated house is becoming more common. The principle is the same as with an industrial or commercial building: a CPU drives the different operations of the home. The homeowner can wake up to brewing coffee or set the temperature of the house to be adjusted to a comfortable range before returning from work or a vacation.

This equipment will be included in more security systems—the CPU will drive all of a home's systems. Building automation is not a thing of the future; it is with us today, commercially ready and the number of installations will continue to grow.

Photovoltaic systems

Currently, sustainable energy supplies are most often used for equipment in isolated areas. This is because the cost of building fossil fuel and atomic power grids is immense. In areas like the United States, where a large power infrastructure has been built, the cost of photovoltaics is not always immediately effective in relationship to available power. The payback can take a decade or so. But this will change rapidly as utility prices rise and the cost of photovoltaic equipment drops.

Photovoltaic cells were first developed for the space program. In the United States, they tend to be used in places where a hookup with the traditional power grid would be expensive, but they are a common sight if one takes a close look. For example, photovoltaic panels often drive courtesy phones beside the freeway. Another common example is the lighting in small, independent structures like the shelter for bus stops. If the costs of a power drop obviously outweigh the price of the solar panels, they are very often the choice. This is true in many instances, such as agricultural pumps in isolated areas.

Once installed, they are very reliable and require little maintenance. And there is no fuel cost related to their operation.

Photovoltaic panels create a flow of electrons from the energy of light; the theory is not new to the security systems contractor. The name comes from the combination of photo that is an offshoot of the Greek, *phos,* meaning light. And volt comes from Allesandro Volta's name; he was one of the early scientists involved with electricity in the early 19th century. In simple terms then, the word means "electricity from light."

This is a simple, excellent source of power. We already use them abundantly in calculators, watches, and in equipment located in areas where grid hookups are expensive. As the cost of new generation plants, fuels, and power plant waste disposal increase, photovoltaics will become more important. Once they are installed, they are an amazing power supply.

New PV products are continually developing. The deep research is in defense and the space programs all around the world, so it is extremely well funded. They are very valuable sources of energy for military requirements, so the development, refining, and breakthroughs will continue rapidly.

Implementation into the private sector is always slower than military and the space program, but major private installations already have been running for several decades. As factories develop in Asia, the prices for PV equipment will continue to drop. As the Chinese implement production, there could be a drastic drop in the price of installation products.

We already have the real estate in place for their installation: tops of commercial buildings are an invaluable resource. The tops of homes, windows, and curtain wall are also excellent areas for the installation of photovoltaic panels.

These spaces are already being utilized. Arrays of PV panels can be seen on the tops of commercial buildings; they are an ideal source of quiet electricity for low-voltage systems. Solar shingles are already being installed on homes. They are a good-looking product and the owner can send power back to the grid, rather than storing it with battery arrays. Electricity-producing windows are being installed in commercial buildings, where there is a great deal of surface area.

Typically, moving into PV installations is straightforward for the security systems contractor. Many companies have already used panels for an installation. Photovoltaics offer some real advantages over batteries, generators, and even the typical utility suppliers. They are very reliable, need no fuel, are clean and environmentally friendly, and have low construction costs.

This is an excellent and growing market for a business that is parallel and can overlap security installations. Because they are so reliable and require so little maintenance, they power all of the satellites that are in orbit. They are most often placed near the equipment they drive so that the cabling is simple and the need for transformers is eliminated. They are quick to install, have low labor overhead, and typically require a lot less permit paperwork than many sources of energy.

Watching the marketplace

As electronic systems and the computer chips and software that drive devices become more sophisticated, the various systems in a building will become more integrated. Thus, in time, computers will become a part of the building, just as telephone equipment and cable television equipment very often stay with the building when new tenants take over. In time, the Central Processing Unit for the building will be several powerful chips in array that will do anything required in the building. The CPU will be mounted in a safety closet and will control all of the systems that take care of and secure the building, as well as all data-processing and communications devices. Simple plug-in devices, as easy to use as phone jacks, will be located anywhere in the structure.

We are already moving in that direction and the progress this century will be faster than it was last century because computers are used for engineering and so many of the functions that bring products to market.

As fast as one company disappears, another takes its place. Companies have shown up to further the automation of homes. The advent of home entertainment systems has provided a major thrust for companies to provide this equipment for the homeowner. The other part of the equation for the new-market thrust of home systems is the home office. The idea of the vendors is that between the ever-expanding home entertainment industry and the growing need for people who work at home, the various components will be integrated home networking and entertainment centers, making it smoother to work at home and to enjoy the huge selection of entertainment devices.

The integration of the local area network within the home and the entertainment hardware will again be based around integrated cables that are run throughout the structure. This is a very straightforward approach to cable and it will eventually become the standard. Like the early days of personal computers, the market will stabilize and manufacturers will zero in on products as prices come down.

The home office will have simple jacks throughout the structure. All of the available I/O (input/output) devices (computers, digitizers, and printers) will be used at a central area or at spread-out workstations, and they will be simply shared, thus avoiding major expenses throughout the home.

This same ability to create workstations and device stations for the data-oriented equipment will, in turn, be available for the distribution of the media devices. Video and audio equipment can be used wherever the person desires. So the computers, stereos, cable, and home-automation devices will be very versatile for their usage. The stations will be moveable without having to rewire the home and it will be simple to move them. This will be a real revolution for the homeowner.

Working with the various power and transfer systems has been a real hindrance in homes. The owner can now have a good deal of control over the use of the home that wasn't available before. Changes in equipment can be welcomed and adjusted to readily.

The security system and the home controls will be run from a central panel, along with everything else. This wiring could be designed to serve the owner for many years and still have the capability of adjusting to each change in technology. When children grow up and leave home, their rooms can readily become offices and other rooms can easily be reused for entertainment.

The future is ripe with opportunity for the security systems contractor. The wise contractor can look at a long-term game plan and use all of this growth and change as an opportunity to realize those dreams.

2

The Field

5

Basic Security System Considerations

The use of security/fire-alarm systems is not only advisable, but is also mandated in most populated towns and cities by national codes and local ordinances. This chapter introduces the basic types of security-system equipment and components.

Signaling Systems

Signaling techniques are not new. Methods were devised more than 5000 years ago to signal individuals and tribes of danger and of on-coming strangers: Indians used smoke signals to communicate with each other; other tribes used drums and animal horns; bells were used extensively during the settling of the United States to announce meetings and warn of fires and other dangers, and, of course, military troops have used flags and horns to communicate for thousands of years.

When electricity was put to practical use around the latter part of the 19th century, methods were devised to use electrical buzzers and bells (such as doorbells, entrance detectors, and manually operated fire-alarm signals) for signaling devices. However, at that time, electrical and electronic devices were usually limited to certain specialized structures such as banks and school buildings.

Today, all apartment buildings and town houses in almost every section of the United States must have an adequate number of smoke detectors installed to warn occupants of fire. Such buildings as nursing homes, schools, hospitals, and hotels are required to have an approved fire-alarm system installed, as well as fire sprinkler systems. The latter is usually designed to operate in conjunction with the fire-alarm system. Banks and similar institutions would not think of using a building without adequate security systems installed on the premises.

The list of applications is endless, and great opportunities await the trained security technician. To verify this, look in the Yellow Pages of any city phone directory under Alarm Systems, Burglar Alarm Systems, or Fire-Alarm Systems and note the number of businesses listed. Each firm probably has a dozen or more employees, and is eager to hire several more trained technicians. And, employees leave companies to start their own companies regularly.

Classification of Signal Circuits

A signal circuit used for a security or fire-alarm system can be classified as open circuit or closed circuit. An *open circuit* is one in which current flows only when a signal is being sent. A *closed circuit* is one in which current flows continuously, except when the circuit is opened to allow a signal to be sent.

All security systems have three functions in common:

- Detection
- Control
- Annunciation (or alarm) signaling

Many systems incorporate switches or relays that operate when entry, movement, pressure, infrared-beam interruption, or other intrusions occur. The control senses the operation of the detector with a relay and produces an output that can operate a bell, a siren, a silent alarm such as telephone dialers to law enforcement agencies, or any other signals. The controls frequently contain ON/OFF switches, test meters, time delays, power supplies, standby batteries, and terminals to connect the system together. The control output usually provides power on alarms to operate signaling devices or switch contacts for silent alarms (Fig. 5.1).

An example of a basic closed-circuit security system is shown in Fig. 5.2. The detection (or input) subdivision in this drawing shows exit/entry door or window contacts. However, the detectors could just as well be smoke or heat detectors, switch mats, or ultrasonic detectors.

The control subdivision for the system in Fig. 5.2 consists of switches, relays, a power supply, a reset button, and related wiring. The power supply shown is a 6-V nickel-cadmium battery that is kept charged by a plug-in transformer unit. Terminals are provided on the battery housing to accept

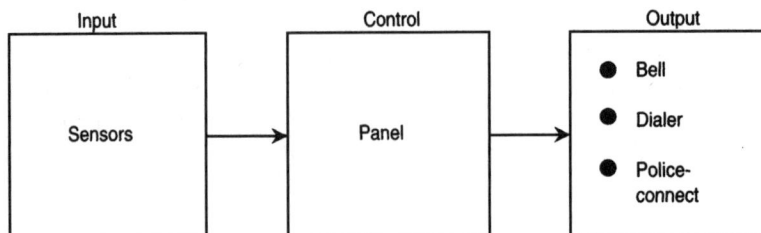

Figure 5.1 Basic subdivisions of an alarm system.

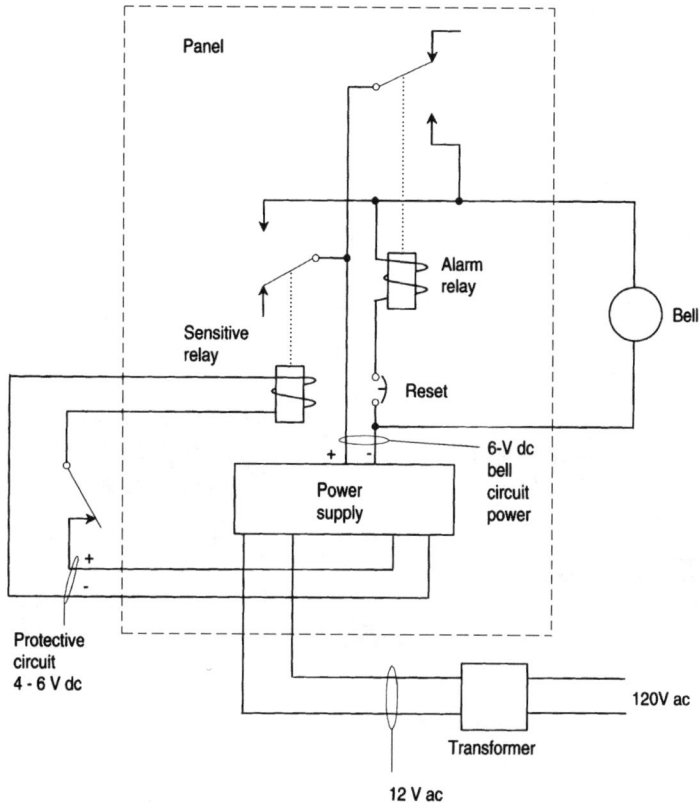

Figure 5.2 Basic closed-circuit security alarm system.

12-V ac charging power from the plug-in transformer, which provides 4 to 6 V for the detection (protective) circuit and power to operate the alarm or output subdivision.

Figure 5.3 shows another closed-circuit system. The protective circuit consists of a dc energy source, any number of normally closed intrusion-detection contacts (wired in series), a sensitive relay (R_1), and interconnecting wiring. In operation, the normally closed intrusion contacts are connected to the coil of the sensitive relay. This keeps the relay energized, holding its normally closed contacts open against spring pressure, the all-clear condition of the protective circuit. The opening of any intrusion contact breaks the circuit, which deenergizes the sensitive relay and allows spring force to close the relay contacts. This action initiates the alarm.

The key-operated switch shown in the circuit in Fig. 5.3 is provided to open the protective circuit for test purposes. A meter (M) is activated when the switch is set to CIRCUIT TEST. The meter gives a current reading only if all intrusion contacts are closed. All three sections of the switch (S_1, S_2, S_3) make contact simultaneously as the key is turned.

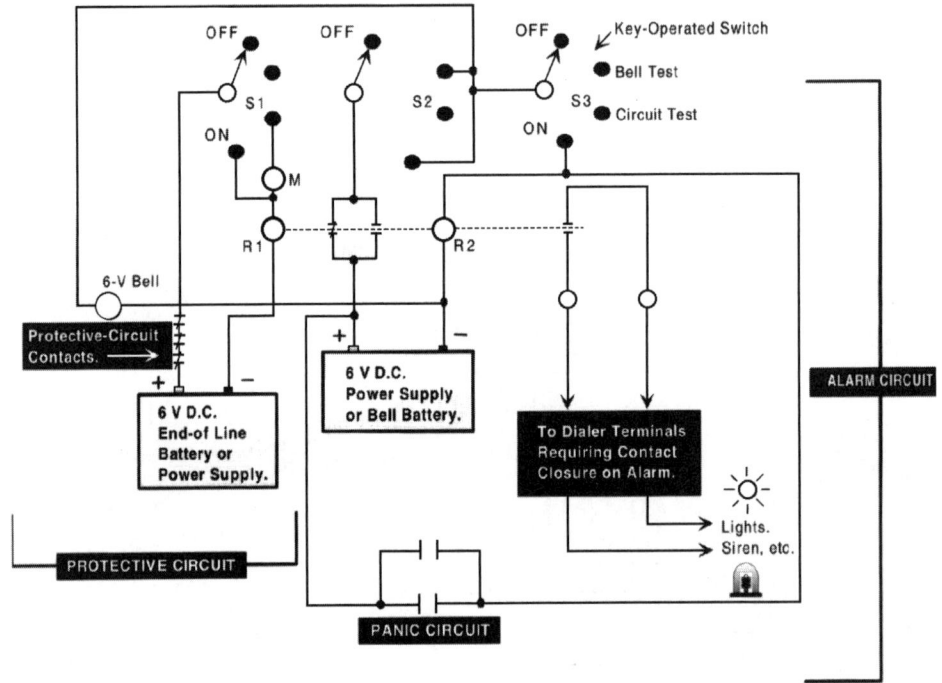

Figure 5.3 Closed-circuit security alarm system.

Opening of intrusion contacts is not the only event that causes the alarm to activate. Any break in protective-circuit wiring or loss of output from the energy source has the same effect. The circuit is broken, which deenergizes the sensitive relay and allows spring force to close the relay contacts, thus sounding the alarm. Any short circuit between the positive and negative wires of the protective circuit also keeps current from reaching the relay coil and causes a dropout, which again sounds the alarm.

Other components of the alarm circuit in Fig. 5.3 include a second energy source, an alarm bell, and a drop relay (R_2). When the keyed switch is at ON, dropout of the sensitive relay (R_1), and closing of its contacts completes a circuit to energize the coil of drop relay (R_2). Closing the drop relay's normally open contacts rings the bell and latches in the drop-relay coil so that R_2 stays energized even if the protective circuit returns to normal and opens the sensitive relay's contacts. As a result, the bell continues to ring until the key switch is turned away from ON to break the latching connections to the R_2 coil.

Drop relays often have additional contacts to control other circuits or devices. The extra contacts in the circuit in Fig. 5.3 are for turning on lights, triggering an automatic telephone dialer, etc. But the main two functions of the drop relay are actuation of the alarm and latching the coil to keep the circuit in the alarm condition.

The majority of burglar systems use a closed-loop protective circuit. Typically, the system consists of an annunciator connected to a contact on each door and window and a relay connected so that when any window or door is opened it will cause current to pass through the relay. The relay, in turn, will operate to close a circuit on a bell, horn, or other type of annunciator, which will continue to sound until it is shut off, thereby alerting the occupants or needed agencies.

The wiring and connections for the open-circuit system are shown in Fig. 5.4. This diagram shows three contacts, but any number can be added. Closing one of the contacts completes the power circuit through the winding of the proper annunciator drops, the constant-ringing switch, the constant-ringing relay, the alarm bell, and the bell-cutoff switch. The current passing through the winding of the constant-ringing relay operates to complete a circuit placing the alarm bell directly across the battery or other power source so the bell continues storing until the cutoff switch is opened. At the same time, current in another set of wires operates a relay that closes an auxiliary circuit to operate other devices, such as lights and an automatic telephone dialer.

Contacts for closed-circuit operation are shown in Fig. 5.5A. The contacts are surface-mounted opposite each other, one on a stationary window or door frame; the other on the movable part of the window or door. When the window is raised, or the door is opened, the contacts break and sound the alarm. Contacts for recessed mounting (Fig. 5.5B) operate the same way as described for the surface-mounted contacts.

A spring-type contact for open-circuit operation is shown in Fig. 5.6. This device is recessed in the window frame or a door jamb so that the cam projects outward. When the window is raised, the cam pivots and is pressed in and makes contact with a spring that is insulated from the plate. The contact is connected in series with the power source and the annunciator; that is, one wire is connected to the plate and the other to the spring.

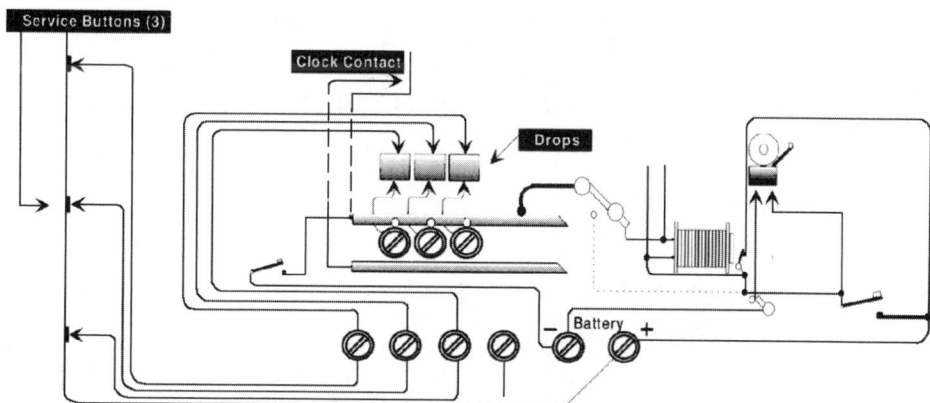

Figure 5.4 Open-circuit security alarm system.

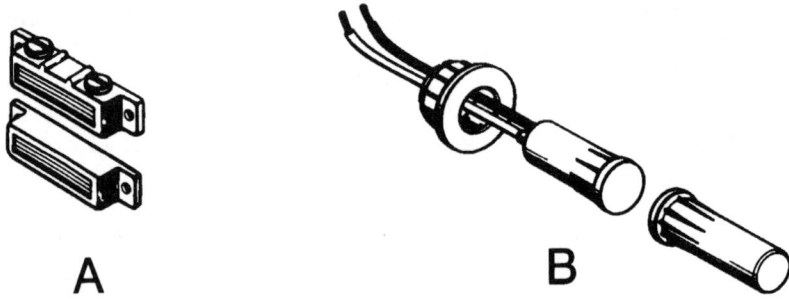

Figure 5.5   Spring-type contact for closed-circuit operation.

Figure 5.6   Spring-type contact for open-circuit operation.

Fire-alarm systems

A fire-alarm system consists of the following:

- Sensors
- Control panel
- Annunciator
- Related wiring

They are generally divided into the following four types:

- Noncoded
- Master-coded
- Selective-coded
- Dual-coded

Each of these four types of alarm has several functional features so designed that a specific system can meet practically any need to comply with local and state codes, statutes, and regulations.

In a noncoded system, an alarm signal is sounded continuously until it is manually or automatically turned off.

In a master-coded system, a common-coded alarm signal is sounded for not less than three rounds. The same code is sounded regardless of the alarm-initiating device activated.

In a selective-coded system, a unique coded alarm is sounded for each fire-box or fire zone on the protected premises.

In a dual-coded system, a unique coded alarm is sounded for each firebox or fire zone to notify the building's personnel of the location of the fire, while non-coded or common-coded alarm signals are sounded separately to notify occupants to evacuate the building.

Figure 5.7 represents a riser diagram of a fire-alarm system. If a detector senses smoke or if any manual striking station is operated, all bells within the building will ring. At the same time, the magnetic door switches will release the smoke doors to help block smoke and/or drafts. This system is also connected to a water-flow switch on the sprinkler system. If the sprinkler valves are activated causing a flow of water through the pipes, the fire-alarm system will again go into operation, energizing all bells and closing smoke doors.

Smoke and fire detectors

Any product of a fire (like aerosols) that changes the ambient conditions in the building is called a *fire signature* and is potentially useful for detection

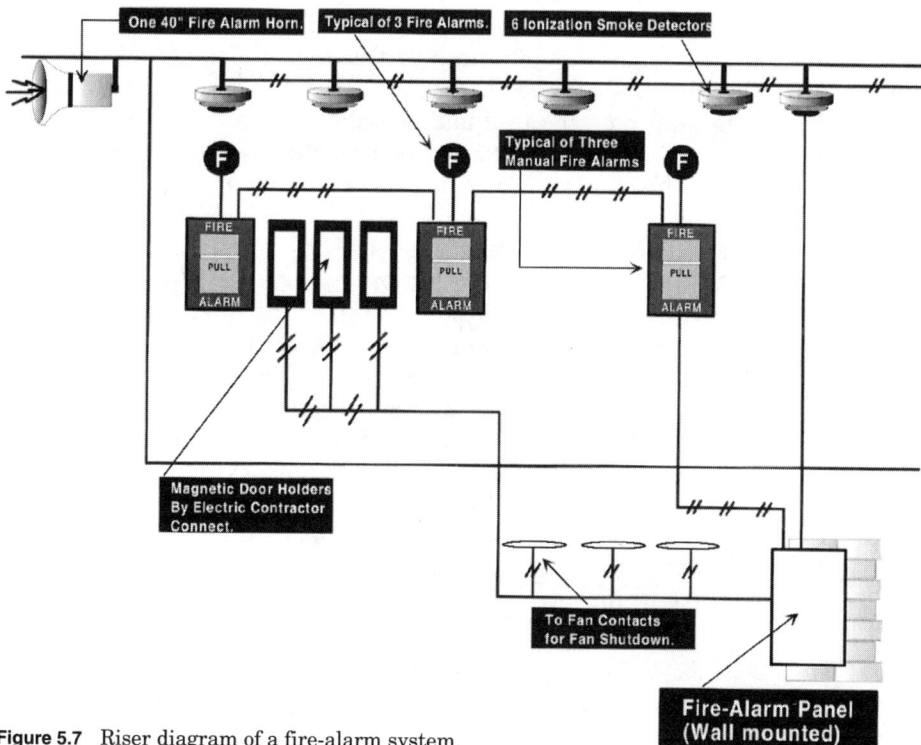

Figure 5.7   Riser diagram of a fire-alarm system.

purposes. The principal fire signature used in residential smoke detectors is aerosol. *Aerosols* are particles suspended in air. The process of combustion releases large numbers of solid and liquid particles into the atmosphere. They can range in size from 10 μm [a micron (μm) is one thousandth of a millimeter] down to 0.001 μm. Aerosols resulting from a fire represent two different fire signatures. Those particles less than 0.3 μm do not scatter light efficiently and are classified as visible. The invisible aerosol signature is usually referred to as the "products of combustion" and the visible aerosol signature as "smoke." Invisible aerosol is the earliest appearing fire signature.

Types of fire-detection devices

Thermal Detectors: Thermal detectors are devices that respond to heat—typically 135°F. These units consist of a bimetallic element that bends to complete a circuit under high heat conditions. Because these units do not detect smoke or products of combustion, they are not recommended for living areas of a residence. However, they do have value for use in attics, unheated garages, and furnace rooms.

Flame Detectors: Flame detectors detect actual flames by sensing ultraviolet emissions. These devices would not be used in residential applications.

Gas Detectors: These units respond to certain gases (propane, carbon monoxide, liquid petroleum, butane, and gasoline vapors) that would not be detected by a smoke and fire detector. Although these detectors do have some uses, they should not be used as a substitute for a smoke and fire detector. They will not respond to aerosols produced by the majority of residential fires.

Ionization Detectors: Inside the ionization chamber, the radioactive source emits radiation, main alpha particles, which bombard the air and ionize the air particles, which, in turn, are attracted by the voltage on the collector electrodes. This action results in a minute current flow. If aerosols, such as products of combustion or smoke, enter the chamber, the ionized air particles attach themselves to the aerosols and the resultant particles, being a larger mass than ionized air, move more slowly, and thus, per unit of time, fewer reach the electrodes. A decrease in current flow, therefore, occurs within the chamber whenever aerosols enter. The decrease in current flow is electronically converted into an alarm signal output (Fig. 5.8).

An ionization type of detector responds best to invisible aerosols, where the particles from burning materials are in the range of 1.0 μm in size down to 0.01 μm. A tremendous amount of these particles are produced by a flaming fire as opposed to a smoldering fire, which produces large and small particles, but, because of low heat, the low thermal lift tends to allow particles to agglomerate into larger particles if the detector is some distance from the fire.

High air flows will affect the operation of this type of unit by reducing the ion concentration in the detector chamber. In fact, with a high-enough air flow,

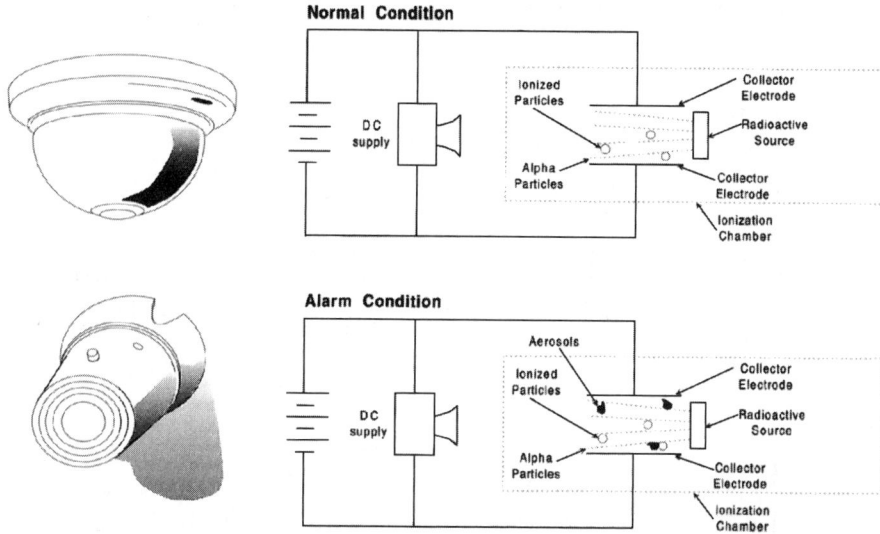

Figure 5.8 Diagrams of ionization detectors. The top diagram shows normal conditions. The bottom diagram shows aerosols, such as products of combustion or smoke, entering the sensor. In the latter condition, the alarm is activated.

the unit will respond and alarm even though a fire does not exist. For this reason, locating ionization detectors near windows, direct air flows from air vents, and comparable areas should be avoided.

Ionization smoke detectors (Fig. 5.9) can be used in place of conventional smoke detectors or can be used in combination with standard smoke detectors. They are more sensitive than the conventional smoke detectors.

Photoelectric Detectors: A beam from the detector's light source is projected across a chamber into a light catcher. The chamber is designed to permit access of smoke, but not access of external light. A photo-resistive cell or light-sensitive device is located in a recessed area perpendicular to the light beam. When smoke enters the chamber, particles will scatter or reflect a small portion of the light beam to the light-receiving device, which, in turn, will provide a signal for amplification to the alarm. Variations in design are sometimes used by manufacturers.

Some photoelectric detectors are adversely affected by dirt films. Any accumulated dirt, dust, film, or foreign matter on either or both lenses of the light source or the photocell will cause an opaque effect and the detector will then become less and less sensitive. Therefore, it will require more smoke in order to respond. Although the latest photoelectric models utilize solid-state light-emitting and receiving devices, which have a longer life than previous light devices, the problem of failure of the light source still exists. Underwriters' Laboratories requires an additional audible alarm in case light failure occurs.

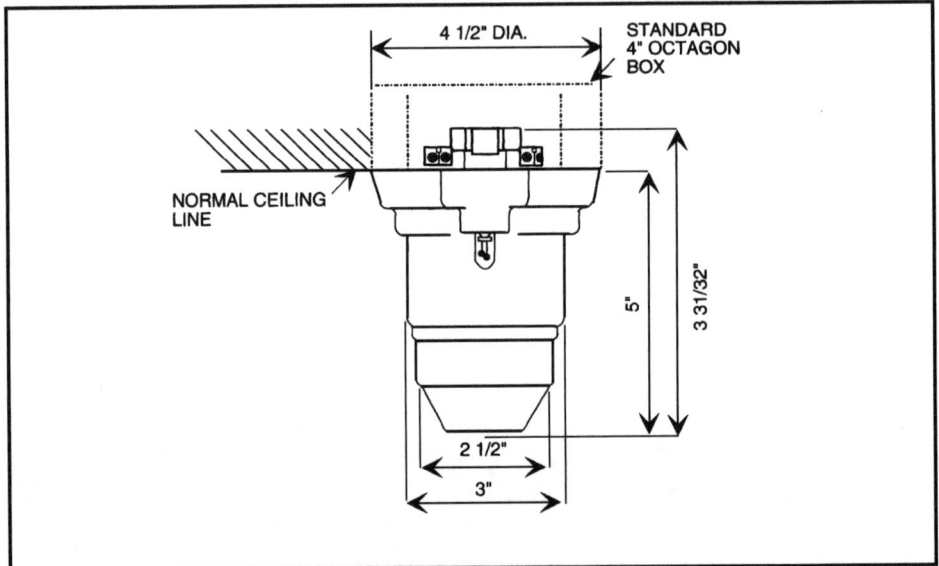

Figure 5.9 Ionization smoke detector.

Photoelectric units respond best to visible aerosols where the particles range from 10 μm down to 0.3 μm. These particles would be given off by a smoldering fire that produces very little heat (Fig. 5.10).

Ionization and Photoelectric Devices: Figure 5.11 can be used to illustrate both types of units—the difference is the use of either an ionization sensor or a photoelectric sensor in the reference chamber and detector portions of the circuit. Under normal conditions, the voltage across the reference chamber and the detection chamber is the same. However, when fire occurs, the detection chamber then functions as described in the previous explanation. Thus, when there is sufficient voltage difference between the two chambers, the alarm is activated through the switching circuit.

Complete descriptive information and practical applications of smoke detectors are covered later in this chapter.

Components of security/fire-alarm systems

Wire sizes for the majority of low-voltage systems range from #22 to #18 AWG. However, in some situations, it might be necessary to use larger wire sizes to prevent excessive voltage drop, for example: where larger-than-normal currents are required for longer distances between outlets. *Note:* Voltage-drop calculations should be made to determine the correct wire size for a given application (see Chapter 2).

Most closed systems use two-wire #22 or #24 AWG conductors and are color-coded to identify them. A #18 pair normally is adequate for connecting bells or

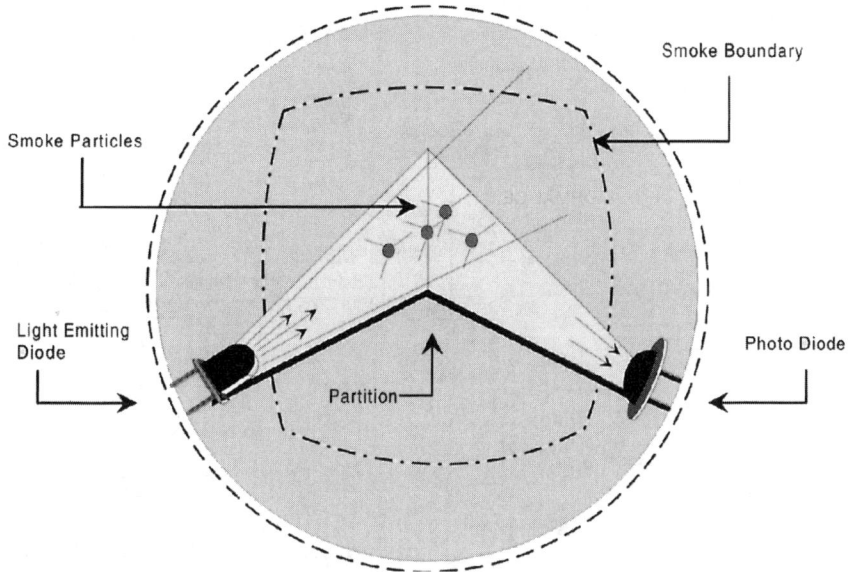

Figure 5.10 Basic operating principles of photoelectric detectors.

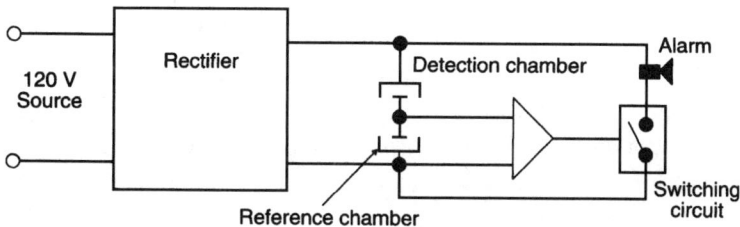

Figure 5.11 Diagram of ionization and photoelectric devices.

sirens to controls if the run is 40 feet or less. However, many installers prefer to use #16 or even #14 nonmetallic cable.

A summary of the various components for a typical security/fire-alarm system is shown in the riser diagram in Fig. 5.12. Notice the varying types of sensors or detectors in this system.

Control stations

The control station is the heart of any security system because the circuitry in control panels senses a broken contact and then either sounds a local bell, a horn, or a silent alarm. Most modern control panels use relay-type controls to sense the protective circuits and regulate the output for alarm-sounding devices. They also contain contacts to actuate other deterrent or reporting devices and a silent holdup alarm with a dialer or police-connected reporting mechanism.

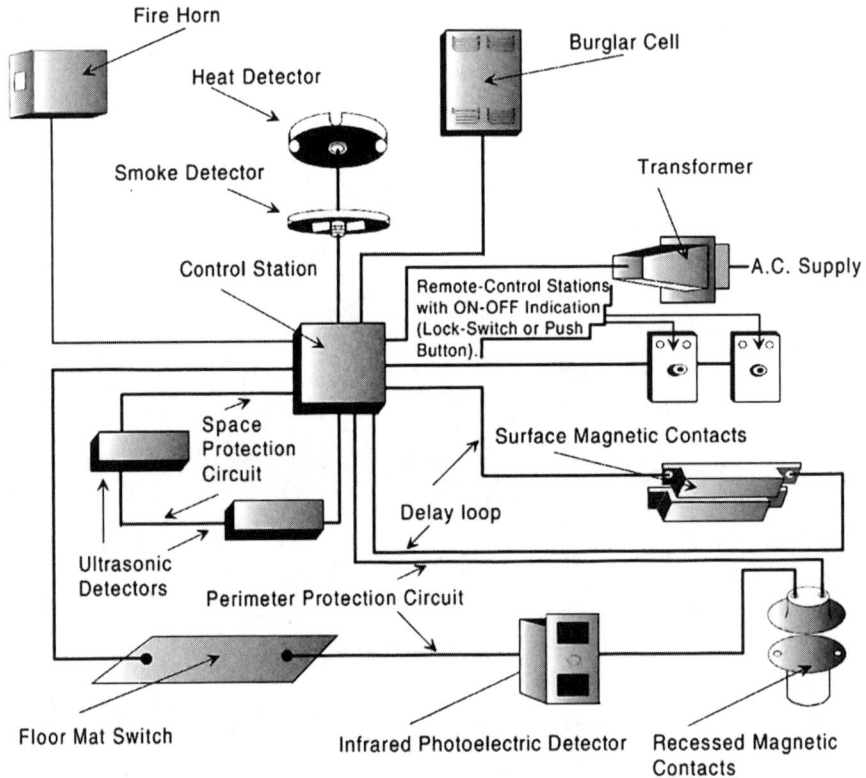

Figure 5.12 Components for a typical security/fire-alarm system.

Power supplies

Power supplies vary for different systems, but, in general, they consist of rechargeable 6-V dc power supplies for burglar alarm systems. The power packs usually contain nickel-cadmium batteries that are kept charged by 12-V ac input from a plug-in or otherwise connected via transformer to a 120-V circuit. High-quality power supplies have the capability of operating an armed system for 48 hours or more without being charged and still have the capacity to ring an alarm bell for 30 minutes or longer. Power supplies are obviously used in conjunction with a charging source and supply power for operation of the alarm system through the control panel.

Contacts and conductive foil

Recessed Magnetic Contacts in Doorways (Fig. 5.13): Holes are drilled in the door and in the casing, one directly across from the other, and a pair of wires from the positive side of the protective circuit is run out through the switch hole. The switch and magnet are then installed with no more than a ⅛″ gap between them.

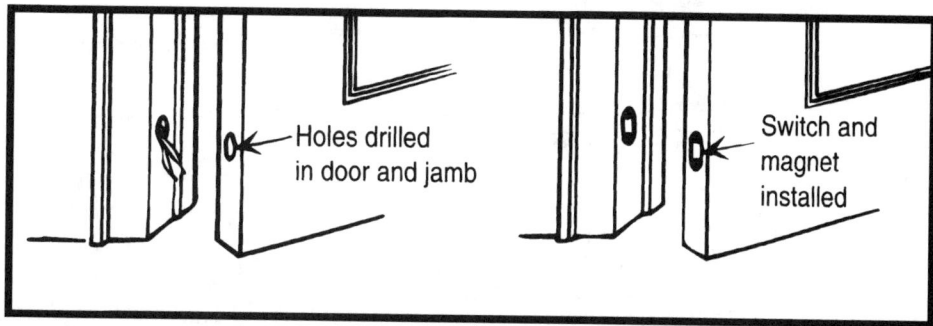

Figure 5.13   Recessed magnetic contacts installed in edge of door and also in door jamb.

Recessed Magnetic Contacts in Casement Windows (Fig. 5.14A): A switch and magnet are installed as in doorways, preferably in the top of the window and underside of the upper window casing, where they will be least noticeable.

Surface-Mounted Magnetic Contacts on Double-Hung Windows (Fig. 5.14B): A switch is mounted on the window casing with a magnet on the window casing and a magnet on the window. As long as the switch and magnet are parallel and in close proximity when the window is shut, they can be oriented side-to-side, top-to-side, or top-to-top.

Conductive Foil on Glass Doors (Fig. 5.15): A self-adhesive foil block (terminator) on the door is connected to a similar unit on the door frame by a short length of flexible cord to allow for door movement. The foil is connected in the positive conductor of the protective circuit and is adhered to the glass parallel to and about 3″ from the edge of the glass by using varnish. Breaking the glass cuts the foil and opens the circuit. To provide more coverage, a double circuit of foil may be taken from the foil block. Coiled, retractable cords are available for use between foil blocks to allow for sliding-door travel.

Complete Glass-Door Protection (Fig. 5.16A): A glass door with a glass transom may be protected by a combination of magnetic contacts and foil.

Surface-Mounted Magnetic Contacts on Door (Fig. 5.16B): Where appearance is not the most important consideration, the use of a surface-mounted switch (on the door frame) and a magnet (on the door) will simplify installation.

Conductive Foil on Picture Windows (Fig. 5.16C): Where a window does not open, a single run of foil is connected to a foil block on the glass, frame, or wall. When the foil crosses over a frame member, a piece of plastic electrical tape should be used to provide an insulated crossover surface for the foil.

Basic Fire-Protection Signaling Systems

All fire alarm systems consist of the following basic parts:

- Control unit

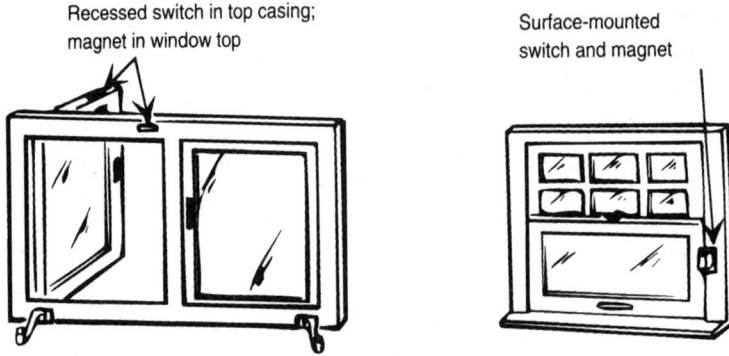

Figure 5.14   Magnetic contacts used on casement windows.

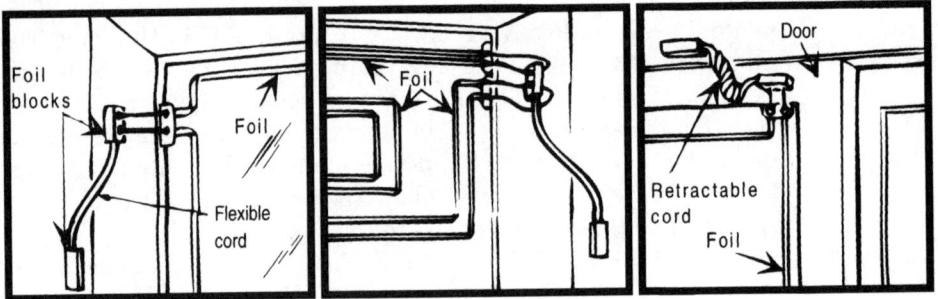

Figure 5.15   Conductive foil on glass doors.

- Initiating device circuits
- Indicating appliance circuits
- Power supply
- Supplementary circuits (optional)

The applications of fire-alarm systems can be broken down into four broad categories:

- Local protective signaling systems
- Auxiliary protective signaling systems
- Remote station protective signaling systems
- Proprietary protective signaling systems

Each of these categories can be defined as follows:

Local protective signaling (fire-alarm) systems: A local system sounds an alarm at the protected premises as the result of the manual operation of a fire-alarm box or the operation of protection equipment or systems, such as water

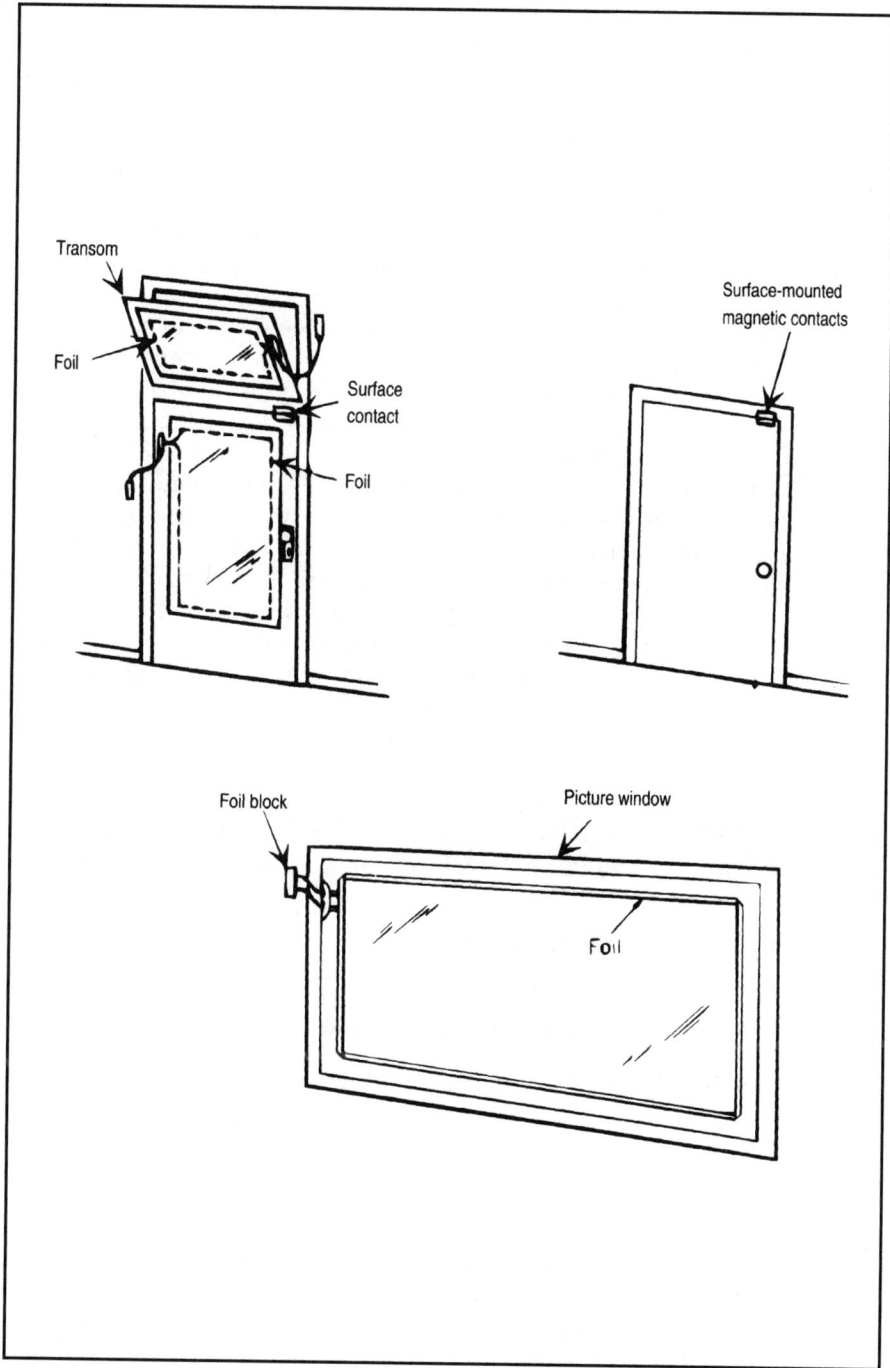

Figure 5.16 Applications of conductive foil.

flowing in a sprinkler system, the discharge of carbon dioxide (CO_2) or Halon, or the detection of smoke or heat.

Auxiliary protective signaling systems: A system utilizing a connection to a municipal fire-alarm box to transmit an alarm signal from a protected building to the municipal communications center.

Remote station protective signaling systems: A system using a supervised direct circuit connection between alarm-initiating devices or a control unit in a protected premises and signal-receiving equipment in a remote station, such as fire houses or police headquarters or other places acceptable to the authority having jurisdiction.

Proprietary protective signaling systems: A system supervised by competent and experienced observers and operators in a central supervising station at the property protected. A proprietary system requires a secondary (standby) source of power and at least one recording device to make a permanent record of signals received.

The specific application of these system types, while sharing common installation requirements, should be selected to meet the requirements of one or more authorities having jurisdiction. In addition, care should be taken to install equipment in accordance with the manufacturer's instructions.

Control unit: The control unit is the brain of the system. It provides power and electrically supervises its circuits. In general, the control unit contains the logic circuits to receive signals from alarm initiating devices and transmit them to alarm-indicating devices and supplemental equipment. Depending on design, the fire-alarm signaling function could provide for one or more of the following:

- Notify all building occupants simultaneously
- Notify occupants in certain portions of the building who are in immediate danger
- Notify key building personnel
- Notify the fire department
- Perform supplementary functions as required

The fire-alarm signaling function falls into two broad categories:

- Coded
- Noncoded

A coded system is one in which not less than three rounds of coded fire-alarm signals are transmitted, after which the fire-alarm system can be manually or automatically restored to normal.

A noncoded system is one in which a continuous fire-alarm signal is transmitted for a predetermined period of time, after which the alarm indicating devices can be manually or automatically restored to normal.

Both coded and noncoded systems are further broken down into several different types as follows:

- Zoned noncoded
- Master-coded system
- March time-coded system
- Selective-coded system
- Zoned coded system

Control unit operation

A fire-alarm control can be a simple, single-zone panel providing for one alarm-initiating circuit and one or two alarm-indicating circuits. However, most control units are modular in construction and are designed to accommodate multiple zones of detection and alarm signaling, as well as the selection of other modular components to perform any of the system functions listed previously.

Fire-alarm systems must perform in an emergency. For this reason, the control panel constantly monitors the integrity of the primary power supply and the installation wires and the connections of the alarm-initiating devices. The control unit will sound a trouble signal to alert operating personnel when a fault condition exists on any of the supervised circuits.

The trouble signal will normally sound to indicate any of the following types of faults, which could interfere with the proper transmission or receipt of an automatic or manual alarm signal:

- Loss of primary power
- An open or ground in a supervised wire
- Loss of an audio amplifier, tone generator, or preamplifier
- Loss of connection between any installation wire and any alarm-initiating device (or indicating appliance) wire or terminal necessary to sense an alarm

The trouble signal is an audible appliance with a distinctive sound. A visible indication (pilot light/LED) can be provided also.

Most problems encountered during the start-up of a fire-alarm system are traced to errors made in the installation wiring or connections, or both. It is, therefore, extremely important to follow manufacturer's instructions when installing a fire-alarm system.

The control panel is usually installed in a wall-mounted cabinet that can be either surface- or flush-mounted, a floor-mounted cabinet, or a desk-type console, depending upon the size of the system and the manufacturer. Normally, a lock is provided for security of the controls.

Alarm-initiating devices

Alarm-initiating devices for fire-alarm systems are either manual stations or automatic detectors and are used to initiate an alarm on a fire-alarm system.

Manual stations can be of the following types:

- Noncoded or coded
- Presignal or general alarm
- Breakglass or nonbreakglass
- Single action or double action

The single-acting devices can be initiated with one motion, and the double-acting devices require two motions. For example, a device that requires lifting the cover and then pulling the handle requires two motions. Double-acting devices are designed to prevent false alarms.

As the name implies, manual stations must be operated by hand, and are seldom, if ever, used alone in a fire-alarm system. Rather, they are used to supplement the automatic detection devices.

Automatic alarm-initiating devices

Automatic alarm-initiating devices are actuated by various factors that could be present as the result of a fire. These factors can be direct fire results: heat, smoke, flame radiation, or combinations of these effects. Devices sensing these direct products of fire are generally referred to as *automatic fire detectors*. Automatic initiating can also be accomplished as the result of detecting flows of water in a sprinkler pipe—either by a vane deflected by the water or a pressure-operated switch mounted on a sprinkler system dry pipe valve. In addition, pressure switches can be mounted on fixed fire-suppression systems that suppress fires by releasing such agents as halon, carbon dioxide, dry chemicals, or foam. These devices are generally referred to by their direct function: *flow switch, pressure switch,* etc.

Automatic fire detectors fall under several different categories:

- By combustion product detected
- By physical arrangement of the detector
- By operating modes

Heat-sensing fire detectors: A heat detector detects abnormally high temperature or rate-of-temperature rise. A fixed-temperature detector responds when its operating element becomes heated to a predetermined level. Its operating element usually consists of a bimetallic, snap-action disc located in the center of the detector, although thermistors and fusible alloys are also used to trigger heat-sensing devices.

Heat-sensitive cables: One type is a line device with a sensing element composed of two current-carrying conductors kept separated by heat-sensitive insulation, which softens at the rated temperature (Fig. 5.17). When the insulation softens, the wires make electrical contact. In another product, a single wire is centered in a metallic tube and the intervening space is filled with a substance which, at a critical temperature, becomes conductive, thus establishing electrical contact between the tube and the wire.

Rate-compensating detectors: A rate compensation detector is a device that will respond when the temperature of the air surrounding the device reaches a predetermined level, regardless of the rate of temperature rise (Fig. 5.18).

Rate-of-rise detectors: A rate-of-rise detector is a device that will respond when the temperature rises at a rate exceeding predetermined ratios, typically 15°F/min.

Pneumatic rate-of-rise tubing: A line-type sensor of small-diameter tubing, usually copper, which is installed on the ceiling or high on the walls throughout the protected area. The tubing is terminated in a detector unit containing diaphragms and associated contacts set to actuate at a predetermined pressure. The system is sealed, except for calibrated vents that compensate for normal changes in ambient temperature (Fig. 5.19).

In referring to Fig. 5.19, note that the copper tubing (A) is fastened to ceilings or walls in a continuous loop and terminates at both ends in chambers (B)

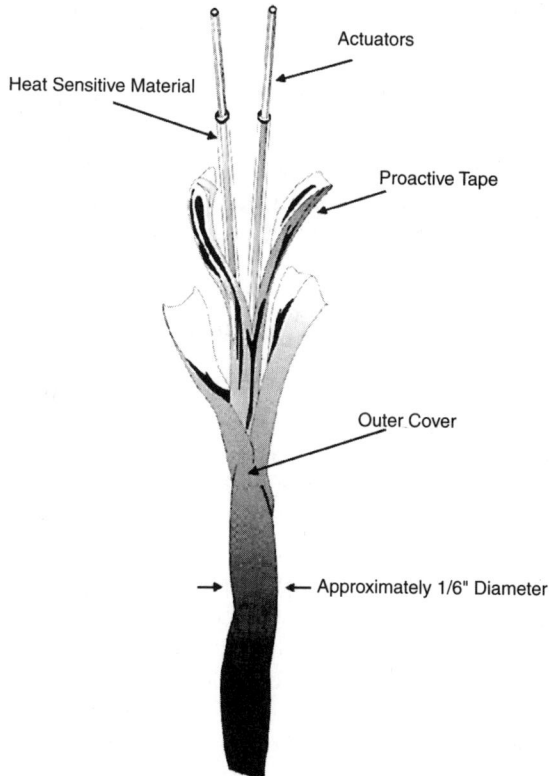

Figure 5.17 Heat-sensitive cable.

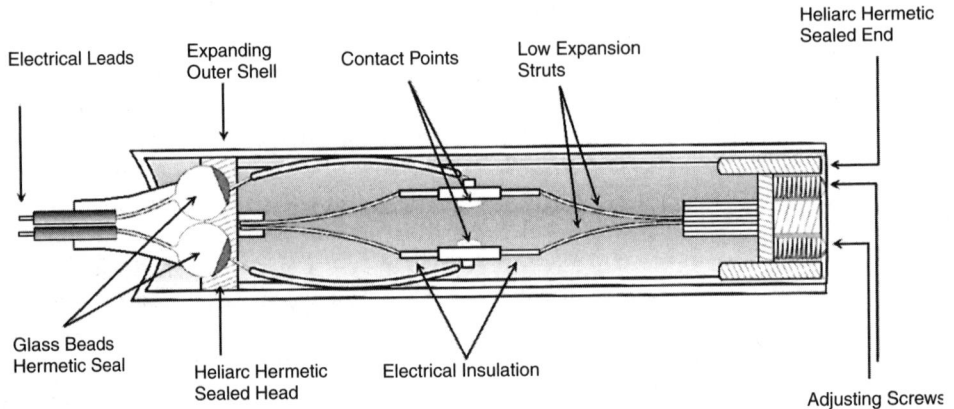

Figure 5.18 Rate-compensation detector.

with flexible diaphragms (C) that control electrical contacts (D). When air in the tubing expands under the influence of heat, pressure builds within the chambers, causing the diaphragms to move and close a circuit to alarm transmitter (E). Vents (F) compensate for small changes in pressure in the tubing brought about by slight changes in temperature in the protected spaces.

Spot-type, rate-of-rise detectors: A device consisting of an air chamber, diaphragm, contacts, and compensating vent in a single enclosure. The principle of operation is the same as that described in (A) in the previous paragraph. Some spot-type, rate-of-rise detectors also incorporate an element using eutectic solder that is arranged to melt at a fixed temperature and cause contacts to close. These detectors are referred to as *fixed temperature* and *rate-of-rise detectors* and combine the two principles of operation in a single unit (Fig. 5.20).

Smoke-sensing fire detectors: Several types of smoke detectors are currently used in fire-alarm systems. As described earlier, a type that is considered to be among the best is the ionization smoke detector. Smoke detectors utilizing the ionization principle are usually of the spot type. An ionization smoke detector has a small amount of radioactive material, which ionizes the air in the sensing chamber, thus rendering it conductive and permitting a current flow through the air between two charged electrodes. This gives the sensing chamber an effective electrical conductance. When smoke particles enter the ionization area, they decrease the conductance of the air by attaching themselves to the ions, causing a reduction in mobility. When the conductance is less than a predetermined level, the detector responds (Fig. 5.21).

Photoelectric light obscuration smoke detectors: Smoke detectors utilizing the photoelectric light obscuration principle consist of a light source which is projected onto a photosensitive device. Smoke particles between the light source and the photosensitive device reduce the amount of light reaching the unit, causing the detector to respond.

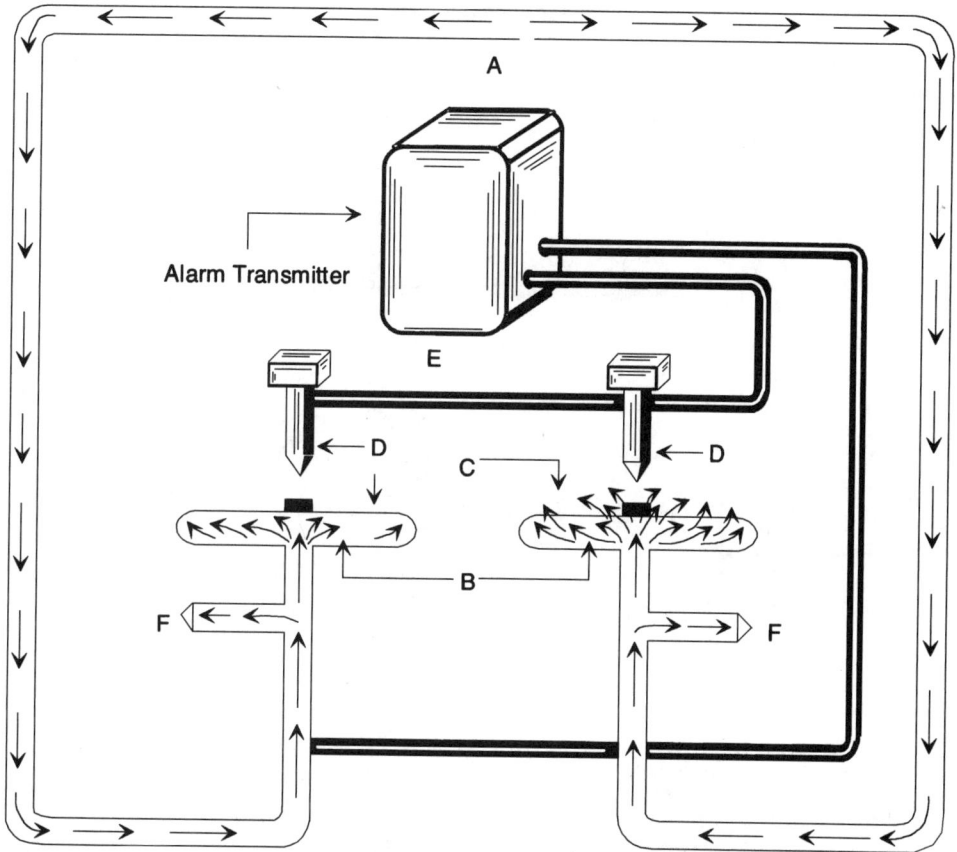

Figure 5.19   Pneumatic rate-of-rise tubing.

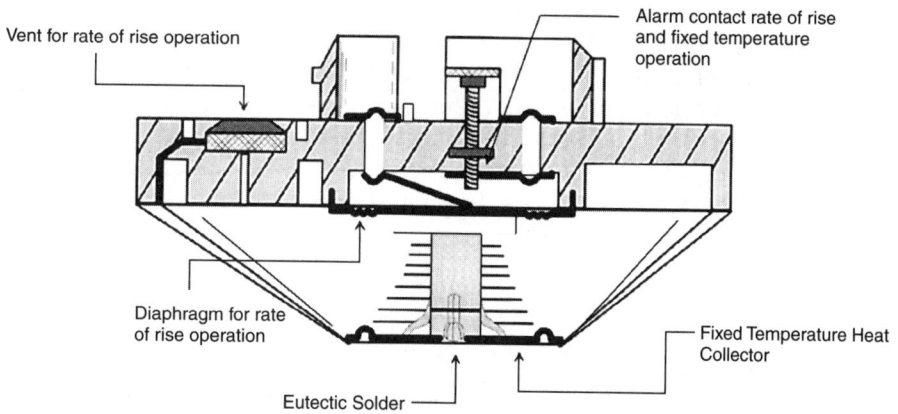

Figure 5.20   Combination spot-type rate-of-rise fixed-temperature detector.

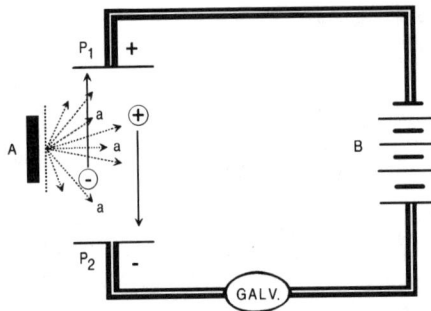

Figure 5.21 Current flow through ionization detector sensing chamber.

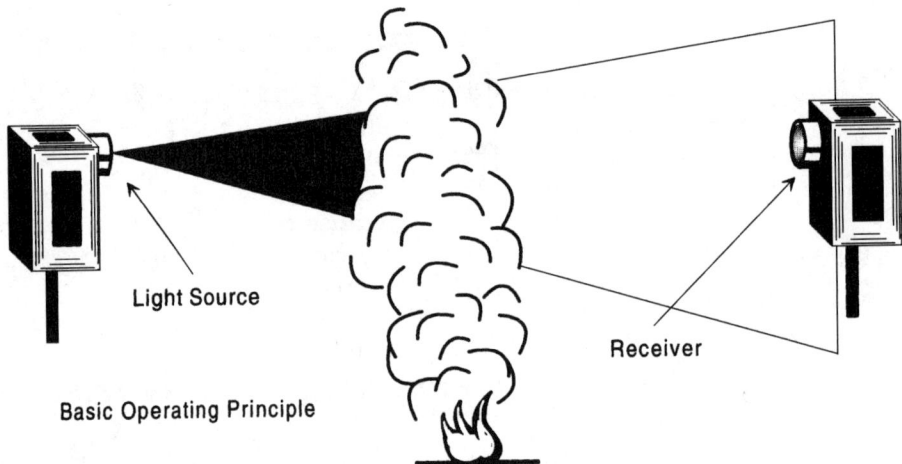

Light Source

Receiver

Basic Operating Principle

Figure 5.22 Projected-beam smoke detector.

Projected beam smoke detectors: A line-type, light-obscuration smoke detector using a light beam projected across the area to be projected (Fig. 5.22).

Photoelectric light-scattering smoke detectors: Smoke detectors utilizing the photoelectric light scattering principle are usually of the spot type. They contain a light source and a photosensitive device so arranged that light rays do not normally fall onto the photosensitive device. When smoke particles enter the light path, light strikes the particles and is scattered onto the photosensitive device, causing the detector to respond (Fig. 5.23).

Wiring installations

The primary rule of installing wiring for fire-alarm systems, and also for making the required connections is:

Follow the manufacturer's instructions

To do otherwise is asking for trouble. This rule cannot be overemphasized because the requirement for fire-alarm circuits and their connections to initi-

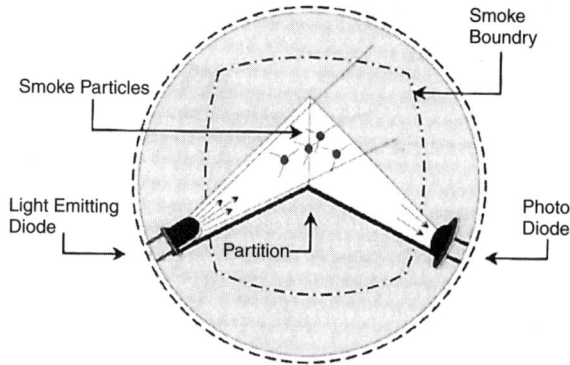

Figure 5.23 Photoelectric light-scattering detector.

ating devices and indicating appliances makes fire-alarm system wiring very different from general wiring.

A manufacturer's installation wiring drawing routes wires and makes connections in a certain manner because of the supervision requirements. Any variance from the drawings might cause a portion of a circuit to be unsupervised and, if an open or short occurred, prevent the circuit from performing its intended function and possibly lead to loss of life.

The rules of supervision are complex. Unless an installer specializes in fire-alarm systems the person might not have a thorough understanding of them. It is possible that hardware that appears to be identical in two different buildings is wired with radical differences at each site.

Fire-alarm system installation drawings take two forms. The first type are produced by a manufacturer of a control panel, or a qualified installer for a particular building. The other type of plan is a generic schematic produced by the manufacturer of the control panel or other components in the system. These are "typical" installation drawings.

An installer that uses the "typical" drawings takes on the responsibility of applying the "typical" drawings in accordance with local code requirements. The manufacturer's drawings will show how the unit is to be connected into a system. However, how to interconnect devices on the same floor, but served by a different riser might not always be shown. Generally, installers using typical drawings should be well qualified in fire-alarm system installation requirements or be under the direct supervision of someone who is very skilled. See Figures 5.24 and 5.25 for samples of correct and incorrect wiring methods.

Alarm-indicating appliances

Alarm-indicating devices provide audible or visible signals, or both. There are several types of alarm signal devices:

- Audible alarm signal

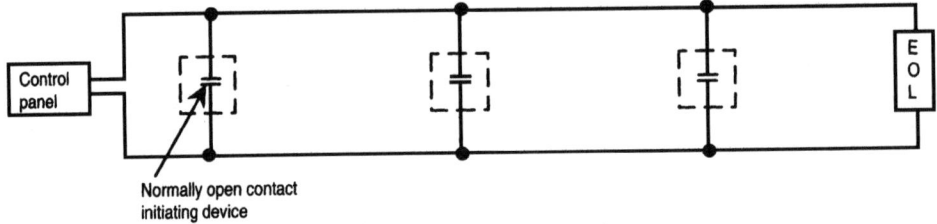

Figure 5.24 Initiating devices incorrectly wired.

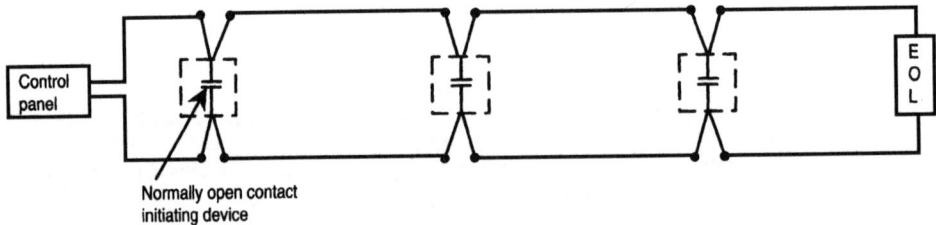

Figure 5.25 Initiating devices correctly wired.

- Visible alarm signal
- Visible alarm signal annunciators
- Audible/visible (combination) alarm signal

All alarm-indicating devices should be installed in the protected property in accordance with local codes, architect's plans and specifications, requirements of authorities having jurisdiction, the manufacturer's recommendations, and the basic requirements of the National Fire Protection Association (NFPA) standards insofar as they apply to the installation.

A wire-to-wire short on an alarm-signal, appliance-installation wiring circuit should sound a trouble signal. A description of some of the audible alarm signal appliances follows (Fig. 5.26).

Bells: Bells may be used for fire-alarm signals where their sound is distinctive and will not be confused with similar audible signals used for other purposes. Single-stroke bells are used to provide audible coded signals. Continuous vibrating types are used primarily for noncoded, continuous-ringing applications. They also may be used to provide coded audible signals.

Horns: Horns are provided for applications requiring louder or more distinctive signals, or both. They may be operated by either alternating or direct current and may be connected in series or parallel. They are usually of the continuous vibrating type and may be used as either coded or noncoded audible alarm signals. They may be surface mounted, flush or semiflush, single projector, or trumpet type in design.

Chimes: Chimes are soft-toned appliances. They are normally used in applications where panic or other undesirable actions might result from the use of

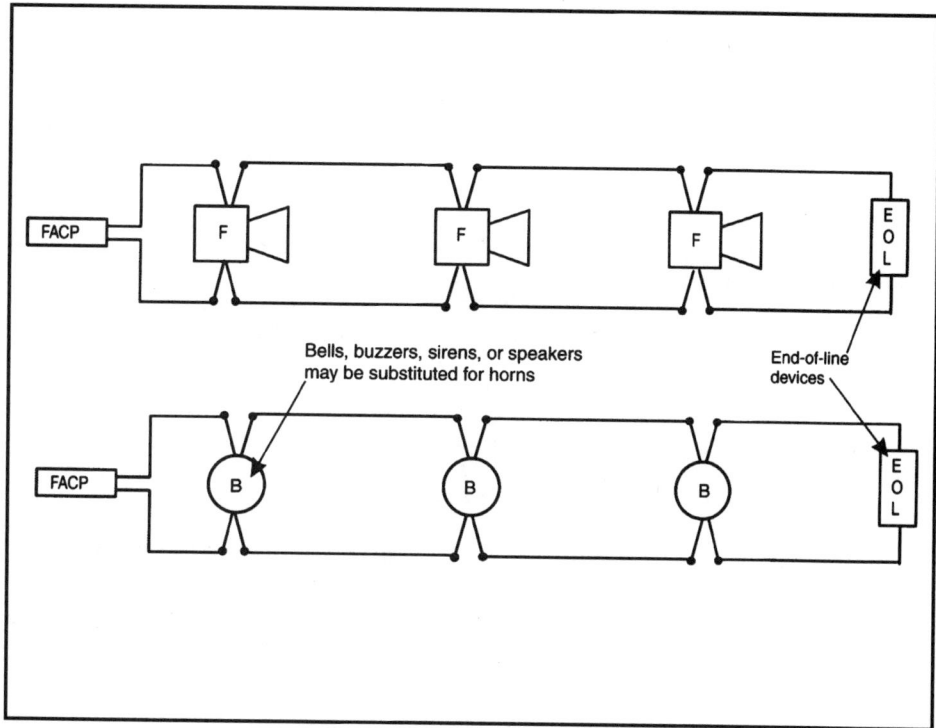

Figure 5.26 Correct wiring method for typical audible signal circuits.

loud, harsh, audible alarm signals, such as in nurses' stations in hospitals. They may be operated by either alternating or direct current.

Buzzers: Buzzers can also be operated from an ac or dc supply. They are used for trouble signals, rather than alarm signals.

Sirens: Sirens are usually limited to outdoor applications but are sometimes used in extremely noisy indoor areas. They are motor-driven and may be either alternating or direct current.

Speakers: speakers are frequently used as fire-alarm signaling appliances. Because they reproduce electronic signals, they can be made to sound like any mechanical signaling device and have the capability of reproducing unique sounds that are not practical on mechanical appliances. The speakers are driven by an electronic tone generator, microphone, tape player, or voice synthesizer and an electronic amplifier. A typical fire-alarm riser diagram appears in Fig. 5.27.

Summary

The design and installation of security and fire-alarm systems include a wide variety of techniques, often involving special types of equipment and materi-

REMOTE STATION PROTECTIVE SIGNALING SYSTEM SCHEMATIC DIAGRAM

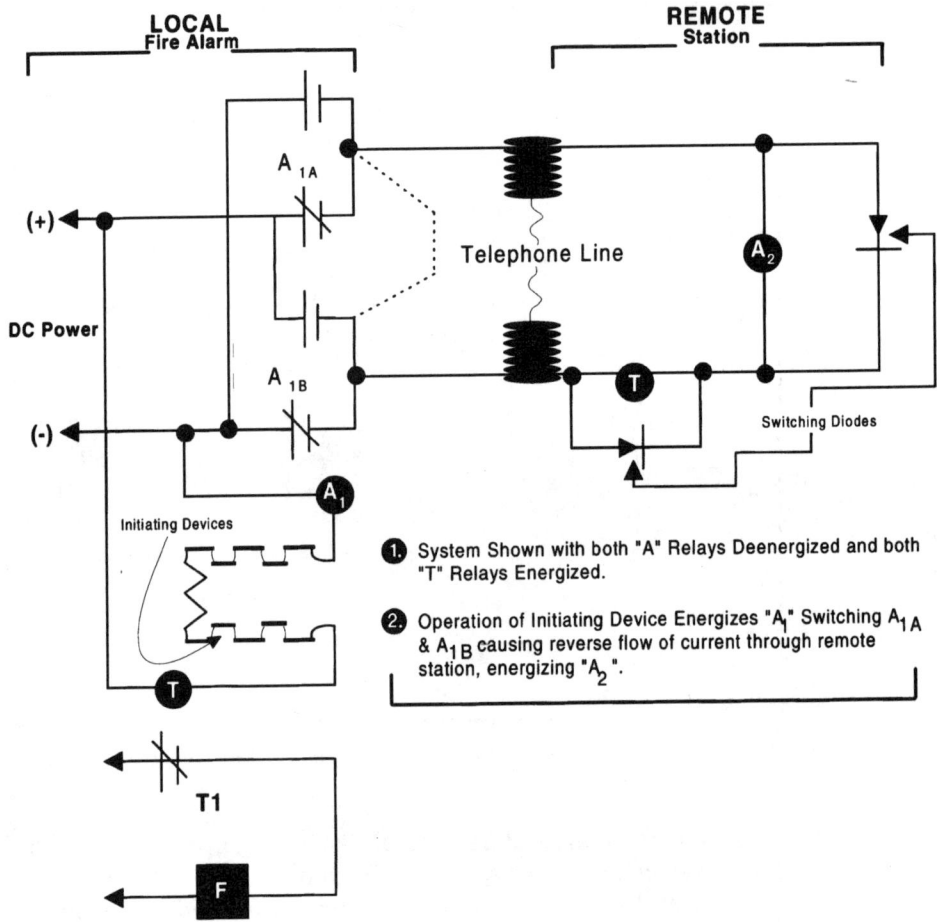

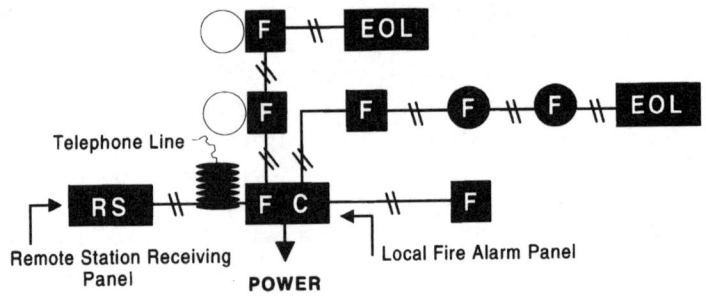

① System Shown with both "A" Relays Deenergized and both "T" Relays Energized.

② Operation of Initiating Device Energizes "A_1" Switching A_{1A} & A_{1B} causing reverse flow of current through remote station, energizing "A_2".

REMOTE STATION PROTECTIVE SIGNALING SYSTEM RISER DIAGRAM

Figure 5.27   Typical fire-alarm wiring diagram and riser diagram.

als designed for specific applications. Many systems operate on low-voltage circuits, but are installed similarly as conventional electrical circuits for light and power. All installations, when used in buildings, must conform to applicable National Electrical Code® (NEC) requirements, local ordinances, and instructions provided by security and fire-alarm system manufacturers and design engineers.

A signal circuit used for a security or fire-alarm system can be classified as open circuit or closed circuit. An *open circuit* is one in which current flows only when a signal is being sent. A *closed-circuit system* is one in which current flows continuously, except when the circuit is opened to allow a signal to be sent.

All alarm systems have three functions in common:

- Detection
- Control
- Signaling

6

Basic Installation Techniques

Before the installation of a security system is started, a sketch of the building or the work area should be prepared or the original working drawings (prints) should be obtained. This sketch should be drawn to scale and should show the location of all windows, doors, chases, and closets, etc. A simple riser diagram showing the various components (such as smoke and heat sensors, control panels, and alarm-signaling devices) should also appear on the sketch. When this is completed the security technician can begin the installation of the security system.

Installation Basics

The installation of a protective security/fire-alarm circuit should always start at the circuit's energy source, as if it were an end-of-the-line battery—a battery remote from the control panel—even though it might actually be a power supply installed in the control panel. A pair of wires are run from this power source to the first contact location, but just the positive wire is cut and connected to the two contact terminals (Fig. 6.1). The neutral or common wire is not cut, but continues on, parallel with the positive wire. The pair is then run to the next contact—be it door, window, or sensor—and again, only the hot (positive) wire is connected to the contacts. This procedure is repeated until all contacts are wired in series. Next the pair of wires is run from the last contact device on the system to the protective-circuit terminals in the main control panel. Although the markings will vary from manufacturer to manufacturer, the terminals for the starting connections will read something like LOOP POWER OUT, and at termination IN or a similar expression will be used.

A simple circuit of the wiring connections just described is shown in Fig. 6.2. Obviously, the system would operate with just a single-wire, positive-leg circuit run from contact to contact, with the negative (common) power-supply terminal connected directly to the negative protective-circuit terminal within the

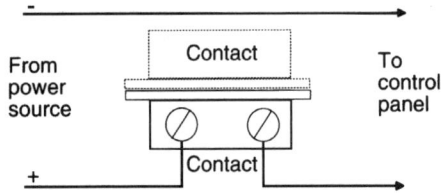

Figure 6.1 Contacts are connected only to the positive wire. The neutral or common remains unbroken.

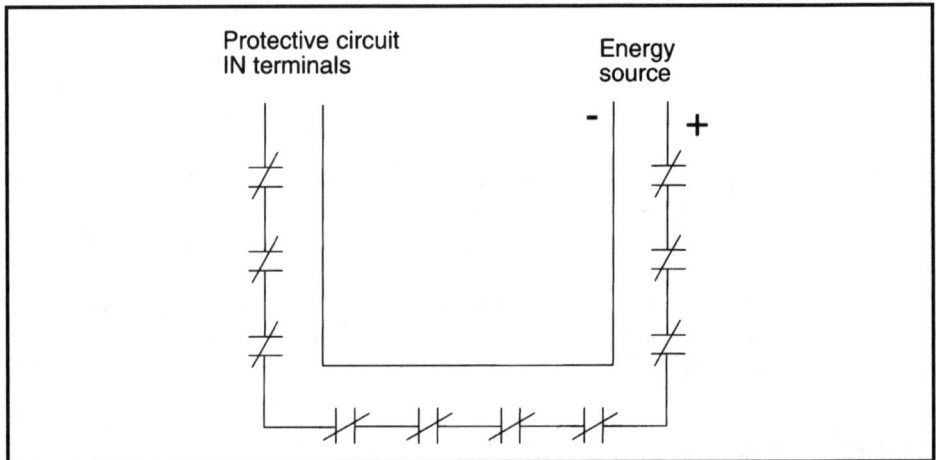

Figure 6.2 Both the negative and positive conductors are run to all contacts—even though the system should operate with just a single-wire, positive-leg wire run from contact to contact.

control panel or cabinet. However, manufacturers of security equipment discourage this practice because troubleshooting a single-wire circuit can be extremely time consuming and the single wire is more vulnerable to defeat by an intruder, with no trouble symptoms occurring to warn the user of the loss of protection.

An exit/entry delay relay is sometimes used on security systems so that authorized personnel can exit and enter (using their door keys) without activating the alarm. However, a shunt switch is used more often (Fig. 6.3). The purpose of the shunt lock is to enable an authorized person with a key to shunt out the contacts on the door used for entry/exit, allowing him or her to enter or leave the premises without tripping the alarm while the system is active. The shunt lock does extend outside the protected premises, however, and it is a potential weak link in the system. Using the following two procedures makes defeat of the shunt lock much more difficult.

- Install the shunt lock at the door that is most brightly illuminated and most readily visible to passersby.

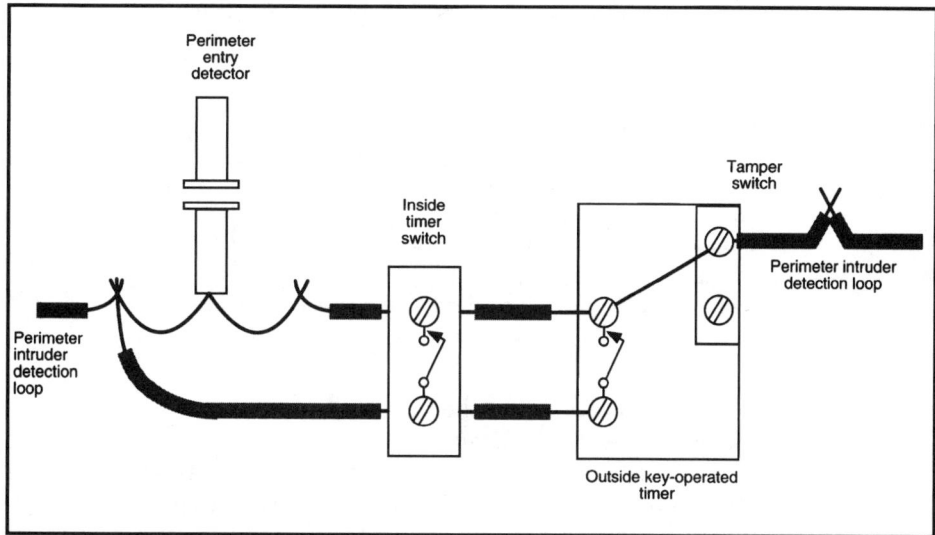

Figure 6.3 A typical shunt switch circuit.

■ Wire the shunt lock switch to the magnetic contact terminals (Fig. 6.4). This arrangement traps the lock so that any attempt to pull it out to gain access to its terminals will break the positive side of the protective circuit and cause an alarm to sound.

Contacts used to signal the opening of doors, windows, gates, drawings, etc., are usually mounted on the frame of the door or window and the magnet unit is mounted on the door or window (moving part) itself. The two units should be positioned so that the magnet is close to and parallel with the switch when the door or window is closed. This keeps the shunt lock actuated, but opening the door or window moves the magnet away and releases the switch mechanism.

As long as the faces of the switch and magnet are parallel and in close proximity when the door or window is closed, they can be oriented side-to-side, top-to-top, or top-to-side. Mounting spacers may be used under the units if necessary to improve their alignment and proximity.

The wiring of any alarm system is installed like any other type of low-voltage signal system; that is, one must locate the outlets, furnish a power supply, and finally interconnect the components with the proper size and type of wire.

Quality of workmanship

Because most security/fire-alarm systems are operated on low-voltage circuits, installers might be tempted not to pay as strict attention to the quality of the workmanship and materials as they would when installing conventional electrical wiring for lighting and power. But security/fire-alarm systems are worthy

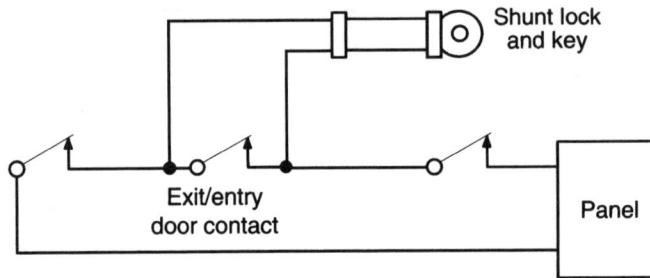

Figure 6.4 Wire the shunt lock switch to the magnetic contacts as shown.

of the best materials and the best workmanship and strict attention to quality work should always be given.

Care must be taken to ensure that all visible components are installed adjacent to and parallel to building lines in order to give a neat appearance. All wiring should be concealed where possible. Wiring that must be exposed should have square corners and should be installed so that it is as inconspicuous as possible.

Only new material of the highest quality should be used and this material should be approved by UL or a similar testing agency. Remember that the protection of the owner's building and its contents are dependent—to a great extent—on the quality of the security system installed.

Wiring Methods

Several types of wiring methods are used for security-system installations. The methods used on a given project are determined by several factors:

- The installation requirements set forth in the NEC and/or by the manufacturer of the security equipment
- Local codes and ordinances
- Type of building construction
- Location of the wiring in the building
- Importance of the wiring system's appearance
- Costs and budget

In general, two types of basic wiring methods are used in the majority of electrical systems:

- Open
- Concealed

In open-wiring systems, the outlets, security devices, and cable or raceway systems are installed on the walls, ceilings, columns, and other surfaces where they are in view and readily accessible. Such wiring is often used in areas where appearance is not important and where it might be desirable to make changes in the security system at a later date. You will frequently find open-wiring systems in mechanical rooms and in interior parking areas of commercial buildings and in almost every manufacturing area of industrial establishments.

Concealed wiring systems have all cable and raceway runs concealed inside of walls, partitions, ceilings, and columns and behind baseboards or molding, where they are out of view and not readily accessible. This type of wiring system is generally used in all new construction with finished interior walls, ceilings, floors, and is the preferred type where appearance is important.

Although most security/fire-alarm systems use low-voltage wiring for perimeter conductors to the various protective devices, the main control panel is nearly always powered with line voltage: 120 to 240 V. Many fire-detection devices also utilize line-voltage circuits to reduce the voltage drop over long circuit runs. Therefore, wiring methods suitable for these voltages are required and all such installations must comply with the latest edition of the NEC, along with all local ordinances.

The following wiring methods describe the majority of all systems that will be encountered by the security technician.

Raceway Systems

A *raceway* is any channel used to hold wires, cables, or busbars and is designed solely for this purpose. Types of raceways include rigid metal conduit, intermediate metal conduit (IMC), rigid nonmetallic conduit, flexible metal conduit, liquid-tight flexible metal conduit, electrical metallic tubing (EMT), underfloor raceways, cellular metal floor raceways, cellular concrete floor raceways, surface metal raceways, wireways, and auxiliary gutters. Raceways are constructed of either metal or insulating material. Raceways provide mechanical protection for the conductors that run in them and prevent accidental damage to insulation and the conducting metal. They also protect conductors from the chemical attack of corrosive atmospheres and prevent fire hazards to life and property by confining arcs from flame caused by faults in the wiring system.

One of the most important functions of metal raceways is to provide a path for the flow of fault current to ground, thereby preventing voltage build-up on conductor and equipment enclosures. This feature, of course, helps to minimize shock hazards to personnel and damage to electrical equipment. To maintain this feature, it is extremely important that all metal raceway systems be securely bonded together into a continuous conductive path. Properly connect the metal raceway system to the system ground, which, in turn, should be connected to a grounding electrode, such as a water pipe or a ground rod.

Rigid metal conduit

Rigid steel conduit is used for both exposed and concealed work. Where corrosion is likely to occur, corrosion-resistant rigid metal conduit and fittings must be used. Corrosion-resistant materials include aluminum, silicon bronze alloy, and plastic-coated steel. All conduit installed in wet locations should be mounted with at least $\frac{1}{4}''$ of air space between the conduit and the wall or other mounting surface.

Rigid metal conduit is available in 10-foot lengths (when the coupling is included in the measurement) with a threaded coupling on one end. The actual length of the conduit section alone is approximately 9' 10".

Most technicians prefer to use a hacksaw with a blade having 18 to 32 teeth per inch for cutting conduit (Fig. 6.5). For cutting larger sizes of conduit ($1\frac{1}{2}''$ and longer), a special conduit cutter should be used to save time. Although quicker to use, the conduit cutter almost always leaves an undesirable hump inside the conduit. If a power band saw is available on the job, it is preferred for cutting the larger sizes of conduit. Abrasive cutters are also popular for the large conduit.

Conduit cuts should be made square and the inside edge of the cut must be reamed to remove any burr or sharp edge that might damage wire insulation when the conductors are pulled inside the conduit. After reaming, most experienced electricians feel the inside of the cut with their finger to be sure that no burrs or sharp edges are present.

Lengths of conduit to be cut should be accurately measured for the size needed and an additional $\frac{3}{8}''$ should be allowed on the smaller sizes of conduit for terminations; the larger sizes of conduit will require approximately $\frac{1}{2}''$ for locknuts, bushings, etc. at terminations.

A good lubricant (cutting oil) is used liberally during the thread-cutting process. If sufficient lubricant is used, cuts can be made cleaner and sharper, and the cutting dies will last much longer.

Full threads must be cut to allow the conduit ends to come close together in the coupling or to firmly seat in the shoulders of threaded hubs of conduit bodies. To obtain a full thread, run the die up on the conduit until the conduit barely comes through the die. This will give a good thread length adequate for all purposes. Anything longer will not fit into the coupling and will later corrode because threading removes the zinc or other protective coating from the conduit.

Clean, sharply cut threads also make a better continuous ground and save much trouble once the system is in operation.

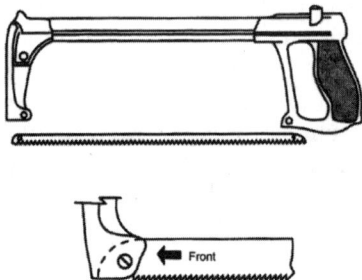

Figure 6.5 Most technicians prefer to use a hacksaw with a blade having 18 teeth per inch for cutting rigid steel conduit and 32 teeth per inch for cutting thinner-walled conduit.

Plastic-coated rigid steel conduit

Plastic-coated rigid conduit has a thin coating of polyvinyl chloride (PVC) over the metal conduit. This conduit is used when an environment calls for the ruggedness of rigid steel conduit and the corrosion resistance of PVC. The following establishments frequently use plastic-coated rigid steel conduit in their electrical wiring installations:

- Chemical plants
- Refineries
- Fertilizer plants
- Paper mills
- Waste-water treatment plants

This type of wiring protection should be used even for low-voltage security wiring when the systems are installed in these establishments.

Aluminum conduit

Aluminum conduit (Fig. 6.6) is typically used when chemical resistance to wet environments and some chemical environments is necessary. Aluminum conduit generally requires less maintenance in such installations as sewage plants and installations around salt water because of its special characteristics.

Aluminum conduit, being made of a nonmagnetic metal, can reduce voltage drop in conductors. Where ac current is involved, voltage drop in conductors

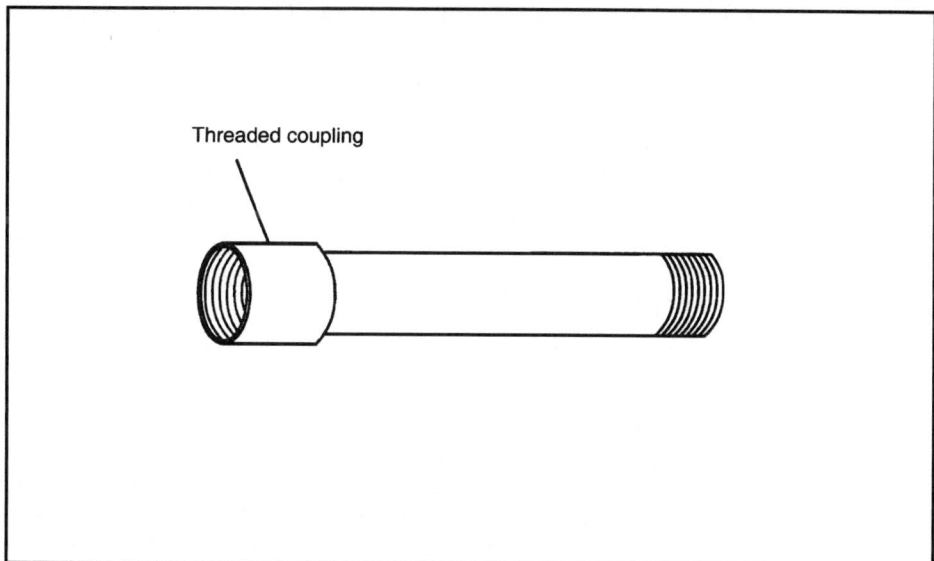

Figure 6.6   Aluminum conduit and threaded coupling.

placed in aluminum conduit can be as much as 20 percent less than that of conductors in steel conduit.

Aluminum conduit is much lighter than steel conduit. A 10-foot section of 3" aluminum conduit weighs about 23 lb. compared to the 68-lb. weight of its steel counterpart. This difference in weight provides some installation advantages. Prices of steel and aluminum vary regularly. However, in general, the price of aluminum conduit is about a third more than that of rigid steel conduit. Because aluminum conduit can be installed easily, the labor costs on some conduit installations will be lowered. Therefore, using aluminum conduit could actually cost less than using steel conduit.

Note
An anti-seize compound should be used on aluminum threaded fittings.

Intermediate metal conduit

Intermediate metal conduit (IMC) was developed to reduce the cost, but retain the high strength of a rigid steel conduit system. IMC has a slightly larger internal diameter than galvanized rigid conduit (GRC). That is, the wall thickness of IMC is less than that of GRC. And size for size, IMC weighs approximately two-thirds less than GRC. The external diameter of a given size of IMC is the same as that of the same size GRC. Therefore, GRC fittings can be used with IMC. Because the threads on IMC and GRC are the same, no special threading tools are needed. Although the internal diameter of a given size of IMC is larger than that of the same size GRC, the conductor fill for both IMC and GRC is the same.

Pulling wire in IMC is often easier than pulling the same size and quantity of wire in GRC. That's because of the IMC's slightly larger internal diameter. Some contractors feel that threading IMC is more difficult than threading GRC because IMC is less ductile than GRC. Bending IMC is easier than bending GRC because of the reduced wall thickness. However, bending is sometimes complicated by "kinking," which may be caused by the increased hardness of IMC. Bending IMC is more difficult than bending EMT.

Rigid nonmetallic conduit

Rigid nonmetallic conduit is manufactured from a polyvinyl chloride base material (PVC). For this reason, rigid nonmetallic conduit is often referred to as "PVC." Because PVC conduit is noncorrosive, chemically inert, and non-aging, it is an excellent choice for installation in wet or corrosive environments. The corrosion problems that can occur with steel and aluminum rigid metal conduits do not occur with PVC.

PVC conduit is lighter than steel or aluminum rigid steel, IMC, and EMT. PVC is much easier to handle because it's lighter. And because the PVC joints are made with glue and require no threading, PVC conduit can usually be installed much faster than other types of conduit.

PVC conduit contains no iron or other metal. Because of this characteristic, the voltage drop of conductors carrying ac current in PVC will be less than that of identical conductors in metal conduit. Also, because of the absence of iron, no magnetic flux can be imposed into PVC by an ac current.

Electrical metallic tubing

Electrical metallic tubing (EMT) can be used for both exposed and concealed work, except where it will be subjected to severe damage during use. In cinder concrete, or in fill where it will be subjected to permanent moisture the tubing may be installed a minimum of 18 inches under the fill.

Threadless couplings and connectors are used for EMT installation and these should be installed so that the tubing will be made up tight. Where buried in masonry or installed in wet locations, couplings and connectors, as well as supports, bolts, straps, and screws, should be of a type approved for the conditions.

Bends in the tubing should be made with a tubing bender so that no injury will occur and so the internal diameter of the tubing will not be effectively educed. The bends between outlets or termination points should contain no more than the equivalent of four quarter-bends (360° total), including those bends located immediately at the outlet or fitting (offsets).

All cuts in EMT are made with either a hacksaw, power hacksaw, tubing cutter, or other approved device. Once cut, the tubing ends should be reamed with a pipe reamer to remove all burrs and sharp edges that might damage conductor insulation.

EMT couplings

EMT is joined by couplings. Two types are in common use:

- Set-screw couplings
- Compression couplings

Both types are shown in Fig. 6.7 and a brief description of each follows.

As its name implies, the set-screw coupling relies on set screws to hold the EMT to the coupling. This type of coupling does not provide a seal and is not permitted to be used in wet locations. However, this type of coupling can be embedded in concrete.

Compression couplings provide a tight seal around the conduit, and can be used in some wet locations as stated in the NEC.

Flexible metal conduit

Flexible metal conduit is manufactured in two types, a standard metal-clad tube and a liquid-tight type. The former type cannot be used in wet locations unless the conductors pulled in are of a type specially approved for such conditions.

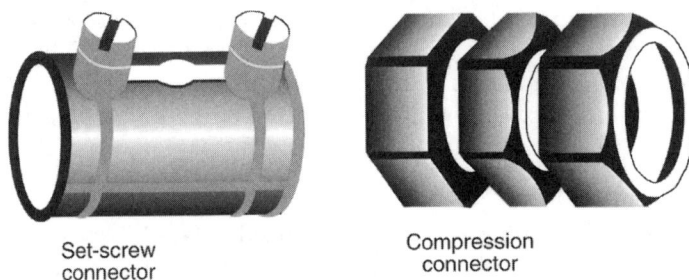

Set-screw
connector

Compression
connector

Figure 6.7 Two types of EMT couplings in common use.

Neither type can be used where they will be subjected to physical damage or where any combination of ambient and/or conductor temperature will produce an operating temperature in excess of that for which the material is approved. Other uses are fully described in Articles 350 and 351 of the NEC.

When this type of conduit is installed it should be secured by an approved means at intervals not exceeding $4\frac{1}{2}$ feet and within 12″ of every outlet box, fitting, or other termination points. In some cases, however, exceptions exist. For example, when flexible metal conduit must be finished in walls and ceilings, securing the conduit at these intervals would not be practical. Also, where more flexibility is required, lengths of not more than 3 feet can be utilized at termination points.

Flexible metal conduit can be used as a grounding device where both the conduit and the fittings are approved for the purpose. In lengths of more than 6 feet, it is best to install an extra grounding conductor within the conduit for added insurance.

Surface metal molding

When it is impractical to install the wiring in concealed areas, surface metal molding (Figs. 6.8 through 6.11) is a good compromise. Although it is visible, the proper painting of surface metal molding to match the color of the ceiling and walls makes it inconspicuous. Surface metal molding is made from sheet metal strips drawn into shape and comes in various shapes and sizes with factory fittings to meet nearly every application found in finished areas of commercial buildings. A complete list of fittings can be obtained at your local electrical equipment supplier.

The running of straight lines of surface molding is simple. A length of molding with the coupling is slipped in the end and out enough that the screw hole is exposed, and then the coupling is screwed to the surface to which the molding is to be attached. Next, another length of molding is slipped on the coupling.

Factory fittings are used for corners and turns or the molding can be bent (to a certain extent) with a special bender. Matching outlet boxes for surface mounting are also available, and bushings are necessary at such boxes to prevent the sharp edges of the molding from injuring the insulation on the wire.

INSTALL 2000B BASE ON SURFACE.
Starting with feed section, mount entire run off
2000B base with No. 8 flat head screw through
screw piercings and knockouts. Cut base to
length at corners and end of run, as required.

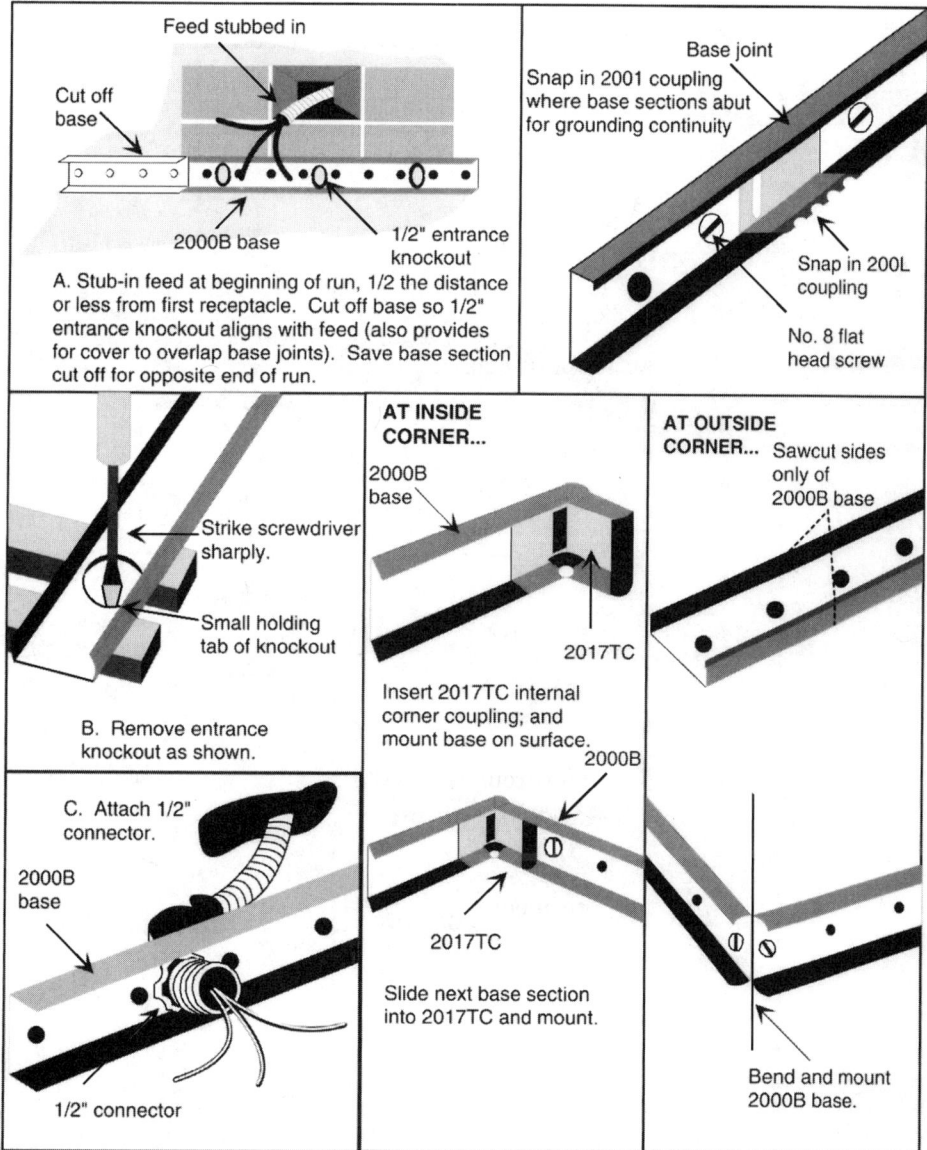

BRING FEED INTO 2000B BASE.
Back-feed connection shown.

Feed stubbed in

Cut off
base

2000B base

1/2" entrance
knockout

A. Stub-in feed at beginning of run, 1/2 the distance
or less from first receptacle. Cut off base so 1/2"
entrance knockout aligns with feed (also provides
for cover to overlap base joints). Save base section
cut off for opposite end of run.

Base joint

Snap in 2001 coupling
where base sections abut
for grounding continuity

Snap in 200L
coupling

No. 8 flat
head screw

Strike screwdriver
sharply.

Small holding
tab of knockout

B. Remove entrance
knockout as shown.

**AT INSIDE
CORNER...**

2000B
base

2017TC

Insert 2017TC internal
corner coupling; and
mount base on surface.

2000B

2017TC

Slide next base section
into 2017TC and mount.

**AT OUTSIDE
CORNER...** Sawcut sides
only of
2000B base

Bend and mount
2000B base.

C. Attach 1/2"
connector.

2000B
base

1/2" connector

Figure 6.8   Installation methods of Wiremold 2000 Snapicoil.

CONNECT SNAPICOIL
TO FEED.

Lay out Snapicoil along entire run of base so that receptacles are not located over feed or in corners. Connect to feed wires with W30 pressure type wire connectors (for common connection of 2, 3, or 4 No. 12 or No. 14 solid conductors); NOT TO BE USED to connect equipment grounding conductors. Insert only conductors of same color in a connector.

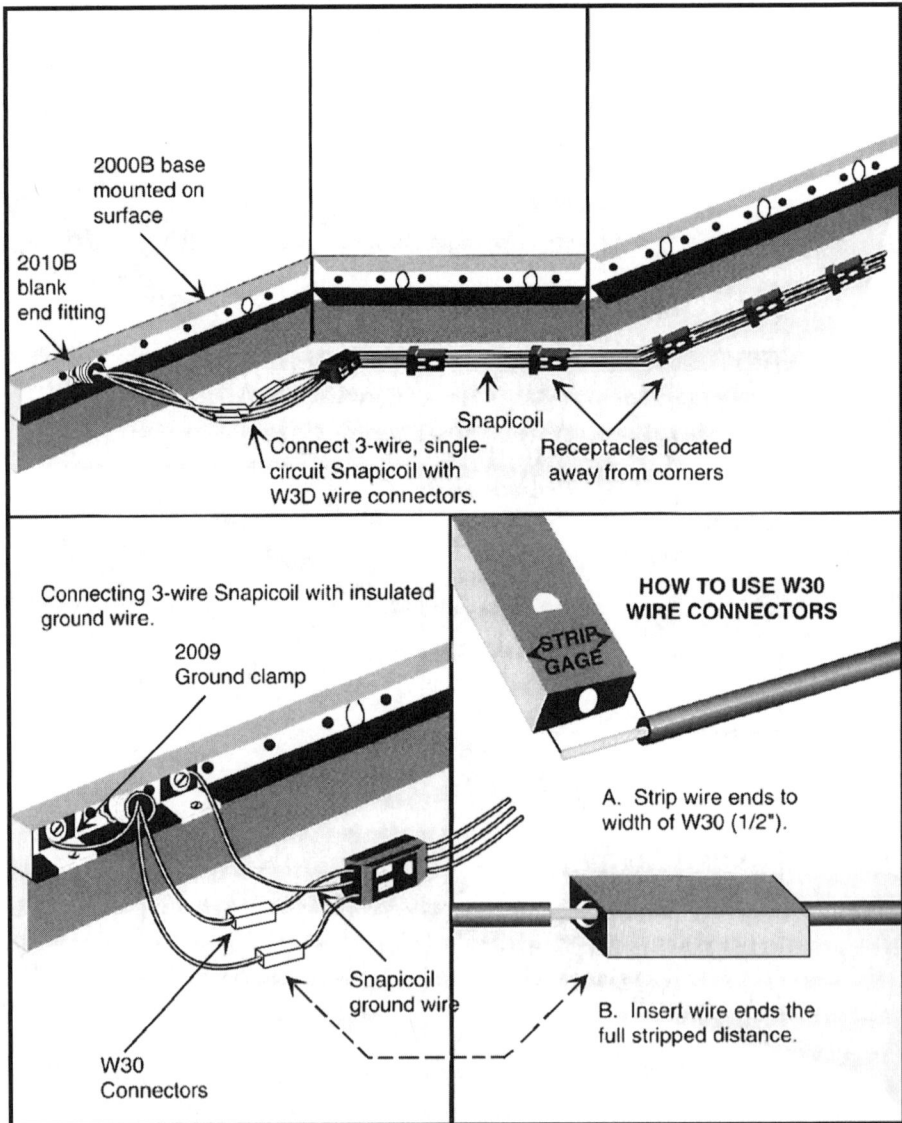

Figure 6.9 Installation methods of Wiremold 2000 Snapicoil (*Cont.*).

ASSEMBLE SNAPICOIL AND 2000C COVER IN 2000B BASE.

Cover sections should overlap base joints for rigidity and better ground continuity.

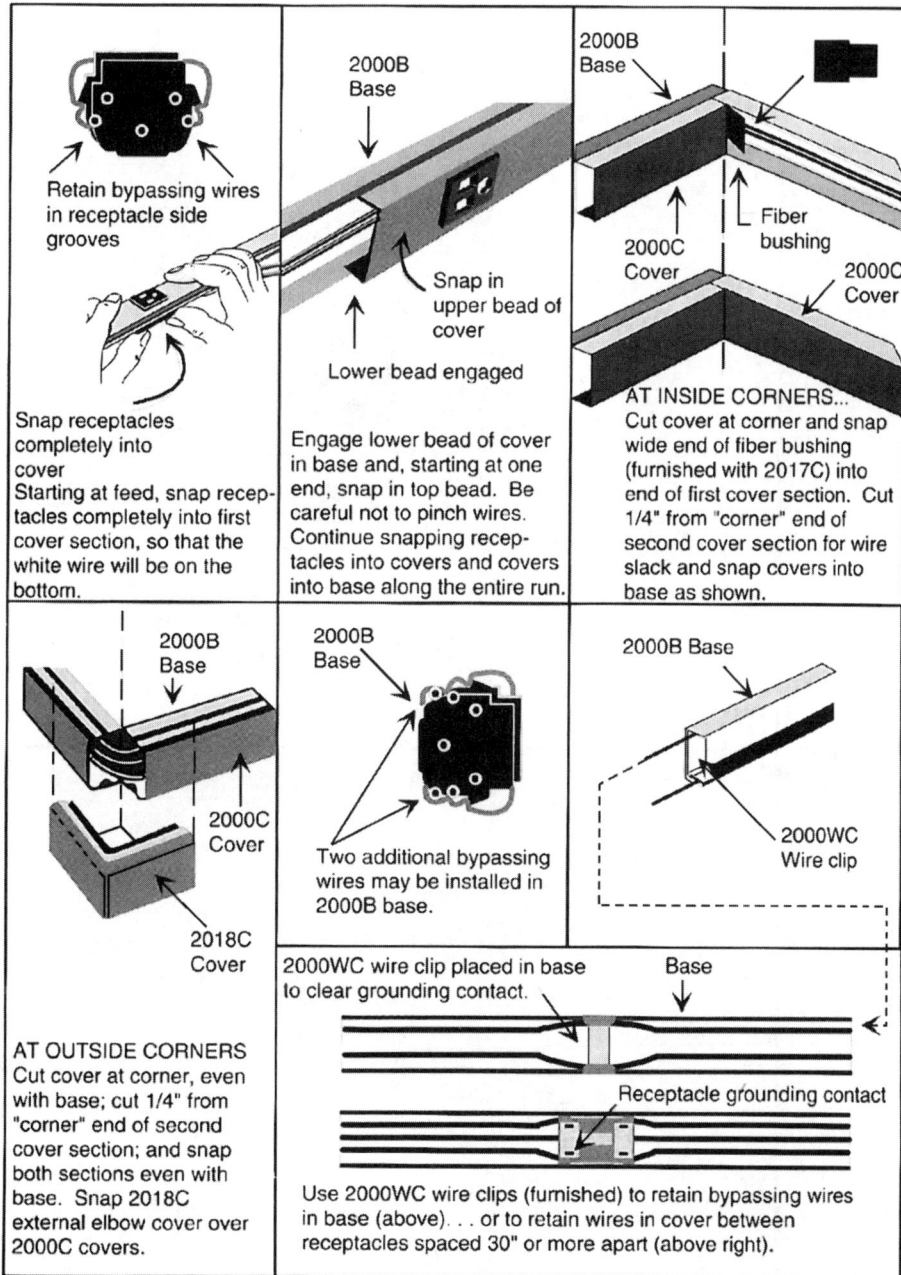

Retain bypassing wires in receptacle side grooves

Snap receptacles completely into cover
Starting at feed, snap receptacles completely into first cover section, so that the white wire will be on the bottom.

2000B Base

Snap in upper bead of cover

Lower bead engaged

Engage lower bead of cover in base and, starting at one end, snap in top bead. Be careful not to pinch wires. Continue snapping receptacles into covers and covers into base along the entire run.

2000B Base

Fiber bushing

2000C Cover

2000C Cover

AT INSIDE CORNERS...
Cut cover at corner and snap wide end of fiber bushing (furnished with 2017C) into end of first cover section. Cut 1/4" from "corner" end of second cover section for wire slack and snap covers into base as shown.

2000B Base

2000C Cover

2018C Cover

AT OUTSIDE CORNERS
Cut cover at corner, even with base; cut 1/4" from "corner" end of second cover section; and snap both sections even with base. Snap 2018C external elbow cover over 2000C covers.

2000B Base

Two additional bypassing wires may be installed in 2000B base.

2000B Base

2000WC Wire clip

2000WC wire clip placed in base to clear grounding contact.

Base

Receptacle grounding contact

Use 2000WC wire clips (furnished) to retain bypassing wires in base (above). . . or to retain wires in cover between receptacles spaced 30" or more apart (above right).

Figure 6.10   Installation methods of Wiremold 2000 Snapicoil (*Cont.*).

Surface Metal Raceway

NEC Article 352

Must be of such construction as will distinguish them from other raceways

Surface metal raceways must comply with the applicable provisions of NEC Article 300

Must have means for connecting an equipment grounding conductor NEC Section 352-9

The number of conductors installed must be no greater than the number for which the raceway is designed NEC Section 352-4

May extend through dry walls, partitions, and floors, if lengths are unbroken NEC Section 352-5

Where used for both for signaling and for lighting and power, the different systems must be run in separate compartments of the raceway system

Splices are permitted if they are accessible after installation NEC Section 352-7

No conductor larger than that for which the raceway is designed shall be installed NEC Section 352-3

Multiout assembly

A type of surface or flush raceway

designed to hold conductors and receptacles,

assembled in the field or at the factory.

May extend through, but not in, dry partitions

Must not be used:

Where concealed

Where subject to physical damage

Where voltage is 300 V or more

In hoistways

Must be used only in dry locations

NEC Article 353

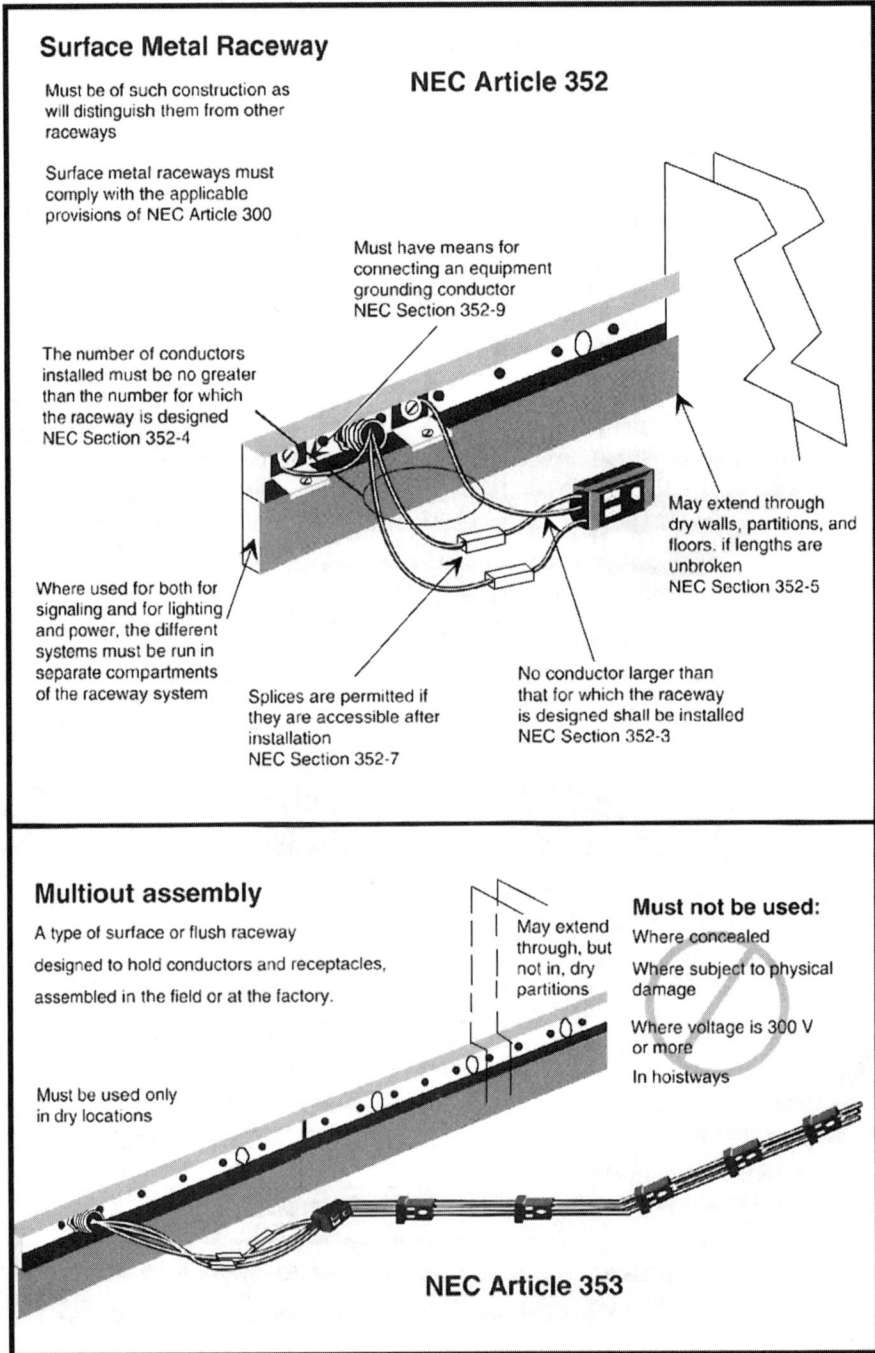

Figure 6.11   Installation methods of Wiremold 2000 Snapicoil (*Cont.*).

Clips are used to fasten the molding in place. The clip is secured by a screw and then the molding is slipped into the clip, wherever extra support of the molding is needed, and then fastened by screws. When parallel runs of molding are installed, they can be secured in place by means of a multiple strap. The joints in runs of molding are covered by slipping a connection cover over them. Runs of molding should be grounded the same as any other metal raceway, and this is done by use of grounding clips. The current-carrying wires are normally pulled in after the molding is in place.

The installation of surface metal molding requires no special tools unless bending the molding is necessary. The molding is fastened in place with screws, toggle bolts, etc., depending on the materials to which it is fastened. All molding should be run straight and parallel with the room or building lines (baseboards, trims, and other room moldings). The decor of the room should be considered first and the molding made as inconspicuous as possible.

It is often desirable to install surface molding not used for wires in order to complete a pattern set by other surface molding containing current-carrying wires, or to continue a run to make it appear to be part of the room's decoration.

Wireways

Wireways are sheet-metal troughs with hinged or removable covers for housing and protecting wires and cables and in which conductors are held in place after the wireway has been installed as a complete system. They can only be used for exposed work and shouldn't be installed where they will be subject to severe physical damage or corrosive vapor, nor in any hazardous location, except NEC Class II, Division 2.

The wireway structure must be designed to safely handle the sizes of conductors used in the system. Furthermore, the system should not contain more than 30 current-carrying conductors at any cross section. The sum of the cross-sectional areas of all contained conductors at any cross section of a wireway shall not exceed 20 percent of the interior cross-sectioned area of the wireway.

Splices and taps, made and insulated by approved methods, can be located within the wireway provided that they are accessible. The conductors, including splices and taps, shall not fill the wireway to more than 75 percent of its area at that point.

Wireways must be securely supported at intervals not exceeding 5 feet, unless specially approved for supports at greater intervals, but in no case shall the distance between supports exceed 10 feet.

Wireway fitting

Wireways are constructed of four major parts:

- The main housing or trough
- End cap

- U connector
- Trough cover

Also, many different shapes of connectors are designed to suit practically any application, for example, the trough cross, 90-degree internal elbow, and T connector.

The wireway fittings are attached to the wireway duct with slip-on connectors. All attachments are made with nuts and bolts or screws. Usually, it is best to assemble sections of the wireway system on the floor and then raise the sections into position by hand. Both the fittings and the duct come with screw or hinged covers to permit conductors to be laid in or pulled through.

As you can see in Fig. 6.12, this type of raceway is also manufactured in a raintight enclosure for outdoor use.

Busways (Duct Systems)

Several styles of busways or duct systems are designed for electrical transmission and feeder purposes. Lighting duct, trolley duct, and distribution bus duct are just a few. Each one is designed for a specific purpose, and the security technician should become familiar with all types before an installation is laid out.

Lighting duct, for example, permits the installation of an unlimited amount of footage from a single working platform. As each section and the lighting fixtures are secured in place, the complete assembly is then simply transported to the area of installation and installed in one piece.

Trolley duct is widely used for industrial applications. Where the installation requires a continuous polarization to prevent accidental reversal, a polarizing bar is used. This system provides polarization for all trolleys, permitting standard and detachable trolleys to be used on the same run.

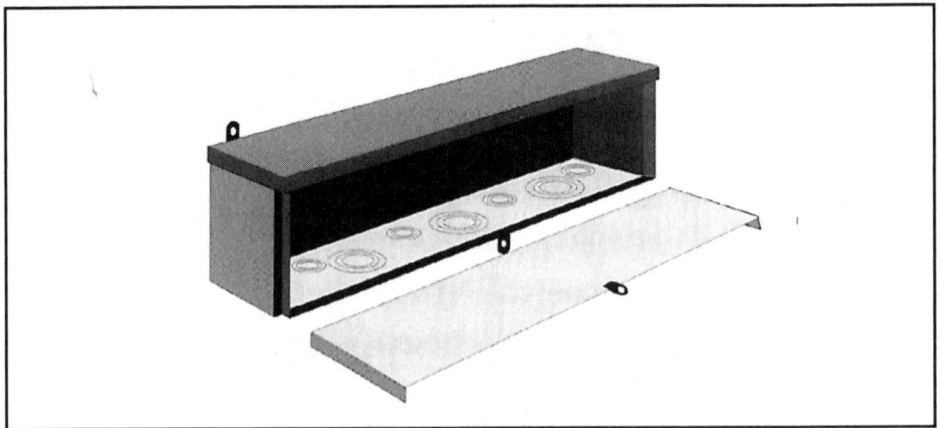

Figure 6.12　Raintight wireway.

Plug-in bus duct is also widely used for industrial applications, and the system consists of interconnected prefabricated sections of bus duct so formed that the complete assembly will be rigid in construction and neat and symmetrical in appearance.

Cable Trays

Cable trays are used to support electrical conductors. They are mainly used in industrial applications, but are sometimes used for communication and data-processing conductors in large commercial establishments. The trays themselves are usually made up into a system of assembled, interconnected sections and associated fittings, all of which are made of metal or other noncombustible material. The finished system forms into a rigid structural run to contain and support single, multiconductor, or other wiring cables. Several styles of cable trays are available, including ladder, trough, channel, solid-bottom trays, and similar structures (Fig. 6.13).

A box or fitting must be installed at:

- Each conductor splice point
- Each outlet, switch point, or junction point
- Each pull point for the connection of conduit and other raceways

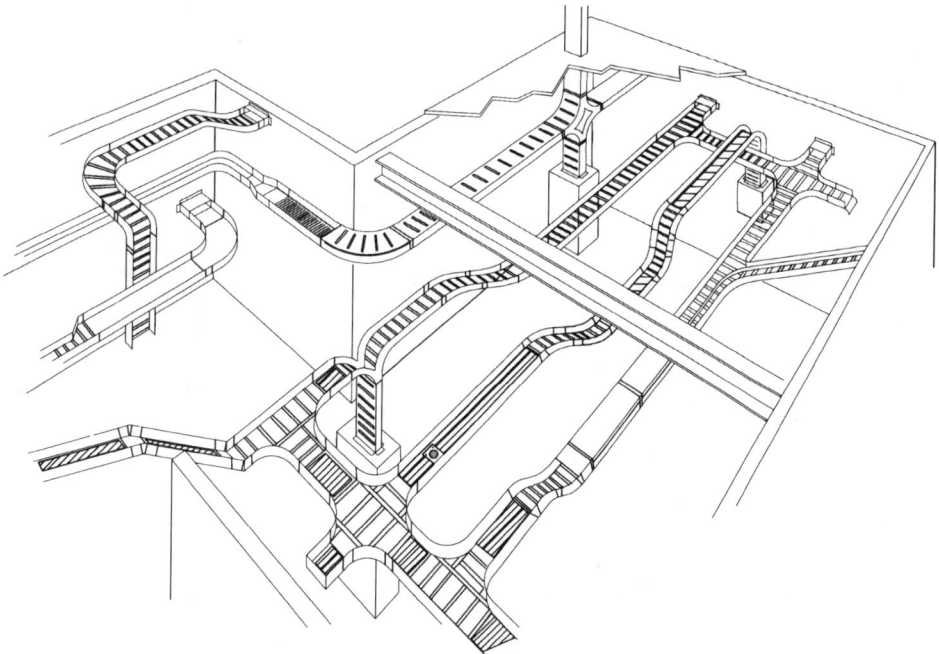

Figure 6.13 Typical cable tray system.

Furthermore, boxes or other fittings are required when a change is made from conduit to open wiring. Electrical workers also install pull boxes in raceway systems to facilitate the pulling of conductors.

In each case (raceways, outlet boxes, and pull and junction boxes), the NEC specifies specific maximum fill requirements; that is, the area of conductors in relation to the box, fitting, or raceway system.

Cable systems

Several types of cable systems are used to feed security equipment and also to provide perimeter-protection circuits. We have already discussed some types of low-voltage cable in other chapters, so a description of this type of cable is not repeated here.

Instrumentation Control Wiring: This type of cable links the field-sensing, controlling, printout, and operating devices that form an electronic instrumentation control system for sophisticated security systems in large commercial buildings. The style and size of the control wiring must be matched to a specific job.

Control wiring usually has two or more insulated conductors (Fig. 6.14). An outer layer (called the *jacket*) protects the pairs of conductors inside. The number of pairs in a multiconductor cable depends on the size of the wire used. A multipair cable might consist of as many as 60 (or more) pairs of conductors.

Shields are provided on control wiring to protect the electrical signals traveling through the conductors from electrical interference or noise. Shields are

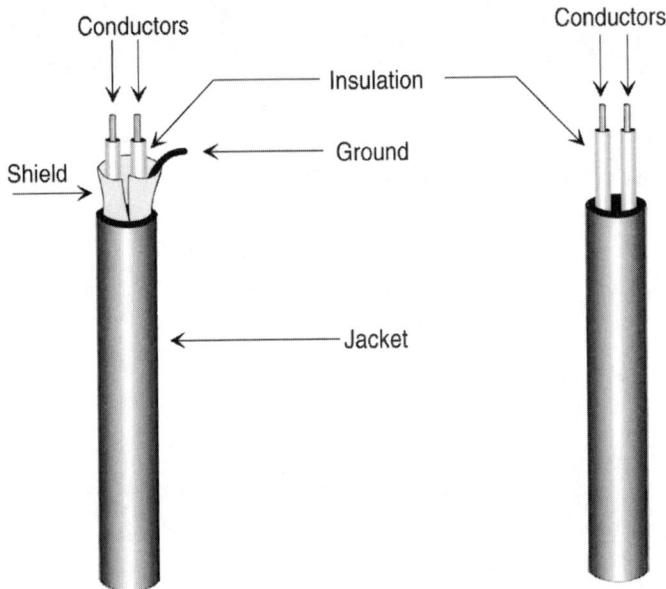

Figure 6.14   Two-conductor instrumentation control cable.

usually constructed of aluminum foil bonded to a plastic film. If the wiring is not properly shielded, electrical disturbances could cause erratic or erroneous control signals, false alarms, and improper operation of security devices.

A *ground wire* is a bare copper wire used to provide continuous contact with a specified grounding terminal. A ground wire allows connections of all the instruments within a loop to a common grounding system. In some electronic systems, the grounding wire is called a *drain wire.*

In most cases, instruments connected to the system are not grounded at both ends of the circuit. This is to prevent unwanted ground loops in the system. If the ground is not to be connected at the end of the wire, do not remove the ground wire. Rather, fold it back and tape it to the cable. This is called *floating the ground.* This is done in case the ground at the opposite end of the conductors develops a problem.

Jackets: A synthetic plastic jacket covers and protects the components within the cable. Polyethylene and PVC jackets are most commonly used. Some cable jackets have a nylon rip cord that allows the jacket to be peeled back without the use of a knife or cable cutter. This eliminates the possible nicking of the conductor insulation when preparing for terminations of the conductors.

Nonmetallic-Sheathed Cable (Type NM): This cable is manufactured in two- or three-wire assemblies, and with varying sizes of conductors. In both two- and three-wire cables, conductors are color-coded: open conductor is black while the other is white in two-wire cable; in three-wire cable, the additional conductor is red. Both types will also have a grounding conductor that is usually bare, but is sometimes covered with a green plastic insulation, depending upon the manufacturer. The jacket or covering consists of rubber, plastic, or fiber. Most will also list the manufacturer's name or trademark, the wire size, and the number of conductors on the jacket. For example, "NM 12-2 W/GRD" indicates that the jacket contains two #12 AWG conductors along with a grounding wire; "NM 12-3 W/GRD" indicates three conductors plus a grounding wire. This type of cable can be concealed in the framework of buildings, or in some instances, can be run exposed on the building surfaces. It may not be used in any building exceeding three floors above grade; as a service-entrance cable; in commercial garages having hazardous locations; in theaters and similar locations; places of assembly; in motion picture studios; in storage battery rooms; in hoistways; embedded in poured concrete, or aggregate; or in any hazardous location, except as otherwise permitted by the NEC. Nonmetallic sheathed cable is frequently referred to as Romex on the job (Fig. 6.15).

Type AC (Armored) Cable: Type AC cable (commonly called *BX*) is manufactured in two-, three-, and four-wire assemblies, with varying sizes of conductors, and is used in locations similar to those where Type NM cable is allowed. The metallic spiral covering on BX cable offers a greater degree of mechanical protection than with NM cable, and the metal jacket also provides a continuous grounding bond without the need for additional grounding conductors.

BX cable can be used for under-plaster extensions, as provided in the NEC, and embedded in plaster finish, brick, or other masonry, except in damp or wet

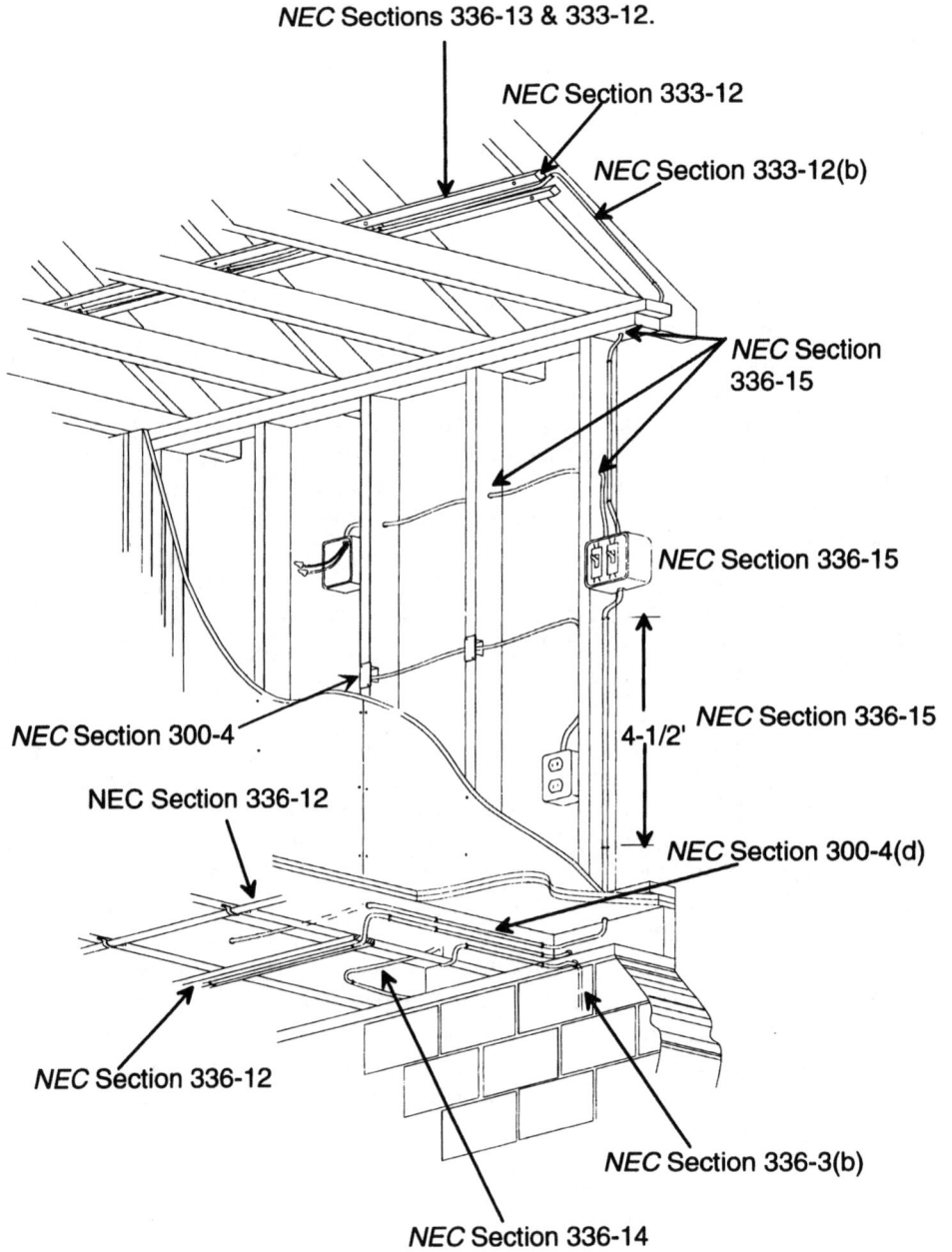

Figure 6.15   NEC installation requirements for Type NM cable.

locations. It can also be run or "fished" in the air voids of masonry block or tile walls, except where such walls are exposed or subject to excessive moisture or dampness, or are below grade. This type of cable is a favorite for connecting line-voltage devices above acoustical ceilings in commercial installations (Fig. 6.16).

Type NMC Cable: This type of cable is similar in appearance and used as conventional Type NM cable, except that NM cable is suitable for dry locations only and Type NMC cable is permitted for dry, moisture, damp, or corrosive locations.

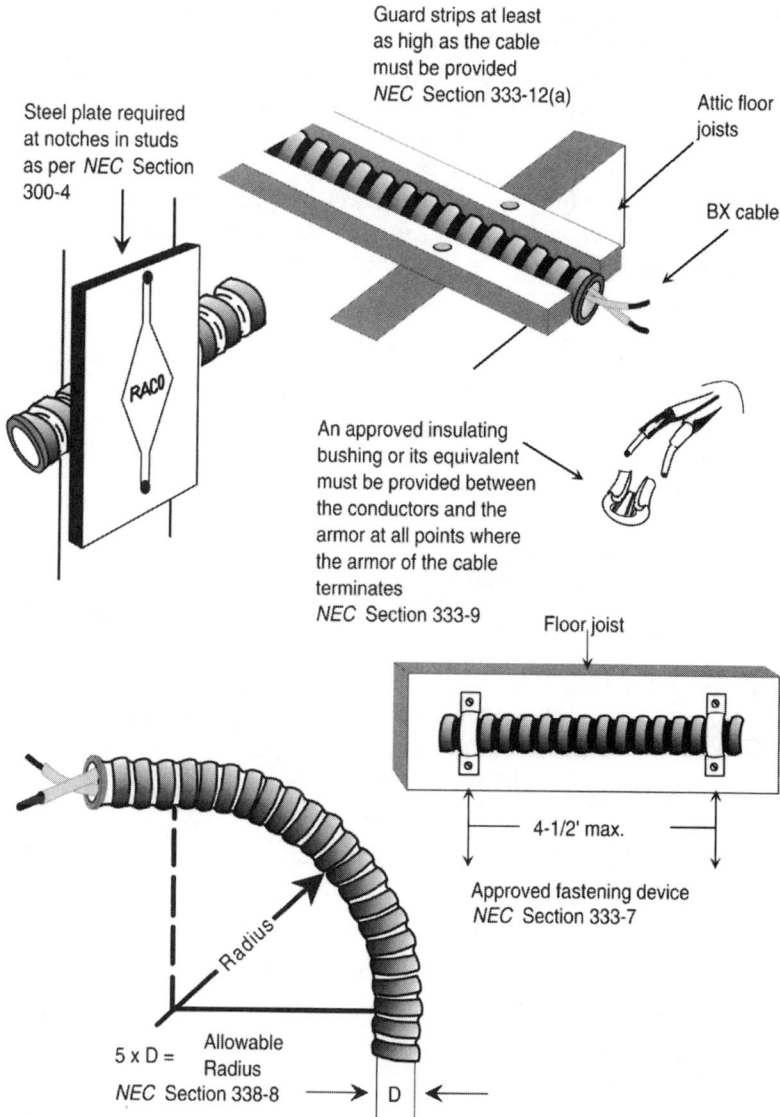

Figure 6.16 *NEC* installation requirements for Type AC cable.

Underground Feeder Cable: Type UF cable can be used underground, including direct burial in the earth, as a feeder or branch-circuit cable, when provided with overcurrent protection at the rated ampacity, as required by the NEC. When Type UF cable is used above grade where it will come in direct contact with the rays of the sun, its outer covering must be sun resistant. Furthermore, where Type UF cable emerges from the ground, some means of mechanical protection must be provided. This protection may be in the form of conduit or guard strips. Type UF cable resembles Type NM cable in appearance. The jacket, however, is constructed of weather resistant material to provide the required protection for direct-burial wiring installations.

Shielded Nonmetallic-Sheathed Cable: Type SNM cable consists of a conductor assembly that is enclosed in a PVC jacket, an overlapping layer of copper tape, a wire shield, and, finally, a nonmetallic sheath. This rugged assembly is as indicated in the description given in NEC Article 337. This cable is intended for use in cable trays and other raceways and is suitable for use in hazardous locations under certain conditions, as prescribed in NEC Articles 500 and 516.

Electrical Circuits

A variety of materials are used to transmit electrical energy, but copper, because of its excellent cost-to-conductivity ratio, still remains the basic and most ideal conductor. Electrolytic copper, the type used for most electrical applications, can have three general characteristics:

- Method of stranding
- Degree of hardness (temper)
- Bare, tinned, or coated

Method of Stranding: Stranding refers to the relative flexibility of the conductor. A conductor might consist of only one strand or many thousands, depending on the rigidity or flexibility required for a specific need. For example, a small-gauge wire that is to be used in a fixed installation is normally solid (one strand), whereas a wire that will be constantly flexed requires a high degree of flexibility and would contain many strands.

- *Solid* is the least-flexible form of a conductor and is merely one strand of copper.
- *Stranded* refers to more than one strand in a given conductor and can vary from 3 to 37, depending on size.
- *Flexible* simply indicates that it has a greater number of strands than are found in normal stranded wire.

Degree of Hardness: Temper refers to the relative hardness of the conductor and is noted as soft drawn-annealed (SD), medium hard drawn (MHD), and hard drawn (HD). Again, the specific need of an installation will determine the required temper. Where greater tensile strength is indicated, MHD would be specified over SD, and so on.

Bare, Tinned, or Coated: Untinned copper is plain bare copper that is available in either solid, stranded, or flexible and in the various tempers just described. In this form, it is often referred to as *red copper.*

Bare copper is also available with a coating of tin, silver, or nickel to facilitate soldering, to impede corrosion, and to prevent adhesion of the copper conductor to rubber or other types of conductor insulation. The various coatings will also affect the electrical characteristics of copper.

Conductor Size: The American Wire Gauge (AWG) is used in the United States to identify the sizes of wire and cable up to and including #4/0 (0000), which is commonly pronounced in the electrical trade as "four-aught" or "four-naught." These numbers run in reverse order as to size; that is, #14 AWG is smaller than #12 AWG, and so on up to size #1 AWG. To this size (#1 AWG), the larger the gauge number, the smaller the size of the conductor. However, the next larger size after #1 AWG is #1/0 AWG, then #2/0 AWG, #3/0 AWG, and #4/0 AWG. At this point, the AWG designations end and the larger sizes of conductors are identified by circular mils (cmil). From this point, the larger the size of wire, the larger the number of circular mils. For example, 300,000 cmil is larger than 250,000 cmil. In writing these sizes in circular mils, the "thousand" decimal is replaced by the letter k. For example, instead of writing *500,000 cmil,* it is usually written as *500 kcmil.*

Properties of conductors

Various NEC tables define the physical and electrical properties of conductors. Electricians and security technicians use these tables to select the type of conductor and the cable jacket or size of raceway to enclose the conductors in specific applications.

NEC tables tabulate properties of conductors as follows:

- Name
- Operating temperature
- Application
- Insulation
- Physical properties
- Electrical resistance
- ac resistance and reactance

NEC Table 310-13 gives the name, operating temperature, application, and insulation of various types of conductors, while tables in NEC Chapter 9 (Tables 5, 6, 7, 8, and 9) provide the physical properties and electrical resistance.

Conductor insulation

Thermoplastic is a popular and effective insulation material used in conductors for security systems. The following thermoplastics are widely used as insulation materials:

Polyvinyl Chloride (PVC): The base material used for the manufacture of TW and THW insulation.

Polyethylene (PE): An excellent weatherproofing material used primarily for insulation of control and communication wiring.

Cross-linked Polyethylene (XLP): An improved PE with superior heat- and moisture-resistant qualities. Used for THHN, THWN, and XHHW wire.

Nylon: Primarily used as jacketing material. THHN building wire has an outer coating of nylon.

Teflon: A high-temperature insulation. Widely used for telephone and security-system wiring in a plenum (where other insulated conductors require conduit routing).

Letter coding

Conductor insulation is coded by letters. The letters generally, but not always, indicate the type of insulation or its environmental rating. Some types of cable have the number of conductors, wire size, and other information inscribed on the cable jacket. The conductor insulation in Fig. 6.17 apply to those used in NEC wire tables.

Low-voltage transformers are used extensively in security-system control circuits to obtain a lower voltage than is available from the main power supply. For example, many control circuits operate at 24 V; normally, 120 V is the lowest voltage rating used in any building's electrical system. Therefore, a control transformer is used to reduce the 120-V circuit to the required 24 V. In selecting such a transformer, Class-2, low-voltage control systems are limited to transformers with a maximum output capacity of 75 volt-amperes (V) (watts). If a control transformer has a smaller output capacity than is required to service the system, it will eventually fail. Therefore, security systems requiring the use of a transformer to reduce voltage in the control circuits should be checked to ensure that the rating of the transformer will not be exceeded when the system is in use.

Electrical line-current capacity

Electrical conductors must be sized according to the NEC as well as the requirements of good wiring practices. Besides the information given in various NEC tables as to the allowable amperes that will safely flow through any given wire size, the wire or conductor should also be sized to limit the voltage drop to a maximum of two percent in any electrical circuit. This ensures efficient operation of both controls and equipment.

Even when sizing wire for low-voltage security systems, the voltage drop should be limited to three percent because excess voltage drop causes:

Letter	Description	Letter	Description
A	Asbestos	RH	Heat-Rated Rubber
B	Braid	RHW	Weather-Rated, Heat-Rated Rubber
E	Ethylene or Entrance		
F	Fluorinated or Feeder	THW	Weather-Rated, Heat-Rated Thermoplastic
H	Heat Rated or Flame Retardant		
I	Impregnated	THWN	Weather-Rated, Heat-Rated Thermoplastic with Nylon Cover
N	Nylon	XHHN	Heat-Rated, Flame-Retardant, Weather-Rated, Cross-Linked Polyethylene
P	Propylene		
R	Rubber		
S	Silicon or Synthetic	USE	Underground Service Feeder
T	Thermoplastic	ZW	Weather-Rated Modified Ethylene Tetrafluorethylene
U	Underground		
V	Varnished Cambric	V	Varnished Cambric
W	Weather Rated	TA	Thermoplastic and Asbestos
X	Cross-Linked Polyethylene	TBS	Thermoplastic Braided Silicon
Z	Modified Ethylene Tetrafluorethylene	SA	Silicon Asbestos
		AVB	Asbestos Varnished and Braided
TW	Weather-Rated Thermoplastic	SIS	Synthetic Impregnated
UF	Underground Feeder	FEP	Fluorinated Ethylene Propylene
FEPW	Weather-Rated Thermosplastic		

Figure 6.17 Letter codes for conductor insulation.

- Failure of controls to activate
- Control contact chatter
- Erratic operation of controls
- Control coil burnout
- Contact burnout

A table that can be used to size low-voltage wire is shown in Fig. 6.18. To use, assume a load of 35 VA with a 50-foot run for a 24-V control circuit. Referring to the table, scan the 50-foot column. Note that #18 AWG wire will carry 20 VA and #16 wire will carry 43 VA, while still maintaining a maximum of three-percent voltage drop. In this case, #16 wire should be used.

When the length of wire is other than listed in the table, the capacity can be determined by the following equation:

$$\text{VA capacity} = \frac{\text{Length of circuit (from table)}}{\text{Length of circuit (actual)}} \times \text{VA (from table)}$$

The three-percent voltage-drop limitation is imposed to ensure proper operation when the power supply is below the rated voltage. For example, if the

AWG	Length of Circuit, One Way in Feet											
Wire Size	25	50	75	100	125	150	175	200	225	250	275	300
20	29	14	10	7.2	5.8	4.8	4.1	3.6	3.2	2.9	2.6	2.4
18	58	29	19	14	11	9.6	8.2	7.2	6.4	5.8	5.2	4.8
16	86	43	29	22	17	14	12	11	9.6	8.7	7.8	7.2
14	133	67	44	33	27	22	19	17	15	13	12	11

Figure 6.18   Table showing length of circuit, one way in feet.

rated 240-V supply is 10 percent low (216 V), the transformer does not produce 24 V, but only 21.6 V. When normal voltage drop is taken from this 21.6 V, it approaches the lower operating limit of most security controls. If it is ensured that the primary voltage to the transformer will always be at rated values or above, the control circuit will operate satisfactorily with more than three percent voltage drop. However, during extremely hot or cold weather, the use of electricity is increased, and many power companies must reduce the voltage (by 10 percent, for example), so that their generators can carry the additional load. You can usually tell when the voltage is reduced on power lines because heating elements and microwave ovens take longer than usual to produce the required amount of heat.

During the installation of low-voltage circuits, it is suggested that one extra line be run for emergency purposes. This extra line can be substituted for any one of the existing lines that might be defective. Also, it is possible to parallel this extra line with the existing line carrying the full load of the control circuit if the length of run affects control operation caused by voltage drop. In many cases, this will reduce the voltage drop and permit satisfactory operation.

Installing Systems in Existing Buildings

Many changes and advances in developing complete security/alarm systems for building operation and protection have taken place in the past few years. Numerous existing buildings are currently having security and fire-alarm systems installed—either to replace their obsolete systems or to provide protection they never had.

The materials used for installing a complete alarm system in an existing building are essentially the same as those used in new structures. However, the methods used to install the equipment and related wiring can vary tremendously and require a great deal of skill and ingenuity. Each structure is unique.

When concealed wiring is to be installed in a finished existing building, the installation must be planned so that a minimum of cutting and patching of exposed surfaces is required. In most cases, this means giving special consideration to the routing of conductors. Unlike the wiring of a new building, where the installer would try to conserve as much material as possible, the amount of material used (within reason) is secondary in existing buildings. The main objective in security/fire-equipment installations in existing buildings is to install the wiring in the least amount of time with the least amount of cutting and patching of the existing finishes of the building.

Prior to any actual work on an existing building, the contractor or the installers should make a complete survey of the existing conditions in the areas where the security system will be installed. If the majority of the work can be done in exposed work areas (as in an unfinished basement or attic), the job will be relatively simple. On the other hand, if most of he wiring must be concealed in finished areas, many problems must be solved. The initial survey of the building (see also Fig. 6.19) should determine the following:

1. The best location for the alarm control panel.

2. The type of construction used for exterior and interior walls, ceilings, floors, etc.

3. The location of any chases that can be used to route the conductors and the location of closets—especially those located one above the other, for possible use in fishing wires.

4. The material used for wall and ceiling finishes: plaster, drywall, paneling, etc.

5. Location of moldings, baseboards, etc., that can be removed to hide conductors.

6. Location of decorations or other parts of the building structure that cannot be disturbed.

7. Location of any abandoned electrical raceways that new alarm-system wires might be fished into. Don't overlook similar possibilities. For example, old abandoned gas lines were recently used to fish security-system wires in an old building in Washington, D.C.

8. The location of all doors and windows, coal chutes, and similar access areas to the inside of the building.

As indicated previously, the most difficult task in running wires in existing buildings is the installation of concealed wiring in finished areas with no unfinished areas or to provide access to them in the area in question. In cases like these, the work is usually performed in one of two ways, namely, by deliberately cutting the finished work so that the new wiring can be installed. Of course, these damaged areas must be patched once the wiring is installed. The second way is to remove a small portion of the finished area (only enough to give access to voids in walls, ceilings, etc.) and then fish the wires in. The removed portions are then replaced after the wiring is complete.

Where outlet boxes are used, they should be designed for installation in the type of finish in the area. Means of securing the boxes to some structural

INITIAL SURVEY OF A BUILDING		
✓	SITUATION	NOTES
	The best location for the alarm control panel.	
	The type of construction used for exterior and interior walls, ceilings, floors, etc.	
	The location of any chases used to route the conductors and the location of closets— especially those located one above the other, for possible use in fishing wires.	
	The material used for wall and ceiling finishes: plaster, drywall, paneling, etc.	
	Location of moldings, baseboards, etc., that can be removed to hide conductors.	
	Location of decorations or other parts of the building structure that must not be disturbed.	
	Location of any abandoned electrical raceways that new alarm-system wires might be fished into. Don't overlook such possibilities as abandoned gas lines.	
	The location of all doors and windows, coal chutes, and similar access areas to the inside of the building.	

Figure 6.19   Initial survey of a building.

member (such as mounting ears or holding devices) should also be given consideration.

Another method of providing outlets in a finished area is to remove the existing baseboard and run the conductors in the usual groove between the flooring and the wall and then replace the baseboard. This method requires less work (cutting and patching) than most other methods when the finished area must be disturbed. Also, a type of metal baseboard on the market can be installed along the floor line and used as a raceway. Most types are provided with two compartments for wires: one for power and one for low-voltage wiring. Using this metal baseboard provides a simple means of routing wires for security/fire-alarm systems with very little cutting or patching. In most cases, wires can be fished from the baseboard up to outlets on the wall—especially if they are less than three feet above the floor. However, if this is not practical, matching surface molding can be installed to blend in very nicely with the baseboard.

When a lot of cutting and patching is required in a finished area, many installers like to hire a carpenter to do the work. The carpenter might know some tricks that will help the alarm-system installers get the system in with the least amount of difficulty. Also, any cutting or patching will be done in a professional manner.

Before doing any actual cutting of an existing building to install security/fire-alarm components, the installer should carefully examine the building structure to ascertain that the wires can be routed to the contacts and other outlets in a relatively easy way (Fig. 6.20). It is possible that a proposed outlet location, for example, could be moved only a foot or two to take advantage of an existing chase. Perhaps a smoke detector or similar component was originally located in a ceiling with insulation, which would make the fishing of cables very difficult. If the detector could be located on a ceiling containing no insulation, the job would be greatly simplified.

Figure 6.20 A flashlight and mirror used in combination are useful for viewing conditions inside of partitions.

When cutting holes in ceilings for outlets, a drop cloth or paper should be spread underneath to catch all dust and dirt. Sometimes an old umbrella can be opened and hung upside down under the spot in the ceiling where the hole is being made to catch the debris and keep it off the rugs and furniture.

Holes for wires and components can be cut through plaster with a chisel, through wood with a keyhole saw after first drilling two or four pilot holes, and in brick or other masonry with a masonry chisel or rotary hammer. To locate the exact spot to cut these openings, it is best to first cut a very small hole in the center of the spot where the larger one will be made. This hole can then be used to locate the area between studs or, in the case of very old homes, the cracks between the plaster lath. It is then possible to shift the mark for the outlet openings so that all obstacles can be avoided and to provide proper anchoring of the outlet box or component.

There are a number of ways to pull and fish wires into walls and openings in finished buildings and, with a little ingenuity and careful thought, workers should be able to solve almost any problem.

When pulling wires into spaces between the studs in walls, a flashlight placed in the outlet box hole is often a great help when feeding the wires in or catching them as they are pushed near the opening. Under no circumstances should a candle or other open flame be used for this purpose. If one must see farther up or down the inside of a partition, a flashlight and mirror used in combination (Fig. 6.20) is a great help. Many installers like to make their own mirror by gluing a small 2″ × 3″ compact mirror on a handle resembling a wooden tongue depressor. Any type of small flashlight can be used.

Where it becomes necessary to remove floor boards during a security/fire-alarm installation, it should be done with the greatest of care so that the edges are not split. Split edges make a poor appearance on the finished job when the boards are replaced. Special saws can cut into floors or other surfaces without drilling holes to start the saw. Then if the tongue (on tongue-and-groove boards) is split off with a thin sharp chisel driven down in the crack between the boards, the board from which the tongue was removed can be pried up carefully without damaging the rest of the floor.

New installation techniques for existing structures

A few years ago, the Diversified Manufacturing and Marketing Co. (Burlington, NC 27215) patented a system that attaches a drill bit to a long flexible spring steel shaft and is known as *D'versiBit*. This system makes it possible to manipulate a drill bit in walls to accomplish complex installation maneuvers in existing buildings. The D'versiBit can be inserted into the wall cavity through a small opening and positioned accurately to drill from midwall to attic or basement, from windows and doorways to basement or attic, etc. The development of this system makes penetration and cable retrieval a much simpler operation than it used to be. The following list of tools are available for use with the D'versiBit system.

Bits: The three types of bits available for this system are shown in Fig. 6.21 on the next page. The auger bit (Fig. 6.21*A*) is for starting and drilling a clean entrance hole, the combination bit (Fig. 6.21*B*) is designed for greater durability, and the masonry bit (Fig. 6.21*C*) has a carbide tip for drilling in cement blocks and plaster. All three of these bits are designed for use with standard electric drill motors.

Alignment Tool: The special alignment tool shown in Fig. 6.22 provides total control of the flexible shaft, and can be used to hold the bit and shaft steady and true toward any desired destination.

Line-Recovery Devices: After the drilling is completed, the system quickly converts to a line-recovery system using the grips shown in Fig. 6.23. These grips attach to holes located in the bit tip or in the shaft end. This feature enables even one person to quickly fish wires or cables through partitions.

Shaft Extensions: The standard lengths of the flexible shaft are 54″ and 72″, but shaft extensions (Fig. 6.24) are available to provide extra distance drilling capabilities. One or more can be attached in special situations, such as from the basement to a smoke sensor in the attic.

The basic shaft is $3/16$″, which will accommodate both $3/8$″ and $1/2$″ drill bits in the three styles mentioned previously. For larger bits (such as $3/4$″ and 1″ sizes) a $1/4$″ shaft is required. This larger shaft reduces the flexibility for complex drilling.

Operation procedures

When drilling with the flexible shaft of the D'versiBit, run the drill motor only when actually drilling. Never run the drill when sliding the bit up or down in the wall cavity as wires—either signal wires or existing electric power wiring—could be cut during the process. Also be certain that the bit is sharp because a dull bit is one of the greatest causes of bit breakage.

If at all possible, a reversible drill motor should be used to withdraw the bit from the wall. The motor should be running only when the bit is actually passing through a wood member. When drilling, force is exerted in one direction. When the bit is being removed, it is removed at a different angle and force is exerted from a different direction. This is why the reverse is used. If

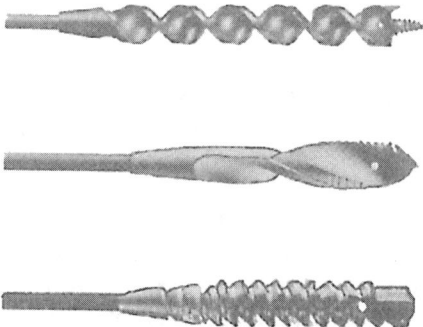

Figure 6.21   Three types of bits available for the D'versiBit system.

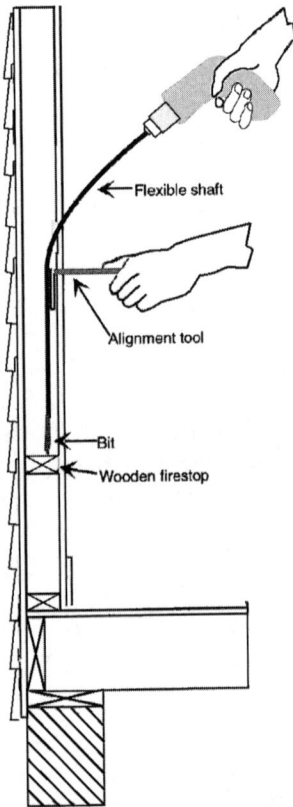

Figure 6.22 The special alignment tool provides total control of the flexible shaft.

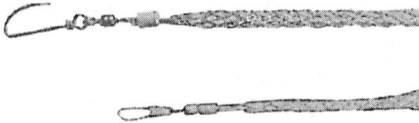

Figure 6.23 Line-recovery devices.

Figure 6.24 Shaft extension.

the flexible shaft is being used with drill motors with no reverse, it would be better to exert force to pull the bit from the hole with the motor running because chances of an easy recovery without damage are much better with the motor running.

When drilling from an attic or crawl space, be certain not to select an area directly above or below a door because this will result in property damage. It is also good to keep a slight tension on the wire when it is being pulled from overhead so that it will not get tangled with the bit and become damaged.

The shaft should not be bowed any more than absolutely necessary to accomplish the job. Excessive bowing will decrease the life of the flexible shaft. Drill motors, of course, should be adequately grounded or have insulated handles.

Practical applications of the D'versiBit

Assume that an outlet box for an infrared photoelectric detector is to be installed above a countertop in a residential kitchen to sense entry of unauthorized persons through the kitchen door. If, upon investigation of the space inside of the partitions, it is found that a wood 2×4 member (fire-stop) blocks the route from the outlet hole to the basement area where the alarm control station is located, an alignment tool must be used.

The flexible shaft containing a drill bit is placed through a cut outlet-box opening and then the special alignment tool is attached to the shaft (Fig. 6.24). By keeping the alignment tool in the same position on the shaft and by lifting the handle, the shaft will bow back toward the operator. As the bit is lowered into the wall cavity, the operator can feel the bit strike the inside wall. When the bit is aligned correctly on the wooden member, the alignment tool is removed while keeping downward pressure on the bit so that it will not slip out of place, and the hole is drilled through the fire-stop. This hole will then act as a guide to drill through the floor plate (Fig. 6.25).

In the case of a wall cavity without fire-stops or purlins, the alignment tool is used to snap the bit back to the inside wall (Fig. 6.26) at which time downward pressure on the drill motor will keep the bit point in place and cause the shaft to bow. Power and pressure are then transmitted from the back wall, which allows proper angle drilling to miss the joist boxing.

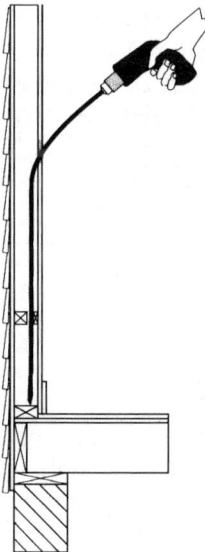

Figure 6.25 The first hole cut acts as a guide for drilling through the floor plate.

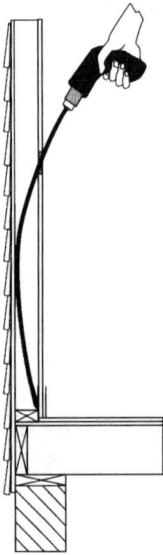

Figure 6.26 Alignment tool used to snap the bit back to the inside wall.

After the bit has penetrated into the basement area (Fig. 6.26), the operator has access to the hole in the drill bit itself for attaching the recovery grip and pulling the wire up to the outlet location—all without damage to the existing finish.

Figure 6.27 shows how the recovery grip is attached to the bit tip eyelet. The swivel located between the cable and the head of the grip prevents the wire or cable from becoming twisted during the fishing process.

Figure 6.28 shows the grip after it has been attached to the bit tip with the line inserted ready for recovery. The installer then operates the drill in reverse (because of the angle of the pull), applies a slight pull, and the wire can be fished through the holes due to the reverse cutting action of the bit. If desired, the drill motor can be removed from the shaft and a recovery grip attached to the chuck end of the shaft for pulling the wires downward toward the basement. This example shows the method of routing wires or cables from an outlet to a basement, but the same procedure would apply for drilling from an outlet opening to an attic space.

To install contacts on windows for a burglar-alarm system, drill from the location of the contact through the casement, lintels, and plates with a $3/8''$ shaft. Attach a recovery grip to the end of the bit, insert the wire to keep the grip from becoming tangled, reverse the drill motor, and bring the wire toward the operator as the bit is being withdrawn (Fig. 6.29).

Burglar alarm contacts or door switches installed at doors are simple projects when one uses the flexible shaft. First cut or drill the entrance hole in the normal manner and then insert the flexible shaft with bit into the entrance hole, slanting the bit as much as possible in the desired direction of travel. Continue by drilling through the door casing and floor jamb into the cavity of

Figure 6.27 Recovery grip attached to the bit tip eyelet.

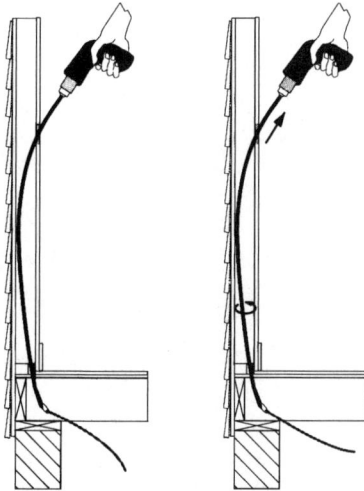

Figure 6.28 Grip attached to the bit tip with the line inserted ready for recovery.

Figure 6.29 Drilling through the window casing.

the wall (Fig. 6.30). The drill is then stopped until it strikes the next stud, which will deflect the bit either up or down, depending on the direction of the drilling. Continue to push the bit until it strikes the top of the bottom plate and then drill through the plate into the basement or attic. The recovery grip is then attached to the bit and the wire or cable can be drawn back toward the operator by reversing the drill motor and keeping a slight tension on the wires as they are being pulled to prevent tangling.

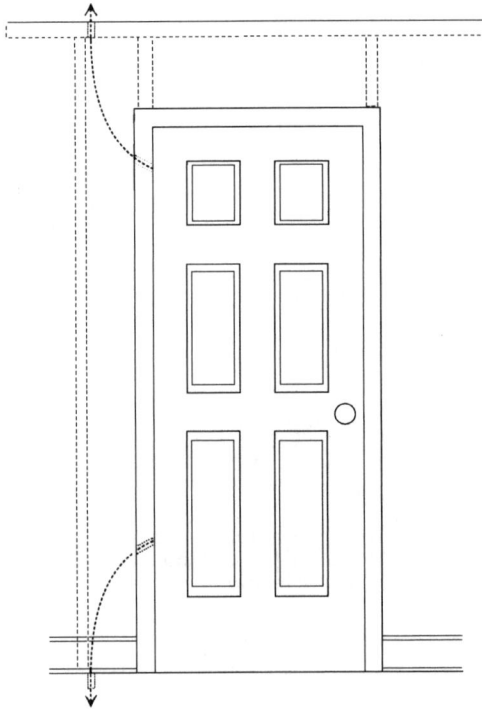

Figure 6.30 Drilling through the door jamb into the cavity of the wall.

With conventional tools, the routing of wires from one outlet to another requires either channeling the wall, using wire mold, or running the wires down to the baseboard, removing the baseboard, and then installing the wires behind it. Instances like these occur when the crawl space is too shallow for workers to crawl into or the house is built on a concrete slab. However, with the flexible shaft, it is possible to drill through the wall horizontally through several studs (Fig. 6.31) and then pull the wires back through the drilled holes to the wall openings.

The installation of an outside annunciator under the eave of a house with an extremely low pitch to the roof would cause several problems in getting wires to the outlet. With the flexible shaft, a hole can be drilled through the boxing (Fig. 6.32). As soon as the bit penetrates the boxing, it is pushed into the attic as far as it will go. A recovery grip is then attached to the bit, the wire or cable inserted, and then pulled backward toward the outlet opening. The outlet box and annunciator (horn, bell, etc.) are installed under the eave and the other end of the cable is connected to the alarm system. Also, because the flexible shaft is more rigid than the conventional fish tape, it will pull through attic insulation if any exists.

If it becomes necessary to install wiring in an attic and run cable from this area to the basement, the installation can be greatly simplified by using a flexible shaft. First drill through the top plate into the wall cavity—ensuring that

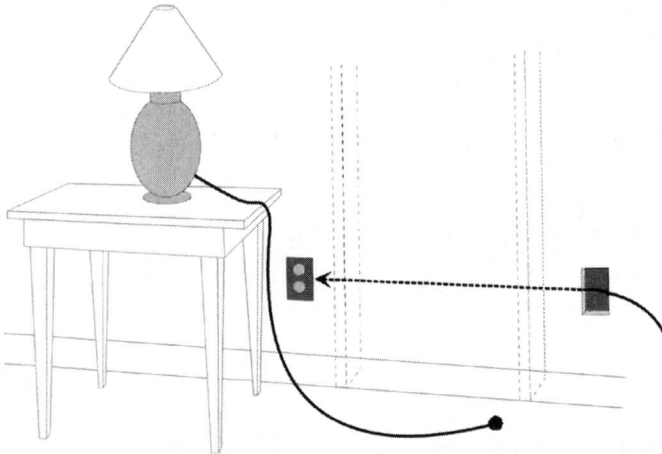

Figure 6.31 With the flexible shaft, it is possible to drill through the wall horizontally through several studs.

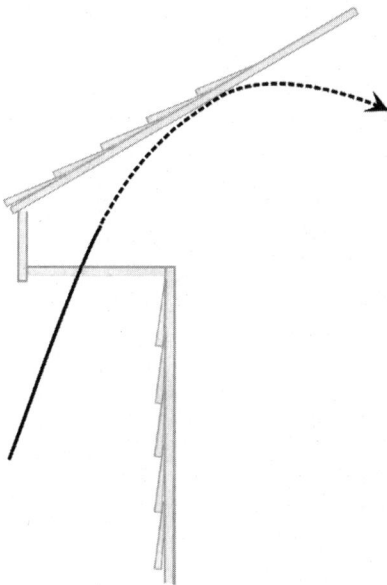

Figure 6.32 Method of drilling a hole through boxing by using a flexible shaft.

the drilling is not being done above a window or doorway or any other obstruction (such as existing wiring, ductwork, etc.) Once through the top plate, the drill motor is turned off and the bit is pushed into the cavity of the wall as far as it will go. If no fire-stops are encountered, the bit is pulled back and an extension is attached to the shaft. With the extension installed, the bit is again lowered into the wall cavity until a fire-stop is encountered. The bit is then positioned and used to drill through the wooden member. Once the wooden member is penetrated, the drill motor is again stopped and the bit is lowered further until the bottom plate is reached. Continue drilling through the bottom plate in the basement or crawl space. Fasten the appropriate recovery grip, insert the wire or cable, and pull up the wire with the flexible shaft. The drill motor should be reversed only when the bit is passing through one of the wooden members.

Those who use this device often are certain to discover many other useful techniques for installing wiring in existing structures.

Residential Security Systems

A wide variety of security systems and accessories are available to take care of almost every conceivable residential application. This fact sometimes poses a problem of weeding out which system is best for the given application. Chapter 7 covers the design of several types of residential occupancies, giving reasons for choosing the various systems.

The Essentials

When planning the design of a security system for residential occupancies, the first decision to be made—and often the hardest—is what the system must accomplish. Will the threat be personal risk encountered in a home because of burglary or psychological harassment? Or will it be loss of property and money because of burglary, or some other danger? The security technician or system designer must decide what type of alarm should be provided. The following list offers the typical uses of residential systems:

- Scare an intruder with an alarm sounder
- Call for help from neighbors
- Summon police or others over telephone lines
- Combination of the above

The allotted budget is another factor to consider. Many residential security systems can be installed for less than $1000 while others may cost several thousand dollars. The best way to determine the amount of protection needed is to consider the threat of personal risk or bodily injury and the possible loss of property. Then discuss how much protection is worth to the homeowner. When this figure is determined, you will have a good guide to how much of an alarm system is needed. Discuss all of these issues with the owner in detail, and do it before any work begins.

With the threat determined and the system goals and a general budget for the system established, the design can begin.

Basic information

A form like the one in Fig. 7.1 should be used by the security system contractor or installer to ensure that all necessary information concerning the design will be obtained. Contractors should have a supply of the form printed with their logo at the top of the page. Manufacturers of security/fire-alarm equipment also furnish survey forms at little cost. The form is a word processing file on your CD. You can modify it to your needs.

Begin filling out the form with the customer's name, address, city, state, and zip code. Be sure to include the actual date that this survey is taken. Continue with the size of family, including all adults and children. Some adults are reluctant to give their age so an estimate in this category is okay: that is, "young couple," "middle-aged," etc. The number and weight of pets might seem ridiculous, but all of this information is pertinent for obtaining and installing the best security system. For example, if pressure-mat detectors are used in the home, they must be calibrated so that a dog or cat walking over the mats at night will not set off the alarm. The same is true for motion detectors. Knowing the type, size, and weight of pets will help you to get the right system for each individual family and premises.

Construction information

The next part of the form (Fig. 7.2) deals with the construction of the building or premises. This information can be collected either by actual measurement or by obtaining existing architectural drawings along with a plot plan, if these are available.

Many security technicians prefer to use graph paper as a sketch pad and then draw a rough sketch (to scale) of the property and floor plan of the building. Each square on the graph paper represents a predetermined size; that is, 1 foot, 2 feet, 10 feet, etc. Consequently, if one wall of the building was, for example, 30 feet in length, a line drawn with a straightedge through 30 squares on the graph paper would then equal 30 feet on the building. The remaining walls are drawn in a similar fashion. Once the outside walls of the building are drawn on the graph paper, the interior partitions are sketched in to complete the drawing.

CAD (computer-aided design) drawings are excellent and can be altered easily. Portable printers and e-mail make accurate presentations much simpler than hand work.

Referring again to Fig. 7.2, note that the age of the home is requested, along with the lot size (width and depth of the lot) and total number of acres or a percentage of an acre. A plot plan (Fig. 7.3) should also be prepared at the same time that the form is being completed. The plot plan should show the relation of the building to the property lines, all nearby roads, alleys, bushes, trees,

Reliable Security Services
4457 Royal Oak Drive
Front Royal, Virginia 22630
(703) 555-1212

SECURITY SURVEY
for

Name_____Date of Survey_____

Address_____

City_____State_____Zip_____

Telephone (home)_____(work)_____

Family Size:
    Adults_____Ages_____

    Children_____Ages_____

    Pets_____Weight_____Height_____

Normal bedtime_____How many in family smoke?_____

Name of closest relative_____Telephone_____

Name of insurance agent_____

    Company_____Telephone_____

Figure 7.1 Cover page used for initial survey of project.

and other pertinent information that can aid in the design of the security system. Also notice any fire hydrants, the closest fire department, and the nearest police station.

Many modern residences have an existing plot plan that can be used by the security technician in his or her design. If the homeowners do not have this plan available, check with the land developer, the architect, or the land surveyor. A

CONSTRUCTION INFORMATION

Age of home:_____ Lot size:_____ Ft. wide:_____ Ft. deep:_____

   Acre(s)_____

Building measurements: _____ Ft. long _____ Ft. wide _____ Ft. High

Walls: Wood _____ Shingle ___ Stone ___ Brick ___ Metal siding ___
   Other: _____

Interior walls: Wood _____ Plaster _____ Plasterboard _____
   Flagstone ____ Brick ____ Fiberboard ____
   Other: _____

Ceilings: Wood ___ Plaster ___ Plasterboard ___ False ___ Acoustic ___
   Other:_____

Roof: Wood shingle ___ Tar paper ___ Tile ___ Asphalt shingle ___
   Rock ___ Slate ___ Pitched ___ Flat ___ Multistory _____
   Other: _____

Floors: Wood ___ Concrete ___ Tile ___ Flat stone ___ Carpet ____
   Other: _____

Attic: Full ___ Partial ___ None ___

Attic crawl space: Good ___ Average ___ Poor ___ None _____

Crawl space under house: Good ___ Average ___ Poor ___ None ___

Basement: Full ___ Partial ___ None ___
   Basement dimensions:_____ x_____

Garage: Attached ___ Free standing ___ Garage dimensions: ___ x_____

Garage walls: Wood ___ Unfinished ___ Plasterboard ___ Brick ___
   Stone ___ Other _____

Type of insulation: Walls_____Floor_____Ceiling_____

Figure 7.2 Form used for construction information on a proposed residential project.

plot plan can also be attached to the property deed in the deed registrar's office for the area. If such a plan already exists, using it will save the security technician much time and usually the original plan will be more accurate than one made by the technician without using surveying instruments.

Once the plot plan is obtained or sketched, trouble spots should be indicated on the drawing. Burglars almost always prefer doors for entrance because doors are usually required to remove such items as television sets from the home. Even if some other means of entry is used, burglars will use doors to exit. Furthermore, they need a concealed approach route and a hiding place for the vehicle to be loaded with the stolen goods. Therefore, back alleys and hidden carports are the obvious vehicle hiding places. All such areas should be clearly indicated on your plot plan. With this knowledge in mind, provide the most vulnerable door in the building with the best physical and burglar-alarm protection. Vulnerable entries should have dead latch and dead-bolt locks on a solid wood or steel door with no windows in it, if possible. If this point of entry is a sliding glass door, care must be taken so that the door cannot be lifted out of the track and the door should be capable of being securely key-locked and pinned into the closed position using appropriate hardware.

Obviously, this most vulnerable entry point must be alarmed with the best equipment available. Similarly, all doors into the home must have some sort of protection.

Windows should be analyzed in a similar manner to doors. Most premises have one or more windows that face a side yard, a fence, or a wall, where no casual observer is likely to see someone making a forced entry. First, protect the window from being forced open and then install a dependable security system.

The form in Fig. 7.4 is designed to furnish security technicians with intrusion security information. Review the information included in this form as you continue with this chapter.

Most security systems use a closed-loop protective circuit where a pair of wires is connected to the alarm control and is then run around the perimeter of a building and finally returned to the alarm control panel. Closed-circuit detectors are connected in series in this loop. A small current flows through the wiring and detectors and any interruption of this current sensed by the detector operation (cutting the wires or shorting the wire pair together) will sound the alarm. Restoring the loop to its original condition, like closing the alarm door after entry, will not stop the alarm condition. Only operating the appropriate control will do this.

Magnetic contacts or switches are by far the most commonly used detection devices for window and door openings. They consist of two pieces: a magnet and a magnetically operated switch enclosed in plastic cases. The magnet is mounted on the edge of the door and the switch section is mounted directly adjacent to the magnet on the door frame. When the magnet is located near the switch section, the switch is turned on and electricity flows through the switch contact. Moving the magnet away from the switch at an opening (such as a door) turns the switch off.

Because this is a closed-circuit system, the current through the loop will cease and the alarm will sound on opening. Magnetic switches are very successful because they are noncrucial in alignment between the magnet and switch section and are extremely reliable in operation. Many switches are

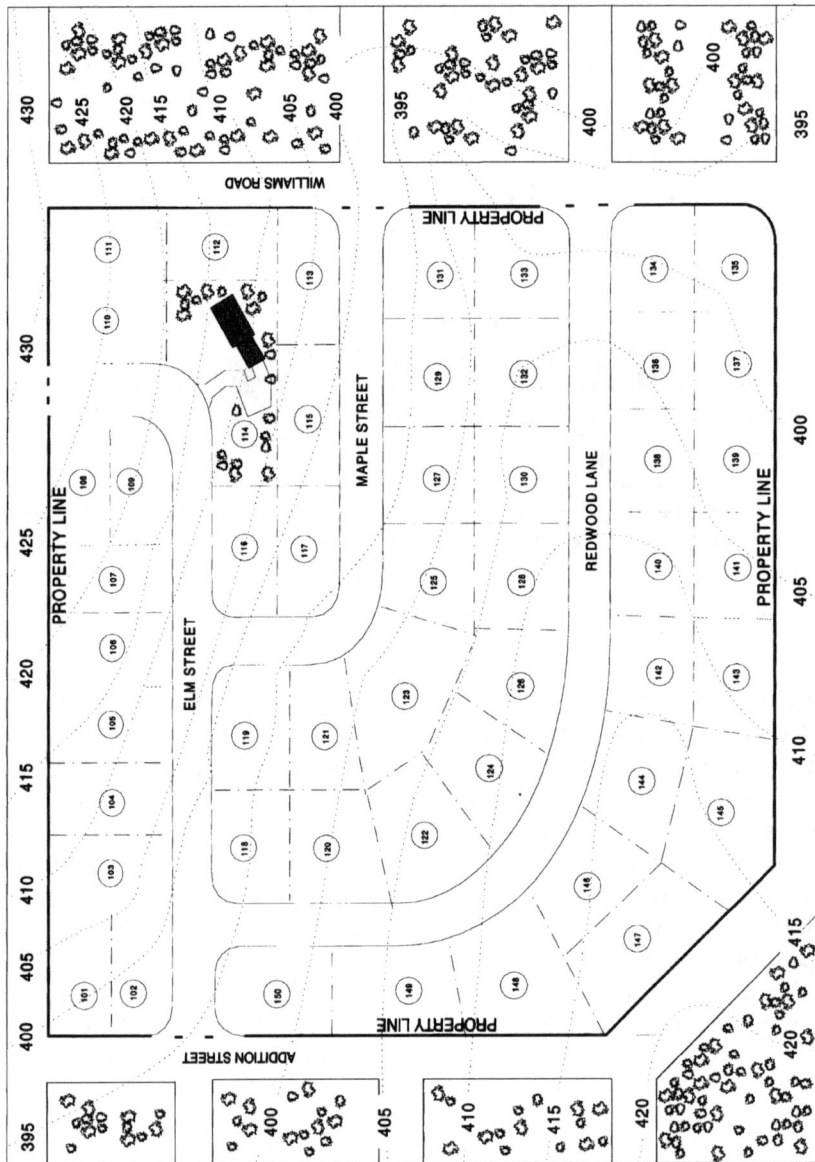

LAND SUBDIVISION TOPOGRAPHY MAP

Figure 7.3   Typical plot plan as prepared by a land surveyor.

INTRUSION SECURITY INFORMATION

Distance to neighbor on:

      Right ____ft. Left ___ft. Front ___ft. Rear ___ft.

Type of public lighting: Street ___ Flood ___ Other _____ None ___

Type of private lighting: Front _____ Side _____ Rear _____
      What lights are left on at night?_____

Visibility onto property from street:
      Good ___ Average ___ Poor ___ Why?_____

Hours home is normally vacant:_____

Number of entrances:
      Doors_____ Sliding glass doors_____ Windows_____ Other_____

Doors: Wood ___ Metal ___ Glass ___ Single ___ Double ___Hollow ___
      Solid:_____ Other:_____

Door frames: Wood ___ Metal ___ Aluminum ___ Other _____

Locks: Single-key ___ Sliding-bolt ___ Double-key ___ Dead-bolt ___
      Padlock ___ Night latch ___ When last rekeyed?_____
      Other: _____

Windows: Sliding ___ Double-hung ___ Single hung ___ Casement ___
      Louver ___ Fixed ___ Other _____

Glass: Single-strength ___ Double-strength ___Plate ___ Frosted ___
      Tempered ___ Thermal pane ___ Other _____

Storm windows: Yes ☐   No ☐

Glass framing: Wood ___ Metal ___ Other_____

Grilles or Screening: Standard screens___ Iron bars ___ Other _____

Garage door: Single ___ Double ___ One-piece overhead ___
      Sectional overhead ___ Swing out ___ Other _____

Outbuildings: Number_____Type _____

Figure 7.4 Form used to obtain intrusion security information.

rated for hundreds of millions or even billions of switch operations. Little mechanical motion takes place in the switch, so replacement will be extremely infrequent, even under the worst of circumstances. By their nature, they are free from false alarms and are easy to troubleshoot and replace if one fails.

Magnetic contacts are also the best method of protecting windows and other openings. To protect glass from breakage in windows or sliding glass doors, a special lead foil is the common means of protection. This foil is put in series with the same burglar circuit that connects the doors and windows. The alloy in the foil is of such composition that any break in the glass will shear the foil and thereby set off the alarm.

It is advisable to include an extra door switch or two on some of the interior doors that are likely to be opened if an intruder somehow penetrates the perimeter circuit. Such doors might include those to a gun closet, fur storage vault, or just between two rooms that have to be traversed to find any valuable property. The intruder is likely to have his guard down at this point and not to be looking for a switch. Motion detectors (ultrasonic, infrared, audio, etc.) are also good insurance for the interior circuit.

Routing the circuit wires around the perimeter in an effective manner is one of the most important parts of a security/fire-alarm system. A pair of either #22 or #24 wires should be run all the way around the home, from the control panel and then back. All detectors are then connected to this perimeter loop.

Wire concealment can be a major problem for the installer. If the house is under construction, the pair of wires can be located at some set distance within the partitions and walls. The installer can then cut into the wall at this distance when the walls are finished to get to the wires for the final connections. For existing construction, much fishing is necessary to route the wires to the various detectors, but Chapter 6 provides several solutions to these problems.

Concealment is important for aesthetic reasons, for making it impossible for the intruder to locate the presence of the system, and for reliability in minimizing damage to the wires.

Residential fire-alarm system

Heat and smoke detectors should be included in any residential security/fire-alarm system. They are generally connected to the system (Fig. 7.5). The fire-detection circuit should be fully supervised as required by UL. The circuit itself should act as a detector in the event of a malfunction; that is, a trouble bell or buzzer should activate in the control unit to alert the occupants of the situation.

The primary location for installing smoke detectors is outside each bedroom area. Because fire travels upward, the top of each stairwell is another important location. The NFPA also recommends that smoke detectors be installed on each living level of a multistory house.

Heat detectors should be installed in each enclosed living area including bathrooms, closets, attics, and basements. Any number of detectors can be used with most fire-alarm systems.

Rate-of-rise heat detectors should be mounted on the ceiling not less than 6 inches from a side wall. Exact detector location can be determined by an evaluation based on good engineering judgment, supplemented if possible by field tests.

Figure 7.6 shows some of the heat/smoke detectors supplied by NuTone. The model number, a description of each component, suggested use, and dimensions are given to assist security technicians.

Although no regularly scheduled maintenance is necessary for most heat/smoke detectors, periodic cleaning of the detection chambers might be required when detectors are located in abnormally dirty or dusty environments.

Residential Security-System Equipment

Figure 7.7 shows a variety of components designed for use in a residential security/fire-alarm system. The following is a brief description of each component and its function within the system.

Surface Magnetic Detector: The surface magnetic detector (Fig. 7.8) is one of the most versatile entry detectors for residential alarm systems and should be considered first as a method of protecting any movable door or window. These detectors can be mounted on wood, metal, and even glass, if necessary. They can be mounted with screws, double-sided tape, or epoxy. Obviously, the tape

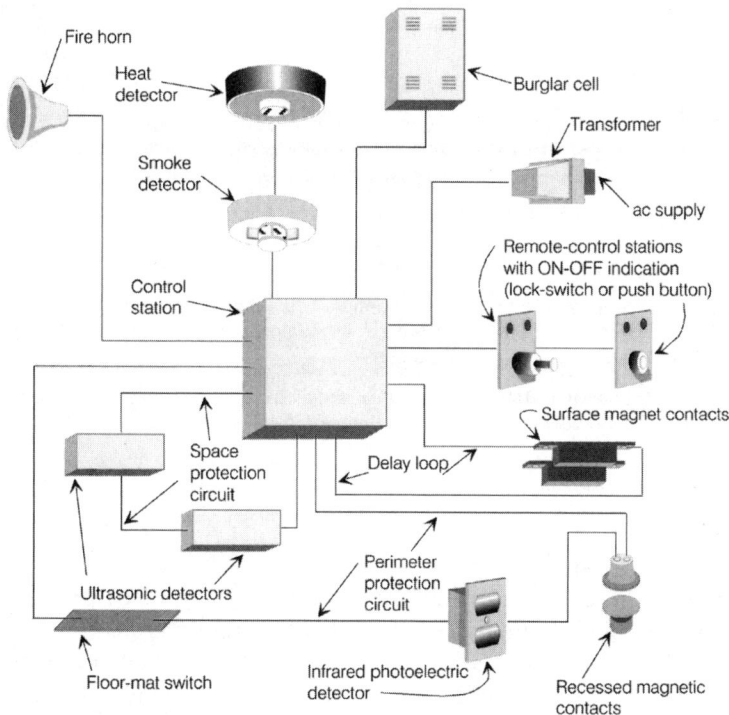

Figure 7.5 Various components of a residential security/fire-alarm system.

Model No.	Description	Suggested Use	Specifications
S-120	135°F fixed-temperature heat detector.	Surface-mount on ceiling in ordinary living areas with normal room temperatures.	1.75 inches in diameter, .75 inch deep. Distance range: 10 ft in all directions. Detector covers area up to 20 x 20 ft.
S-121	200°F fixed-temperature heat detector.	Surface-mount on ceiling in areas where temperatures are higher than normal: furnace or boiler rooms; attics.	Same as above.
S-122	Rate-of-rise/135°F fixed-temperature heat detector.	Surface-mount on ceiling in ordinary living areas with normal room temperatures.	4.5 inches in diameter, 1.75 inches deep. Distance range: 25 ft in all directions. Detector covers area up to 50 x 50 ft.
S-123	Rate-of-rise/200°F fixed-temperature heat detector.	Surface-mount on ceiling in areas where temperatures are higher than normal: furnace or boiler rooms. Note: Use S-121 or S-125 in areas where temperatures consistently exceed 150°F.	Same as above.
SA-124	135°F fixed-temperature heat detector.	Surface-mount on ceiling in ordinary living areas with normal room temperatures.	1.75 inches in diameter, .75 inch deep. Distance range: 15 ft in all directions. Detector covers area up to 30 x 30 ft.
SA-125	200°F fixed-temperature heat detector.	Surface-mount on ceiling in areas where temperatures are higher than normal: furnace or boiler rooms; attics.	Same as above.
S-245	Smoke detector.	Surface-mount on 4-inch square or octagonal wiring box, primarily outside bedrooms.	5.5 inches square x 2⅝ inches deep.
S-245H	Smoke detector with 135°F heat sensor.	Surface-mount on 4-inch square or octagonal wiring box, primarily outside bedrooms.	5.5 inches square x 2⅝ inches deep.
S-240	Smoke-detector supervisory module.	Installs in wiring box at location of last smoke detector in series.	2 inches x 2 inches x 1 inch.

Figure 7.6   Heat-smoke detector selection guide.

and epoxy are useful on glass, aluminum, or any other surface where screws cannot be used. However, when using tape or epoxy, be certain that the surface is clean, dry, smooth, and at least 65°F when applied.

Recess-Mounted Magnetic Detector: Where the appearance of surface-mounted detectors is objectionable, recess-mounted detectors can be used. These detectors are more difficult to install (requiring greater care on the technician's part) but few problems develop if the following precautions are taken:

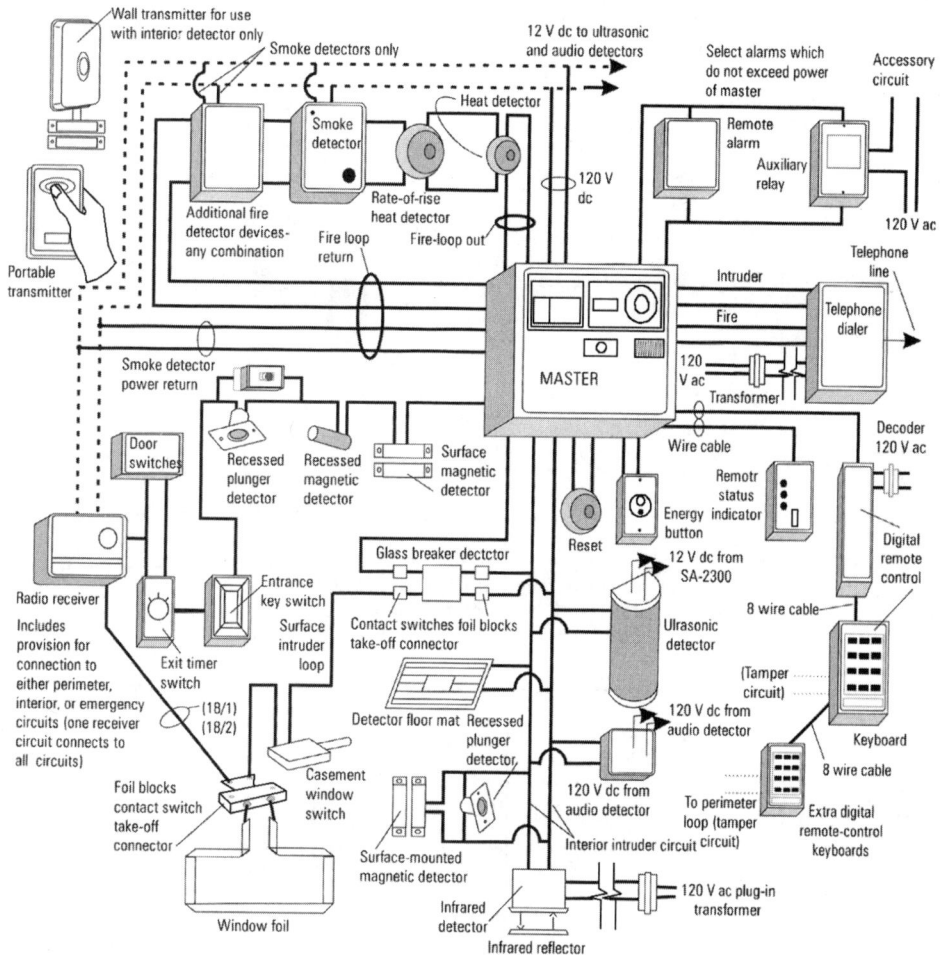

Figure 7.7 Variety of components designed for use in a residential security/fire-alarm system.

- Be careful not to damage or destroy any weatherproofing seal around windows, doors, or other openings.
- If a recessed-mounted entry detector is installed in the window sill, you must prevent water seepage to the switch by applying a sealant under the switch flange and around the switch body.
- When drilling holes to accept each half of the detector, be sure the holes line up and that there is no more than $1/4''$ space between the two sections of the detector.
- Be certain that there is enough space between the window and its frame (or door and its frame) when each is closed; that is, there must be enough space (usually equaling $1/16$ inch) for the protrusion of both sections when they meet.

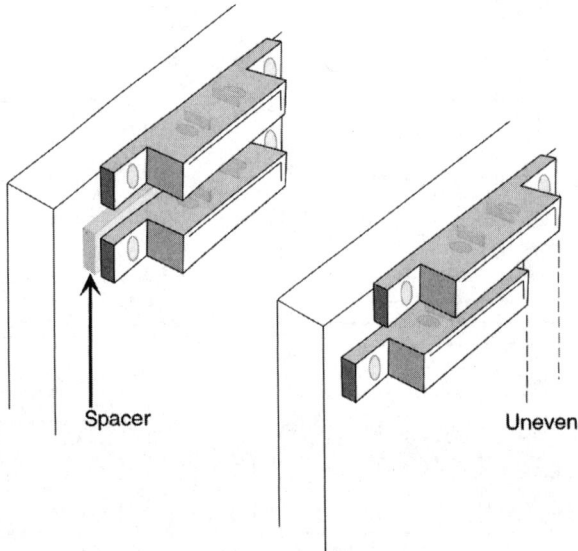

Figure 7.8 The surface magnetic detector is one of the most versatile entry detectors for residential alarm systems.

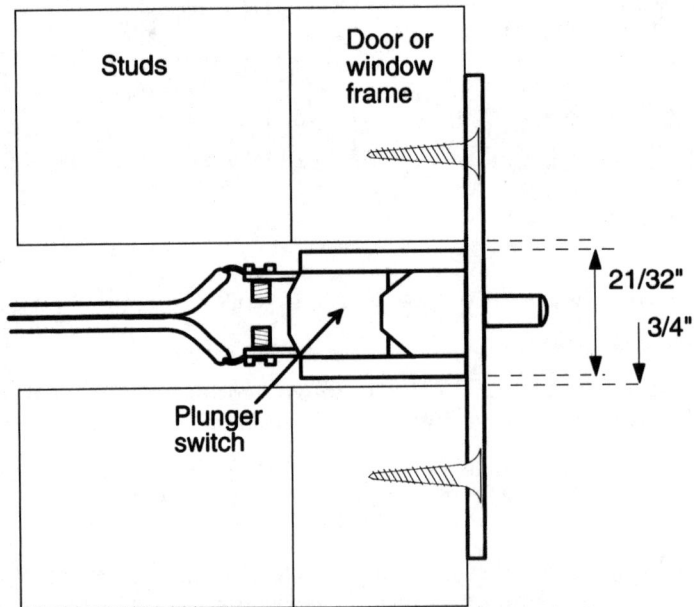

Figure 7.9 Cross-sectional view of a recess-mounted entry detector.

■ If the window frame is not thick enough to accept the magnetic section of the detector, it can be mounted in the side frame.

The recessed plunger detector shown in Fig. 7.9 is mounted so that the door or window will contact the plunger at the tip and push the plunger straight in. Therefore, the area of the window or the door that depresses the plunger should have no slots, cutouts, or step-downs into which the plunger might slip. The surface area should also be hard and free of rubber or vinyl that might be weakened by the plunger and consequently allow the plunger to open. For protecting doors, plunger-type detectors should only be mounted in the door frame on the hinge side of the door.

In cases where it is difficult to protect a window or door by mounting any of the direct-type detectors, the area directly inside the door or window can be protected with interior "space" detectors, such as a floor-mat detector (Fig. 7.10) or an ultrasonic motion detector (Fig. 7.11).

Floor-mat detectors are easily concealed under the rugs at doors, windows, tops or bottoms of stairways, or any other area onto which an intruder is likely to step. A light pressure on the mat triggers the alarm. Obviously, if pets are allowed to roam throughout the house, the mats should be calibrated to withstand the weight of the pets without triggering the alarm. Some types, however, cannot be calibrated. If the mats cannot be calibrated, they will be useless in homes where pets are allowed to roam.

Also, rolls of super-thin floor matting can be cut to any desired length. These rolls can be used on stair treads and in areas near sliding glass doors or other large glass areas, entrance foyers, etc.

Other space detectors include ultrasonic motion detectors, audio detectors, and infrared detectors. Care must be used with any of these units because the

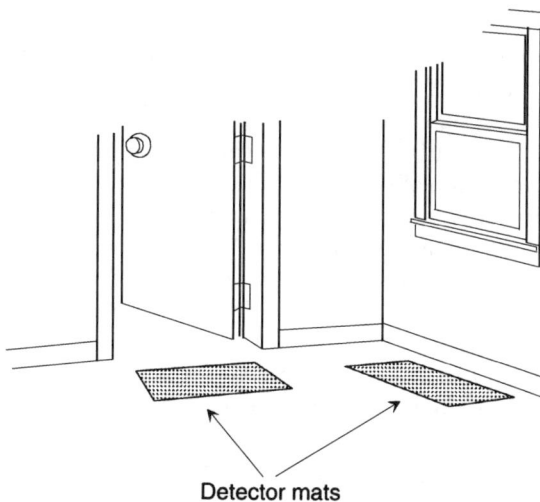

Detector mats

Figure 7.10 Application of floor-mat detectors.

Figure 7.11 Ultrasonic motion detector.

protected area is limited both in width and depth—depending upon the particular unit.

The ultrasonic motion detector can be used in large glass-walled rooms that might otherwise be difficult to protect and in hallways or entries or in virtually any area an intruder would have to pass through when moving about a home. They are especially useful as added protection when conventional detectors are used also to monitor a "valuables" room or area.

Most ultrasonic motion detectors are designed to mount on either the wall or ceiling. They emit inaudible high-frequency sound waves in elliptical patterns that range from 12 to 35 feet to 5 by 20 feet for most residential models. When an intruder moves within the secured area, movement interrupts the established pattern of sound waves and sounds the alarm.

Some designs of motion detectors can be rotated up to 180° for maximum coverage of the area being monitored (Fig. 7.12).

Another type of motion detector is the audio detector (Fig. 7.13). This type senses certain sharp sounds known to be present in forced entry, such as wood splintering or glass breaking. When these sounds are received through the unit's miniature microphone, the detector triggers the control unit to sound an alarm.

Audio detectors are best utilized in areas that are seldom used, such as an attic, garage, or closed-off wing of the house. They can be used in other areas, but when such areas are subject to much daytime activity, it is recommended that the detector only be armed at night when the family retires or is away from home.

Infrared detectors are another type of motion detector. A combination transmitter-receiver is used to project an invisible pulsating beam at a special bounce-back reflector on an opposite wall. Any interruption of the beam activates the system alarms. Infrared detectors can be wired to either the perimeter or interior circuit, but for faster response, it is recommended that it be connected to the interior circuit.

Infrared detectors are designed for indoor areas such as entries, hallways, rooms, etc. Most cover a span from 3 to 75 feet, so they can be used in practically any indoor area or room.

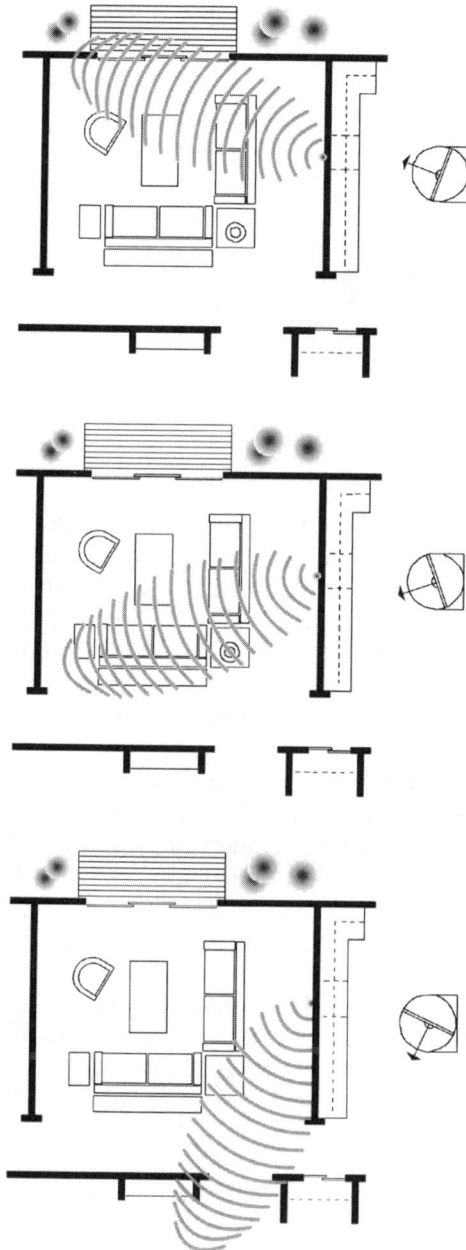

Figure 7.12 Motion detector rotating up to 180° for maximum coverage of the area being monitored.

Perimeter Detectors: Refer again to Fig. 7.7 and note the various detectors available on the perimeter intruder loop. The glass-break detector, for example, is an excellent means to monitor large areas of glass, such as sliding glass doors and picture windows. These detectors, as the name implies, respond only to glass breaks and not to shock or vibrations. Therefore, they seldom produce false

Figure 7.13 Audio detector.

alarms. The size of an area that glass-break detectors will protect varies from manufacturer to manufacturer, but most will average about 10 feet^2 of protection. A small cube like the one in Fig. 7.14 connects to the emergency circuit and the supervised perimeter circuit if they are mounted on movable windows.

Window foil tape is used mostly in commercial and industrial buildings but is sometimes used in residential systems—especially on basement windows. If an intruder breaks the glass, the tape tears, opens the circuit, and causes the alarm to sound.

Where the building construction makes it difficult to install wires, radio controlled intruder detection systems are available. Such systems are also useful for linking outbuildings in a range of 150 feet or more, depending on the type used.

Wall-mounted radio transmitters (Fig. 7.15) are easily mounted behind drapes at windows, above doors, and in similar locations. Any number of transmitters can be used and each can be wired to an unlimited number of detectors as previously described.

When a detector senses forced entry, the transmitter sends a signal via radio waves to the radio receiver. It signals the control unit to sound an alarm—either audible on the premises, or a silent alarm to a local police station.

Practical Applications

To better understand the procedures necessary to design a suitable residential security/fire-alarm system, the floor plan of a single-story residence is shown in Fig. 7.16 on the next page; the legend or symbol list used with this drawing is shown in Fig. 7.17.

The obvious starting place for the design of the system is at the normal entrances, such as the front door opening into the vestibule, the sliding glass door in the dining area, and the kitchen door that opens onto the carport. These types of doors can be protected by several methods, but in this case,

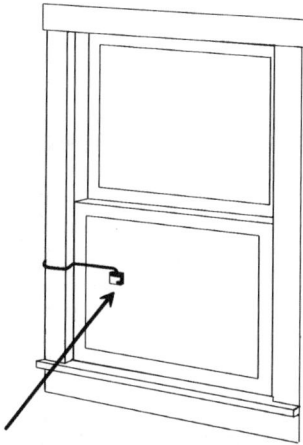

Figure 7.14 Glass-break detector.

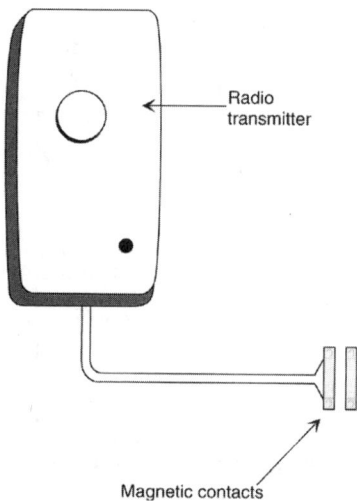

Figure 7.15 Wall-mounted radio transmitter.

infrared photoelectric entry detectors were used at the front door (Fig. 7.18). The transceivers are positioned at the locations indicated by appropriate symbol (1) and the reflectors are located directly opposite the transceivers on the opposite wall. Each of these units resembles a conventional quadruplex receptacle (Fig. 7.18). The centers of these outlets are located approximately 18″ above the finished floor so that an intruder will break the beam as shown.

If pets are allowed to roam in the home, the mounting height should be increased to keep the beam out of reach of pets (for example, 36″ above the finished floor).

The sliding glass doors at the rear of the building could also be protected with photoelectric entry detectors. In fact, magnetic contacts are usually not

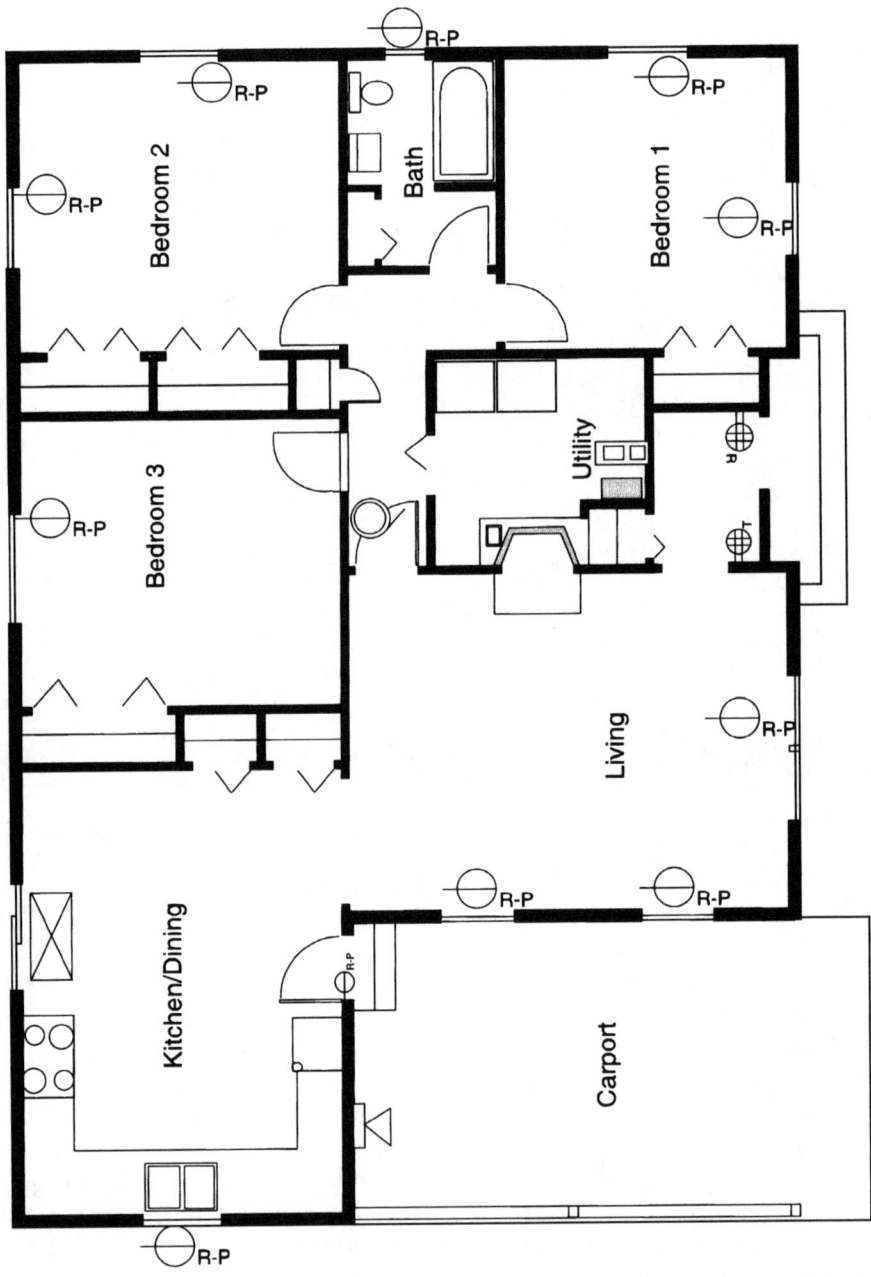

Figure 7.16 Floor plan of a single-story residence to be supplied with a security system.

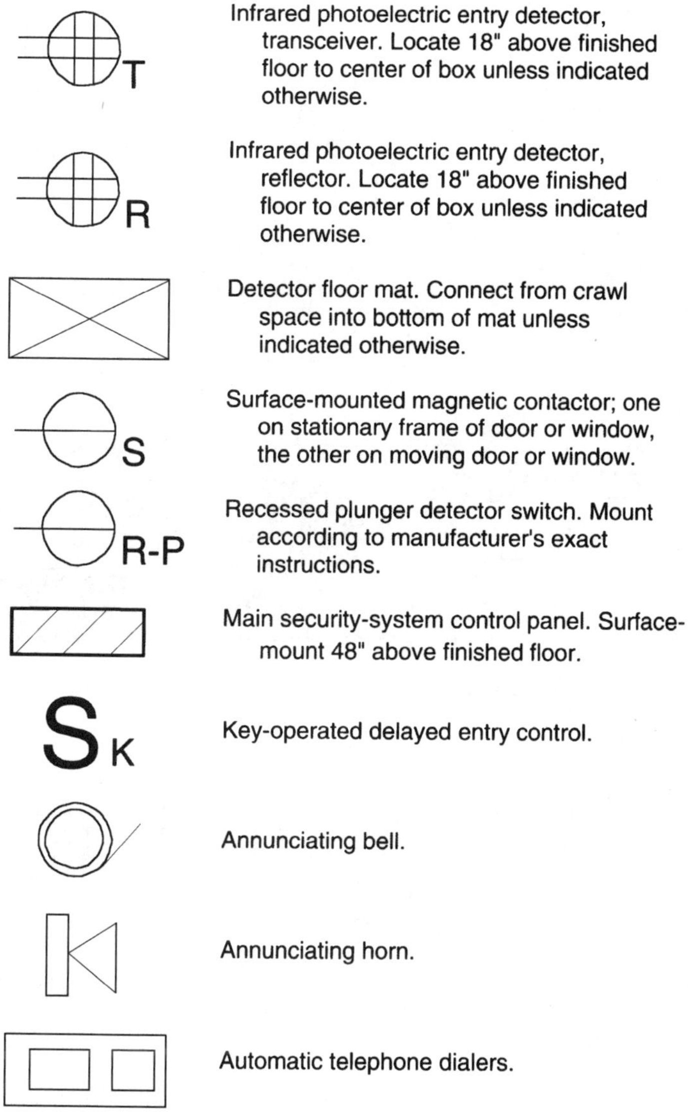

Infrared photoelectric entry detector, transceiver. Locate 18" above finished floor to center of box unless indicated otherwise.

Infrared photoelectric entry detector, reflector. Locate 18" above finished floor to center of box unless indicated otherwise.

Detector floor mat. Connect from crawl space into bottom of mat unless indicated otherwise.

Surface-mounted magnetic contactor; one on stationary frame of door or window, the other on moving door or window.

Recessed plunger detector switch. Mount according to manufacturer's exact instructions.

Main security-system control panel. Surface-mount 48" above finished floor.

Key-operated delayed entry control.

Annunciating bell.

Annunciating horn.

Automatic telephone dialers.

Figure 7.17 Security symbols used with the floorplan drawing in Figure 7.16.

recommended for sliding doors because the doors often jar out of alignment, and the contacts will not function properly. In this case, however, the owners wanted a floor-mat entry detector (Fig. 7.19).

A scatter rug is a good medium for concealing floor-mat entry detectors, but there is always the problem of concealing the connector cord. In the case of the home under consideration, the scatter-rug connection was made in the crawl

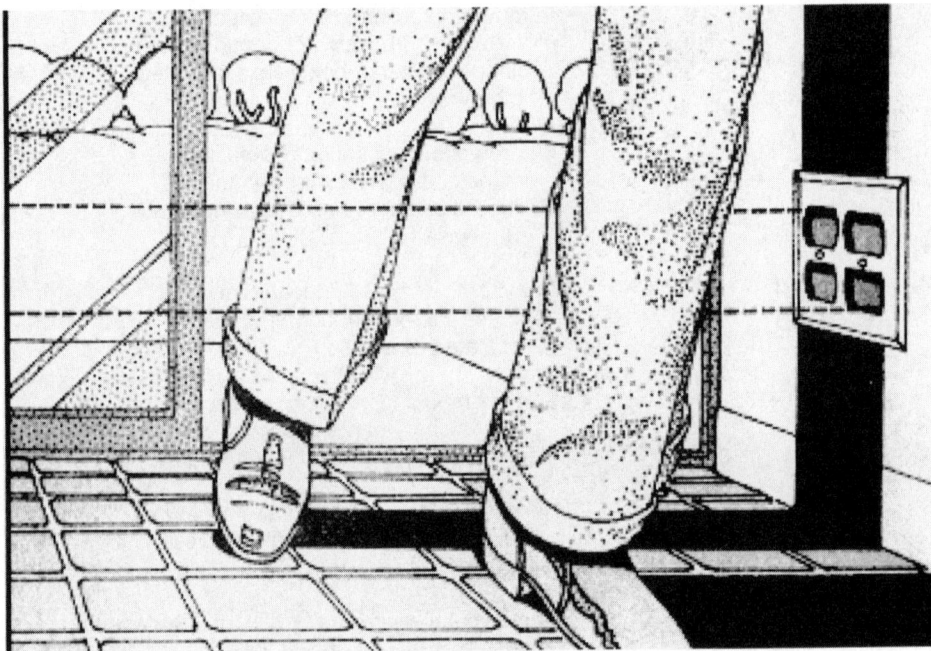

Figure 7.18 Transceivers and reflectors are located approximately 18 inches above the floor so that an intruder will break the beam.

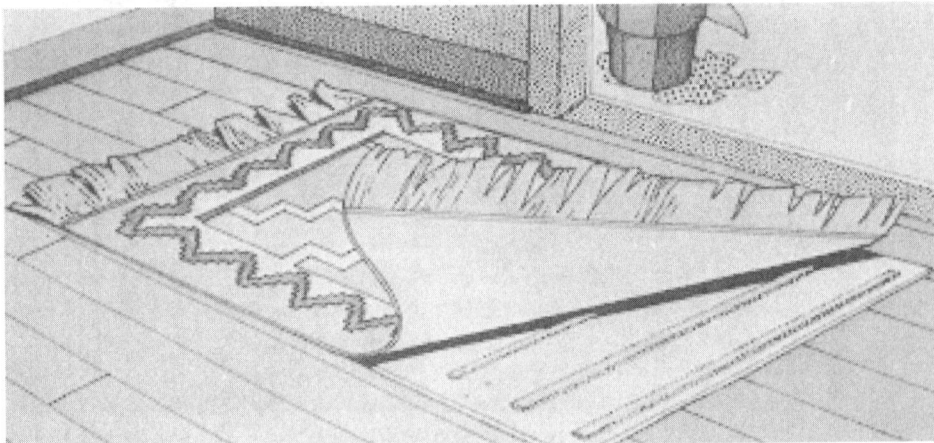

Figure 7.19 A scatter rug is a good medium for concealing floor-mat entry detectors.

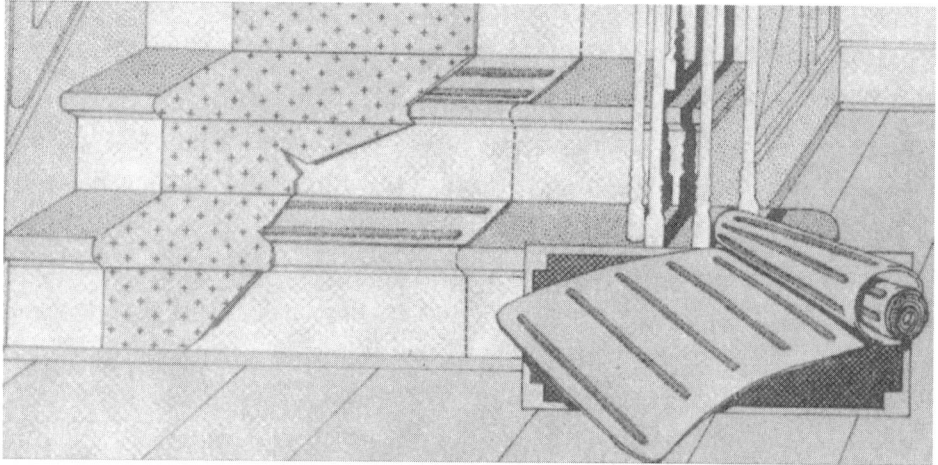

Figure 7.20   Floor-mat detectors can also be located on stairways.

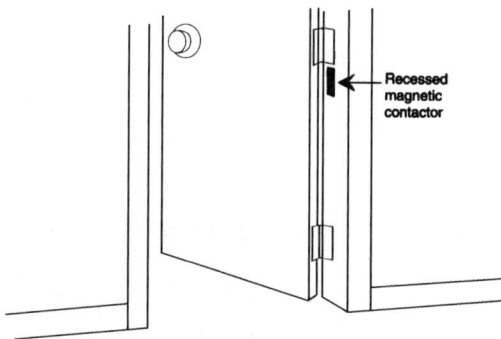

Recessed
magnetic
contactor

Figure 7.21   Recessed plunger-type detectors should
be located on the hinge side of the door.

space beneath the floor—running the security circuit up through the floor and connecting directly to the floor-mat entry detector.

Floor-mat entry detectors can also be used on stairways (Fig. 7.20). Furthermore, they are recommended for use in other interior locations that are likely to be used by intruders.

The kitchen door leading onto the carport is protected by conventional recessed magnetic contacts (Fig. 7.21). Again, either infrared photoelectric entry detectors or a floor-mat entry detector could have been used instead.

The items described thus far could be termed *interior protective devices.* Now, perimeter protection must be provided to ensure an intruder-proof home. Window foil is recommended for use on all windows, but many homeowners

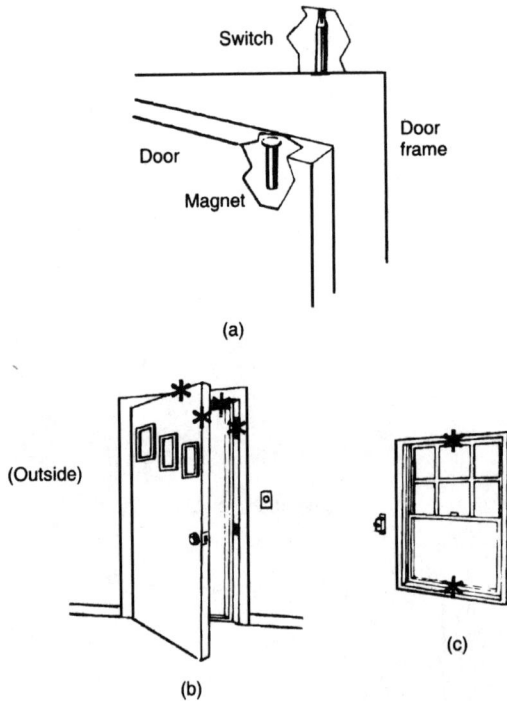

Figure 7.22   Some of the possible locations for recessed magnetic entry detectors.

object to its appearance and prefer to go with magnetic detectors (preferably the recessed plunger type) on windows.

Figure 7.22 shows some possible locations for recessed magnetic detectors on doors and windows. These magnetic detectors are preferred for new construction. However, when existing buildings are to be protected, the surface-mounted magnetic detectors are easier to install.

Of course, the system will need a delayed-entry control (Fig. 7.23) and some means of sounding an alarm. A bell (Fig. 7.24), horn (Fig. 7.25), or telephone dialer (Fig. 7.26) can be used.

If the security/fire-alarm system is operated by conventional house current, you should install a battery backup system. Also, the security technician should consider some possible causes of false or unwanted alarms:

- Severe electrical storms
- Faulty smoke detector
- Faulty wiring, wire connections, staple cutting through insulation, insulation broken by severe bending, closely spaced bare wires that could touch if jarred by vibration of refrigerator, washer, dryer, furnace, etc.
- Electrical transients from heavy-duty appliances, such as refrigerators, relays, etc.

Figure 7.23 A key-operated delayed entry control.

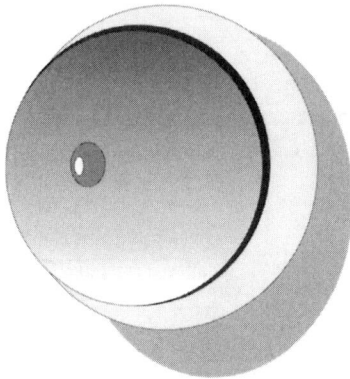

Figure 7.24 A bell can be used as a means of sounding the alarm.

- The use of low-temperature heat detectors in a high-temperature environment, such as an attic or furnace room
- Concentration of sunlight on a heat detector or smoke detector
- Accidental activation of an intruder detector (opening protected door or window, exerting 80 pounds or more of pressure on floor mat, depressing an emergency alarm pushbutton)
- A momentary activation of an entry detector switch on the perimeter or interior detection circuit, caused by a severe vibration
- Amateur or citizens band (CB) transmitter operating with excessive output power near the home

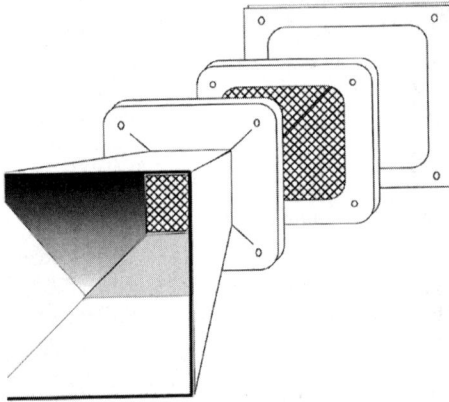

Figure 7.25   A horn is often used to alert neighbors that an intruder is on the premises.

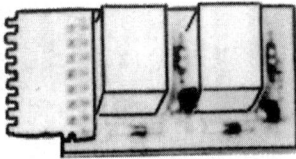

Figure 7.26   A telephone dialer is one means of notifying law-enforcement agencies without scaring off the intruder.

8

Commercial and Industrial Security Systems

The design of commercial and industrial systems is very similar to residential systems, except that heavier-duty equipment is normally used. And, the requirements for these structures are somewhat different from residential demands. This chapter covers the basic techniques necessary to design and calculate the material requirements for typical commercial and industrial occupancies.

The floor plan of a small commercial building is shown in Fig. 8.1. A burglar-alarm panel and a telephone dialer (Fig. 8.2) are located in the storage area. The relay-type control panel has one protective circuit, output for 6-V dc alarm-sending devices, and a silent holdup alarm with telephone dialer to call emergency numbers and deliver voice messages.

Glass on the front door is protected with window foil connected to foil blocks that are then connected to the protective circuit wiring in the alarm system. Door cords are used to provide a flexible connection from the foil blocks on the door and window to a solid contact point adjacent to the door (Fig. 8.3). Flexible door cords also can be used on moving windows and money clips can be installed in cash drawers.

The large display window is again protected with foil connected to foil blocks, and the office area is protected by an ultrasonic motion detector (Fig. 8.4). A surface-mounted door contact (Fig. 8.5) is used to protect the rear door of the building. Legal entry is permitted by use of a key lock.

Security Equipment for Commercial Applications

Magnetic contacts are used on doors and windows in closed-protective circuits, in direct-wire systems, and also in open-circuit applications. Movable elements within the protective switch unit of the magnetic contacts usually consist of a single flexible contact arm that provides a solid metal circuit path from the terminal

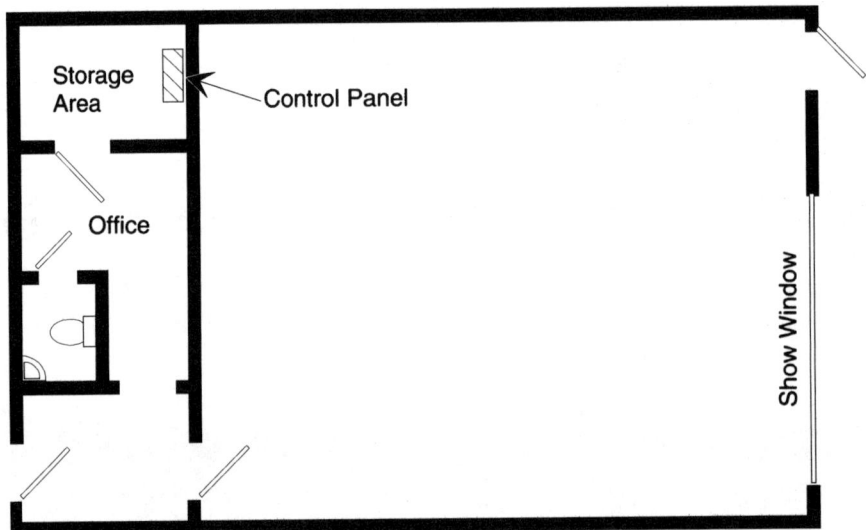

Figure 8.1 Floor plan of a small commercial building.

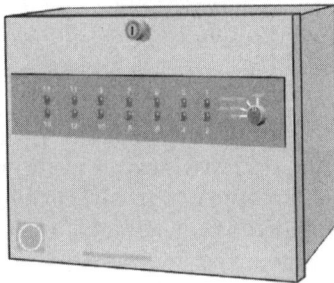

Figure 8.2 Relay-type control panel utilizing one protective circuit and having output for alarm-sounding devices and other detection/alarm capabilities.

screw to the end of the contact point. The circuit continuity should not depend upon conduction across a hinge joint or through a coil spring.

When magnetic contacts are mounted on either noncoplanar or ferromagnetic surfaces, magnet and/or switch units should be held away from their respective mounting surfaces to:

■ Bring switch and magnet into close proximity when the door, window, etc., is closed.

■ Reduce the shunting effect of ferromagnetic materials so that positive switch pull-in occurs when the magnet approaches to within $1/4''$ of the switch.

Mechanical contacts are used as emergency, panic, or fire-test switches. Ball contacts and plunger contacts (Fig. 8.6) are used in both closed- and open-circuit applications.

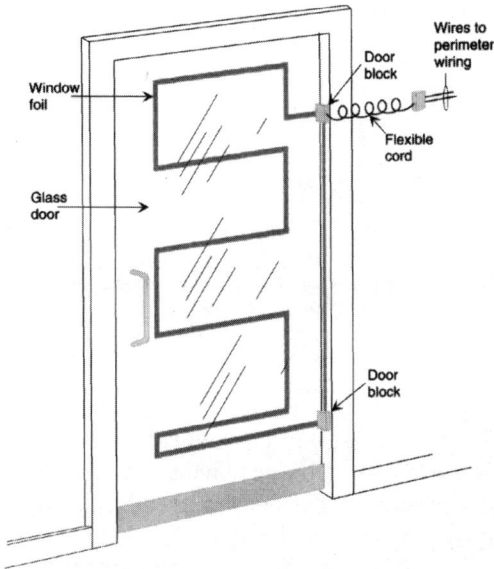

Figure 8.3 Glass door with security devices installed.

Figure 8.4 Ultrasonic motion detector.

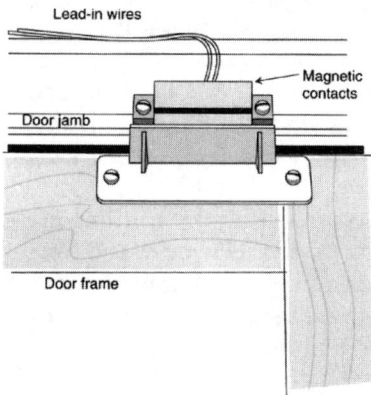

Figure 8.5 Surface-mounted door contact.

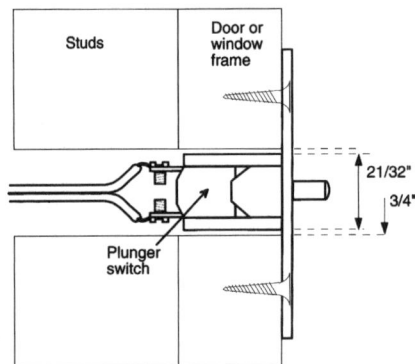

Figure 8.6 Cross-sectional view of plunger contacts.

Mercury contacts are sometimes used in low-energy alarm or signal systems to detect the tilting of horizontally hinged coverings: windows, doors, and access panels. At horizontal openings it is best to install mercury contacts that can be adjusted for sensitivity after installation.

For combined detection of either opening or breakthrough, cord-mounted contacts with foil connected to takeoff terminals should be used. Wiring diagrams of mercury contact connections are shown in Fig. 8.7.

Holdup switches are usually installed under counters or desks in banks or stores, where an employee observing a holdup might be able to signal for help.

In banks and other places where large amounts of money are exchanged, a money-clip alarm device is sometimes used. They automatically trigger an alarm when all bills are removed from a cash drawer. A bill inserted in the clip holds its switch in the normal position. Additional bills on top of the clip keep it concealed. Bills can be added or removed as required for normal business operations so long as one remains in the clip. However, the removal of all bills trips the clip switch to signal an alarm.

Money-clip alarm devices should be installed in the largest bill compartment of cash drawers and connected to the building alarm system by means of a retractable door cord. If exceptionally busy working conditions create the possibility of a false alarm because the bill in the clip might be accidentally removed, two money clips should be used at each station and wired so that both must be emptied to cause an alarm.

Window foil is used extensively in commercial applications. For fixed windows, the connections to the building alarm system are usually made through foil blocks. For movable windows and doors, a retractable door cord must be used, or plunger contacts can be installed (Fig. 8.8).

Ultrasonic motion detectors for commercial applications are essentially the same as those described for residential use. However, the range of detection is sometimes extended on the units designed for commercial use. For example, a typical coverage pattern of a motion detector is shown in Fig. 8.9. Notice the coverage here is 15 feet wide by 30 feet deep.

Terminal block

Connect to
wiring system

Foil takeoff
terminals

Single-pole, single-throw
contact

Terminal block

Connect to
wiring system

Foil takeoff
terminals

Double-pole, double-throw
contact

Mercury
contact
terminals

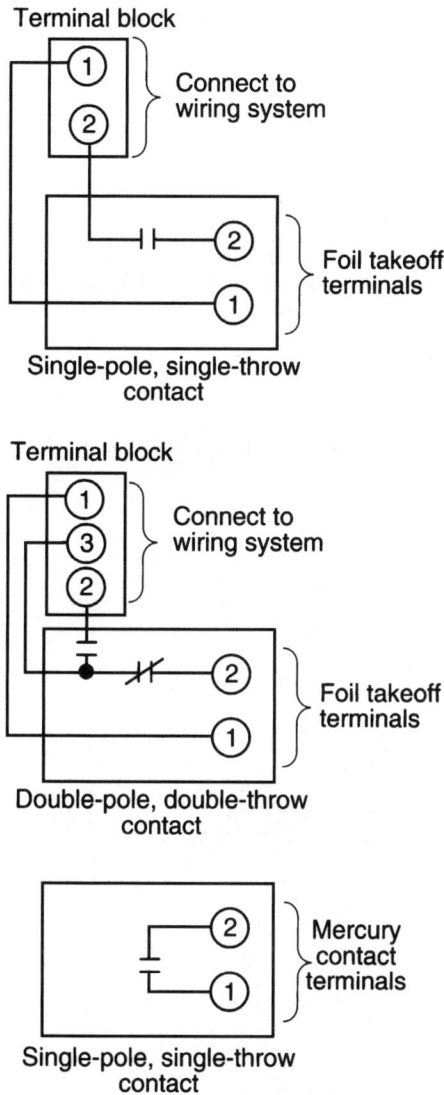

Figure 8.7 Wiring diagram of
mercury contact connections.

Single-pole, single-throw
contact

   Commercial telephone dialers are available that dial emergency numbers
and deliver voice messages. Most distinguish between burglar and fire-alarm
channels. A typical wiring diagram is shown in Fig. 8.10.

Digital alarm transmitters

   Digital alarm transmitters are relatively popular in large commercial establishments. They can be programmed on computer chips to meet the exact
requirements of any business. In a typical model, each line corresponding to

Recessed plunger
contact

←Magnetic
contacts

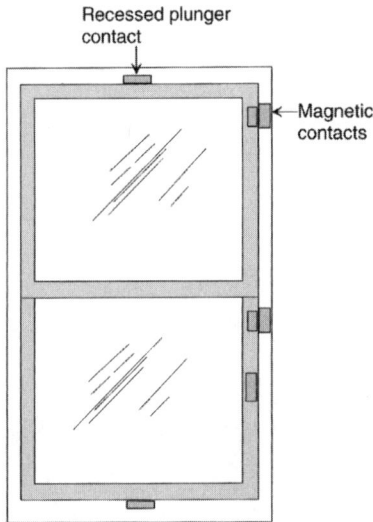

Figure 8.8 Location of plunger
contacts in movable window.

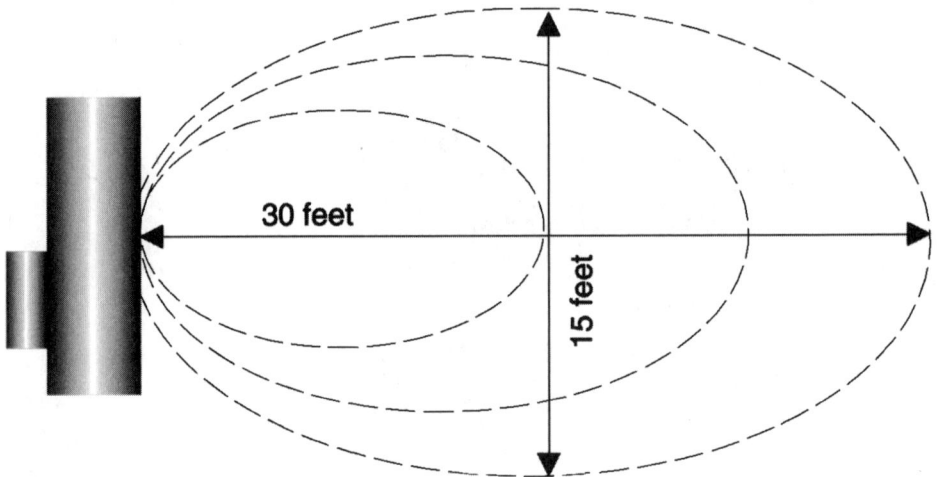

30 feet

15 feet

Figure 8.9 Typical coverage pattern of a motion detector.

an alarm code is limited to13 characters. A blank space between two words on the same line will take up one character.

These units are specifically designed for central-station monitoring of commercial buildings. The unit consists of a transmitter, a special module, and a subscriber control station.

One type of module has three input zones:

- Zone A: Two-wire, closed-circuit input generally used for a foil circuit. This zone can be 24-hour or key-controlled by the subscriber control station.

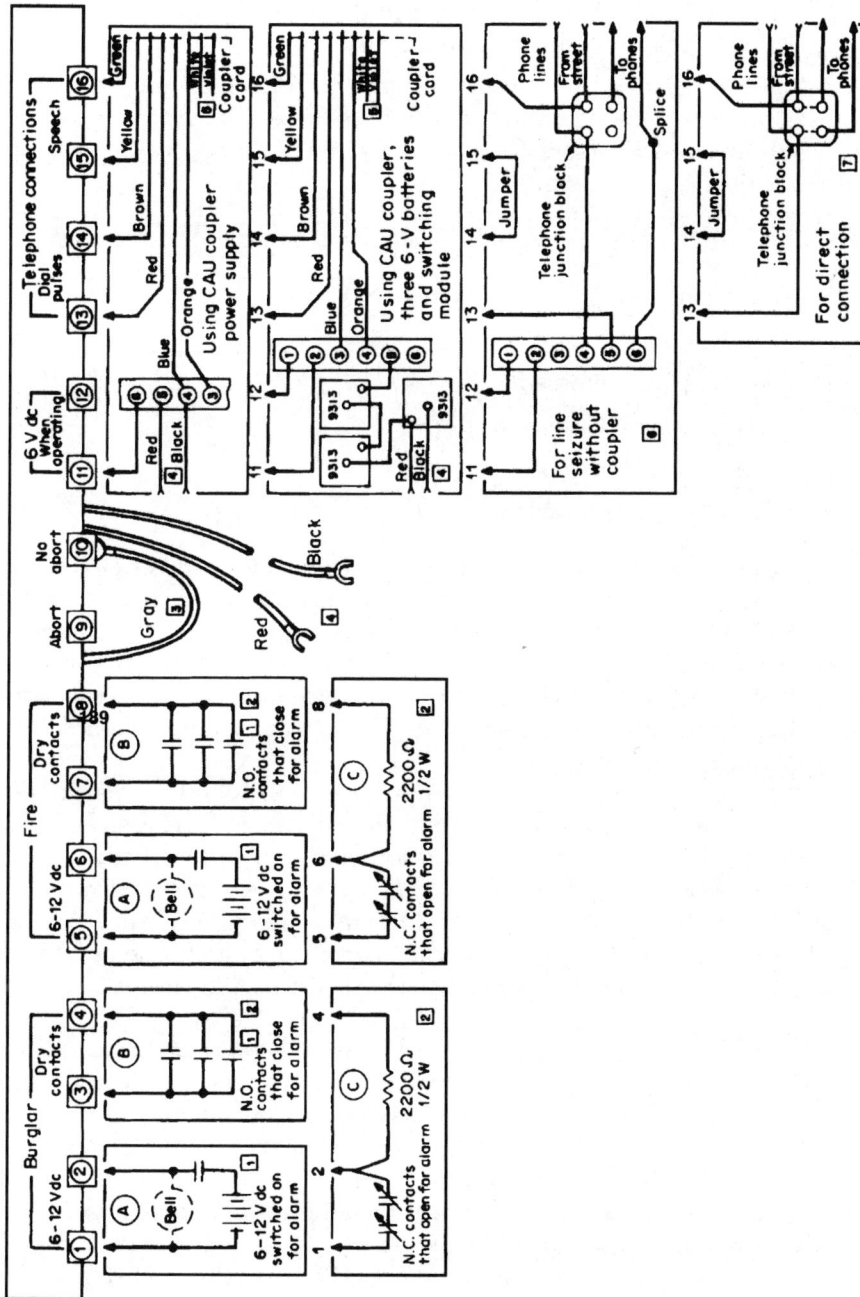

Figure 8.10 Wiring diagram of a commercial telephone dialer.

189

- Zone B: Three-wire, closed-circuit and open-circuit input for perimeter doors and windows. This zone is armed and disarmed by the subscriber control station. It can be connected to the detector contacts so that an open or a crossed circuit will produce an alarm.

- Zone C: Three-wire, closed circuit and open-circuit input for internal protection devices, such as ultrasonic motion detectors, infrared beams and passive infrared detectors. They can be wired so that either an open or a crossed circuit will produce an alarm.

Six output leads are provided from the module to trip the control:

1. Zone A open
2. Zone B open or crossed
3. Zone C open or crossed
4. System armed (closed—night)
5. System disarmed (open—day)
6. Alarm circuit restored

Zones A and B can be connected to the same channel of the transmitter, thereby using only five channels for the module. If this is done, a daytime foil break will be reported as "perimeter alarm—day." The sixth channel may then be used for 24-hour reporting of fire alarm, holdup alarm, or equipment supervisory monitoring. A holdup alarm can also be connected to the same channel of the transmitter as Zone C. If this is done, a holdup would be indicated as "interior alarm—day."

When the subscriber leaves, an interior alarm and the perimeter door circuit will both be tripped and they will be reported. A restore signal will be reported only when all three zones have returned to normal.

The subscriber control station consists of a key switch and two LEDs mounted in a stainless steel plate. One LED indicates that all three zones are good. When the subscriber turns the key to arm the system, the second LED comes on, indicating that the transmitter is reporting the closing signal. This LED goes out when the signal is received at the central station.

Because the transmitter can report multiple signals on the same call, it is practical to locate the subscriber control station near the exit door. Consequently, if the subscriber arms the system and leaves immediately, the interior alarm, perimeter alarm, and restore signal will all be transmitted on a single call. This will take about 30 seconds, including dialing time. The receiver at the central station will be tied up for about 15 seconds after it answers the call.

Surveillance cameras are being used extensively in banks and stores to prevent holdups, pilferage, and burglaries. Because thieves are notoriously camera-shy, the presence of a surveillance camera is often sufficient to make a would-be robber change his or her mind. If a business should be robbed, a surveillance camera provides strong evidence to aid police and courts.

Most surveillance cameras can be adjusted to take individual still pictures at preset intervals to keep a continuous eye on the premises. The Super 8 Kodak surveillance camera, for example, provides continual recorded surveillance for up to 180 hours with each 100-foot roll of film. There are up to 7200 individual photographs to assist in positive identification and apprehension. This unit can be activated by a switch on the camera, by remote control, or automatically by relay from the external alarm system.

Video tapes have been substituted for film in recent years, but the principle is basically the same.

Miscellaneous security equipment

Vibration detectors are often used on buildings to detect vibrations caused by forced entry. Such detectors have been used on a variety of construction materials, such as hollow tile, plaster and lath, brick, concrete, metal ceilings, wood, and concrete block with brick veneer. Once mounted, they can be adjusted with a set screw for the desired sensitivity.

Some establishments maintain a security fence equipped with fence-guard detectors. This type of detector will detect climbing, cutting, or any other penetration of the fenced area. Most of these detectors operate on standard closed-circuit controls.

Fence-guard detectors use a vertical-motion detector that is sensitive to movement created by climbing or cutting the fence. Normal side motions such as wind or accidental bumping do not affect the detector and cause false alarms. They are normally mounted about midway up the fence, and at every 10-foot length of fence. Most of these devices set off the alarm if they are tampered with or if the wire is cut. They may be connected to a control panel and the alarm will "sound" in the form of a bell or horn, or it will silently dial the local law-enforcement agency.

Outdoor microwave detectors are used to protect large outdoor areas, such as parking lots, construction sites, and building perimeters. In operation, a solid circular beam of microwave energy extends from a transmitter to the receiver over a range of up to 1500 feet. Any movement inside of this beam (Fig. 8.11) will activate the alarm.

Thermistor Sensors

The continuous, linear, thermal sensor is a small-diameter coaxial wire that is capable of sensing temperature changes along its entire length. The sensor is made up of a center conductor and an outer stainless steel sheath. The center conductor is electrically insulated from the outer sheath by a ceramic thermistor material.

The term *thermistor* is derived from the words "thermal resistor." Thermistors are actually thermally sensitive semiconductor devices and consist of two basic types:

- NTC (negative temperature coefficient)
- PTC (positive temperature coefficient)

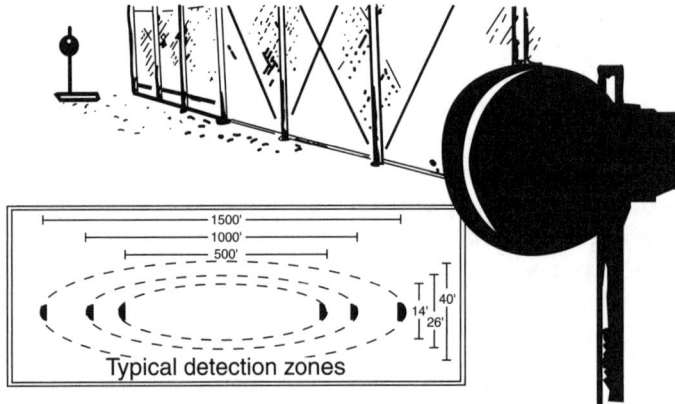

Figure 8.11 A solid circular beam of microwave energy extends from a transmitter to the receiver over a range of up to 1500 feet.

An NTC will decrease its resistance as the temperature increases. The PTC thermistor will increase its resistance as the temperature increases. Of the two types, the NTC thermistor is the most widely used.

Thermistors are highly nonlinear devices and are therefore difficult to use to measure temperature. Devices utilizing thermistors must be calibrated for the particular type of thermistor being used. If the thermistor is ever replaced, it has to be an exact replacement or the circuit will no longer operate correctly.

Thermistors are used as set-point detectors, as opposed to actual temperature measurement. A set-point detector is a device that activates some process or circuit when the temperature reaches a certain level. Because the thermistor has a negative coefficient of resistance, the electrical resistance between the center wire and the outer sheath decreases exponentially as the surrounding temperature increases. The changing resistance is monitored by one of several control panels which actuate extinguishing systems or any other electrically controlled devices.

The sensors have a diameter of approximately 0.080″, and this small mass permits them to sense changes in temperature rapidly. They can react to temperatures from 70°F up to 1200°F, if the thermistor material is properly selected.

Because electrical resistance is measured across two wires (center to sheath), the sensor has the ability to detect a high temperature on a short wire as well as a lower temperature on a longer one.

The element is mounted by clamps spaced along its length and the detectors, being all solid state, have only two electrical failure modes: open circuit and short circuit. Both of these conditions can be caused only by mechanical means and are minimized by rigid mounting. Figure 8.12 shows the construction and mounting details.

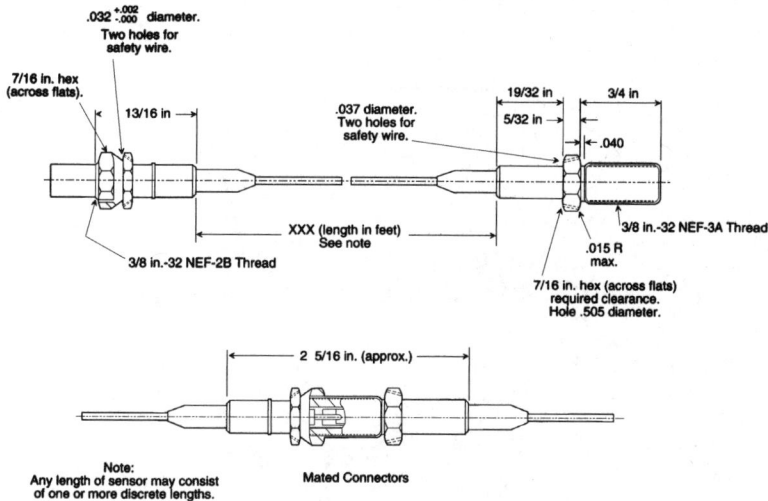

Figure 8.12 Using connectors to supply desired length of sensor cable.

Ultraviolet-radiation devices

Ultraviolet-radiation fire detectors combine large-scale integration circuit techniques with an ultraviolet detection assembly to form a simple, yet flexible, fire-detection system.

The basis of this type of system is a gas-detection tube using the Geiger-Mueller principle to detect radiation waves of a certain length. Notice that visible radiation does not extend into the detector's sensitive area. Similarly, radiation from artificial lighting sources does not extend into the detector's sensitive area.

Welding arcs and lightning strikes, however, will generate radiation to which the detectors are sensitive and precautions must be taken to minimize the effects of large surges.

The ultraviolet-radiation detector's focus of sensitive points is a 60-degree spherical cone with an apex at the detector tube. Figure 8.13 indicates the relationship between viewing angle and relative sensitivity. The sensitivity of the detector tube is a characteristic of its cathode material and is fixed, but its voltage-pulse output rate is directly proportional to flame size; that is, it increases when larger flame fronts are presented to the detector. The pulse output rate is also inversely proportional to the distance of the flame front from the detector tube—the pulse output rate decreases as the distance from the detector tube to the flame front increases.

To illustrate, a 1-ft^2 hydrocarbon fire will cause a pulse output rate of 3 pulses per second at a viewing distance of 30 feet. This same fire will cause a tube pulse output rate of 20 pulses per second at a viewing distance of 20 feet. In a like manner, a 1-ft^2 flame front must be located at a distance of 5 feet to create

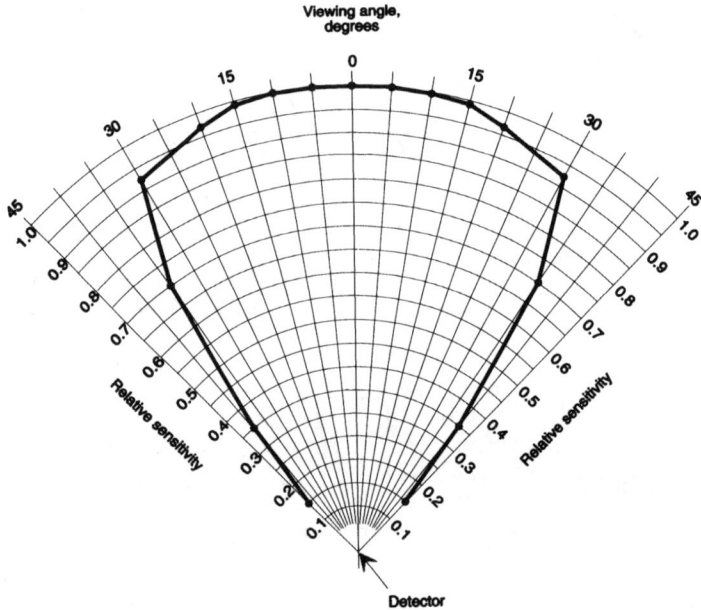

Figure 8.13   Viewing angle of an ultraviolet detector.

a pulse output rate of 30 pulses per second, a 16-ft^2 fire will create the same pulse output rate at a distance of 25 feet.

Firestat

A *firestat* is a safety device mounted in the fan compartment of a heating, ventilating, and air-conditioning (HVAC) system to stop the fan if the return air temperature reaches about 160°F. Most firestats are bimetal actuated and utilize a normally closed switch that must be manually reset before the fan can operate.

The reason for stopping the fan when the high return air temperature exists is to prevent agitation of any open flame in the building, which in turn helps to prevent the spreading of any fire that may be present. Firestats are required on all residential and commercial HVAC systems; in some areas, other safety devices are also required.

Security-System Controls

A typical burglar/fire-alarm control panel is shown in Fig. 8.14. This particular panel is designed for combined burglar-, fire-, and panic-alarm systems. This panel, and most others, operate on low-voltage, alternating current from a plug-in transformer. Many systems also have a rechargeable or dry-cell battery pack for backup power should the ac source fail.

A wiring diagram for a rechargeable 6-V dc power supply is shown in Fig. 8.15. Notice that the transformer is connected to a 120-V ac supply for its primary connection; the transformer then supplies 12-V ac from its secondary. One terminal from the transformer connects to the charging circuit (1) while the other lead connects to one side of the battery (B1). Fuse F1 and resistor R2 offer 6-A circuit protection in this case. R2 also provides short-circuit protection.

The protective-circuit contacts from terminal 5 utilize a 100-W, 2-W resistor in the positive feed to each circuit to keep a cross or short circuit on any one zone from affecting other zones.

Fire-alarm control panels

The heart of any fire-alarm system is the master control panel. To this panel are connected various detector and alarm circuits (Fig. 8.16). In this case, the primary power is taken from an unswitched single-phase, three-wire, 120/240-V ac distribution line. The initiating and alarm circuits are connected to the

Figure 8.14 Typical burglar/fire-alarm control panel.

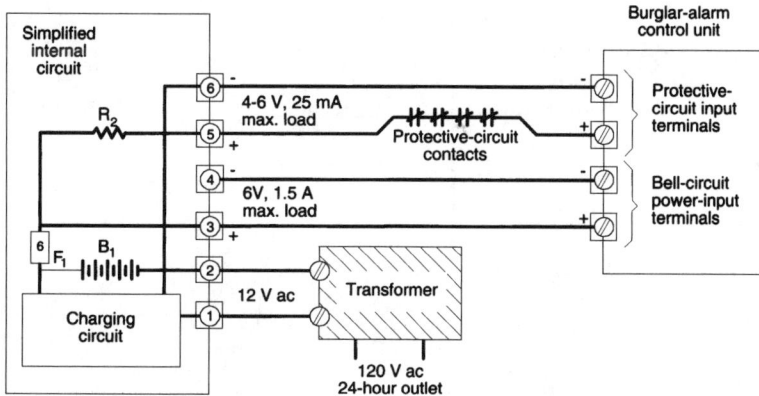

Figure 8.15   Wiring diagram for a rechargeable 6-V dc power supply.

neutral ground and one leg of the main or ungrounded side of the circuit. The trouble-indicating circuits are connected to the neutral ground and to the ungrounded or opposite leg of the circuit.

When an automatic detector or manual station is activated, the contacts close to complete a circuit path and apply 120 V ac to the alarm control circuits in the main panel. On some systems this includes a synchronous motor, which immediately operates cam assemblies that cause the alarm circuit switch contacts to make and break in a code sequence (if a code sequence is used). Additional cam-controlled switches stop the motor and alarm signals after a predetermined time lapse and actuate the alarm buzzer on the main panel.

Most control panels contain a supplementary relay control for connection to an external auxiliary circuit providing its own electrical power. The relay usually has a single-pole, double-throw contact, which operates in step with the master code signal. The circuit can be used to activate other auxiliary alarms or controls, such as a city fire-department connection, fan shutdown, or door release.

A schematic drawing of a day-night police panel is shown in Fig. 8.17; key-switch operating sequences are depicted in Fig. 8.18. In general, any DAY circuit contact opening sounds the buzzer in the panel, but does not ring the alarm bell or disturb police. A holdup contact closure sends a silent police alarm.

On the BELL TEST circuit, the bell can be rung for a test to check the power source and wiring without disturbing the police, but the holdup circuit remains armed during this test.

During the CIRCUIT TEST sequence, the holdup circuit remains armed and the meter shows the current through DAY and NIGHT circuits combined when all contacts are closed. A reading on this particular circuit should be from 2 to 6 mA.

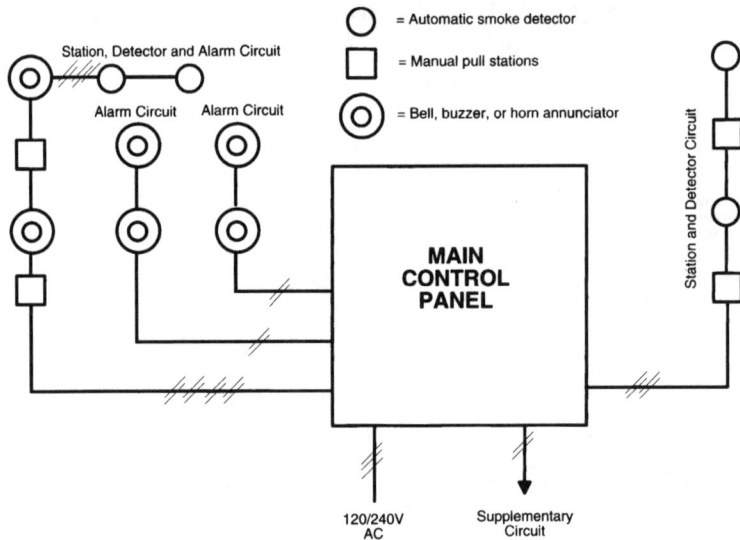

Figure 8.16 Wiring diagram of a fire-alarm main control panel with its connecting detector circuits.

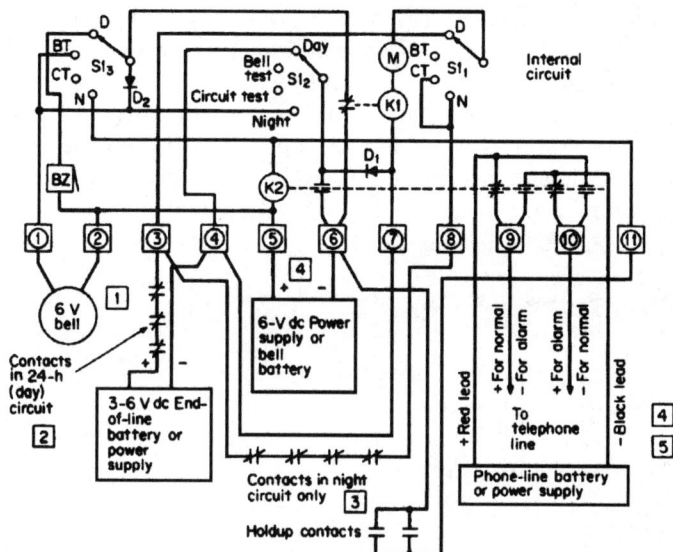

Figure 8.17 Schematic drawing of a day-night police control panel.

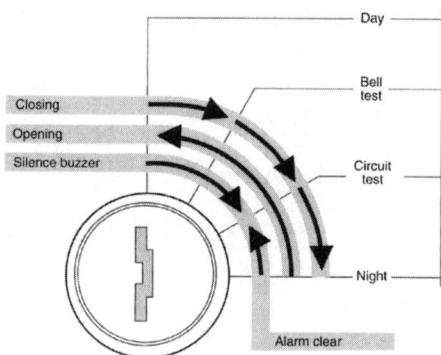

Figure 8.18 Connection detail of entry/exit delay module.

Any contact opening (or cross) in the DAY or NIGHT circuits rings the alarm bell and sends the police alarm. This alarm latches on until the key switch is turned back to CIRCUIT TEST or beyond.

Entry/exit delay module

Solid-state entry/exit delay modules eliminate the need to install a shunt lock across any entry/exit door contacts in a security system. Door contacts are connected to the module, which, in turn, is wired into the protective circuit. Separately adjustable exit and entry delay periods allow the user to turn the system on and leave and then enter and shut the system off without causing alarms.

The module is installed in the alarm system control cabinet (Fig. 8.19). It operates on current from the system's bell battery or power supply and is controlled by the switch functions available in any conventional control unit. It works like a normally closed contact in the negative side of the protective circuit, with all the protective contacts, except the entry/exit door contacts wired into the positive side of the circuit. Opening of any positive-side contacts causes an instant alarm, but the module opens the negative side to cause an alarm only when one of the following occurs:

1. Door contacts have opened once and are still open when the exit delay expires.
2. Door contacts open after the exit delay expires when there was not an exit during the exit delay period.
3. Door contacts open after a proper exit and the system is not shut down before the entry delay expires.

Ultrasonic motion detectors

Ultrasonic detectors work by flooding an area with ultrasonic energy and monitoring the "sound" that returns to the unit from the covered area. In the absence of motion, the received sound is all of a single frequency. Movement of an object in the protected space shifts the frequency of some of the reflected sound, changing the output of the receiving transducer. But such frequency shifts can also be caused by certain environmental factors that must be taken into consideration at installation if false alarms are to be avoided.

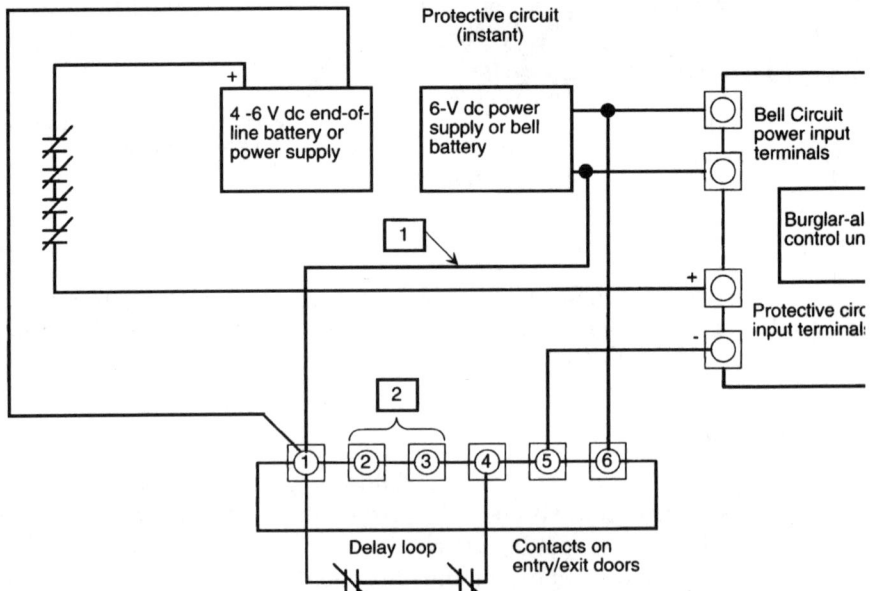

Figure 8.19 Connection detail of entry/exit delay module.

Troubleshooting and Maintenance of Security Systems

Preventive maintenance of a security system is essential whether the system is basic, serving a small installation, or is complex, incorporating supervisory circuits in a large, multifloor building.

Although modern security equipment is highly reliable, components can wear out or corrode in poor environments. Also, circuit parameters may change, and relays can go out of adjustment.

Because security systems are essentially electrical systems, they usually are maintained by electricians. Consequently, anyone involved in the electrical industry should have a working knowledge of fire-alarm systems along with other types of signaling systems installed in all types of building-construction projects.

Figure 9.1, on the following page, shows the parts of a basic fire-alarm system. Reliability of these components is ensured by regular, thorough testing and maintenance programs.

Relays

The two types of relays commonly found in fire-alarm systems consist of sealed plug-in relays and nonsealed plug-in relays.

Maintenance and adjustment of sealed relays (Fig. 9.2) are performed only by the manufacturer. Thus, spare relays should be kept to replace malfunctioning units. Occasionally, relay contacts are burned or pitted by a circuit fault or transient high-current. When this happens, the relay should be sent back to the manufacturer for repair and readjustment. When it returns, it should be kept in reserve as a replacement part.

A major objection to the sealed plug-in relay is the reluctance of many users to maintain an adequate inventory of spare units for emergency use. The chief

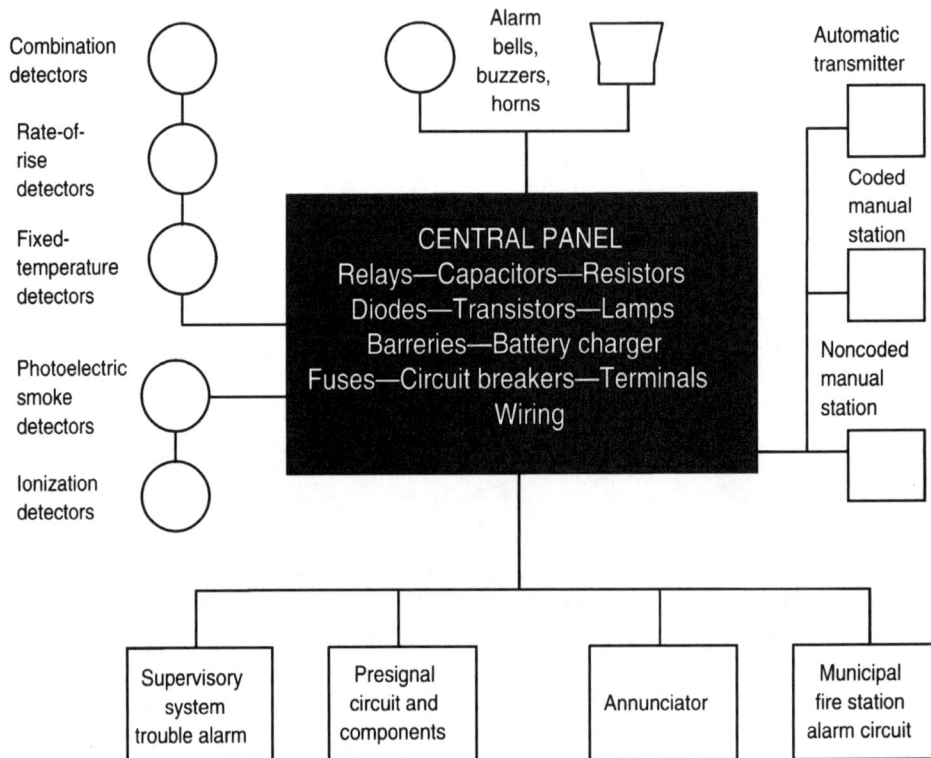

Figure 9.1 Components of a basic fire-alarm system.

advantage of this type of relay is that it is impervious to contamination and it is highly reliable.

Nonsealed plug-in relays have the advantage of permitting adjustment in the field, although they are more susceptible to contamination. In general, relays require readjustment after a period of service, which can range from several months to a few years. The adjustment can be one of armature gap, armature spring tension, contact gap, or contact spring tension.

To clean relay contacts, use a contact burnishing tool, except where excessive pitting or burning has occurred. To remove pits or burns, a fine steel file should be used.

Armature spring tension on some relays can be varied by a screw or knurled nut with or without a locknut; on others by bending a soft brass or steel arm to which a spiral retractile spring is attached. In some instances, it might be necessary to slightly stretch the spring to reduce tension, or cut off a turn or two to increase tension.

Contact springs should be straight. Any excessive belly or waves can be removed by a spring bender or duck-bill pliers. Use a gram gauge to check tension, as recommended by the relay manufacturer. Tension should never be less

Figure 9.2   Sealed relay.

than 6 to 8 grams. The contact gap can be changed on some relays by use of a set screw and locknut; on others, by bending the spring tab that rests on the coil spool head or armature actuator insulator. Set the contact gaps according to the manufacturer's specifications.

Pilot lamps

Where lamps (Fig. 9.3) are only occasionally illuminated, such as in an annunciator, their life can be very long, unless they are subject to considerable vibration from nearby equipment. Where lamps fail often because of vibration, install circuits and lamps rated at 6 or 12 V because these lamps will outlast lamps rated at 24, 32, or 48 V; their filaments are physically stronger.

When pilot lamps are on continuously, as in lighting a colored bullseye lens or a back-lighted panel, it is advantageous to use lamps rated at a higher voltage than the actual supply voltage. This greatly increases lamp life. Typical examples are 32-V lamps supplied at 24 V or 145-V lamps supplied at 120 V. This variation of rated lamp voltage to supply voltage permits ample illumination.

Capacitors

Ceramic capacitors require little maintenance, other than a periodic inspection to see that terminals are free of dirt, grease, or moisture, and to check electrical connections. Failure of this type of capacitor is very infrequent. When a capacitor malfunction is suspected, the capacitor may be tested with an ohmmeter or, at rated voltage, with a voltmeter.

Electrolytic capacitors have a definite life, which varies depending on quality. The best grades should last for at least 7 years. Ambient or working temperature has a direct effect on capacitor life—the higher the temperature the shorter the life.

They can be tested in a manner similar to the dry paper or ceramic capacitors; however, polarity must be observed on full-voltage tests. Terminal areas and electrical connections should be checked when trouble is suspected. And

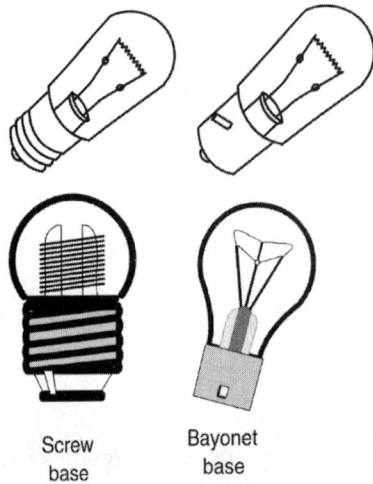

Screw
base

Bayonet
base

Figure 9.3   Basic types of pilot lamps.

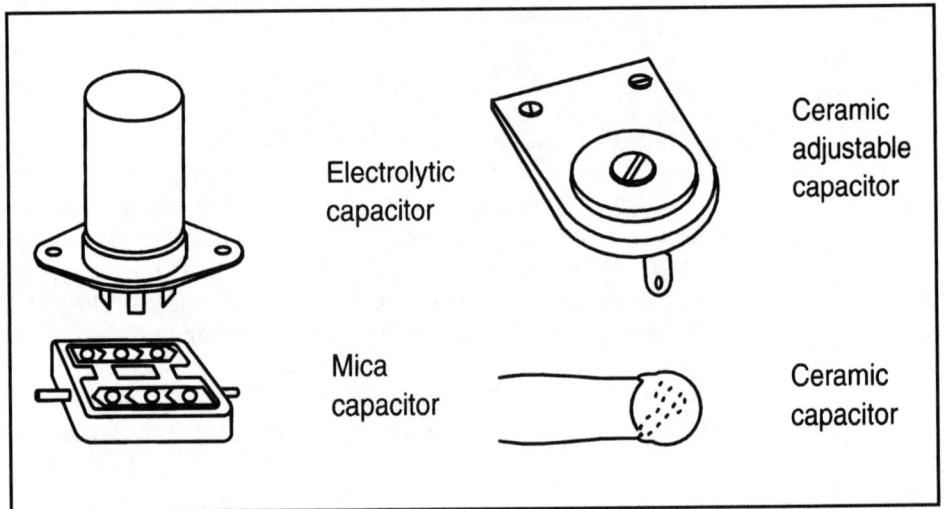

Electrolytic
capacitor

Ceramic
adjustable
capacitor

Mica
capacitor

Ceramic
capacitor

Figure 9.4   Various types of capacitors.

the terminal areas should be kept clean to facilitate inspection and service of connections (Fig. 9.4).

Resistors and rheostats

Resistors and rheostats (Fig. 9.5) require little maintenance. Usually, loose or broken connections or an actual broken resistor will be the cause of trouble. During inspection, terminal-screw connections should be checked for tightness.

Slide bands on resistors and the wiping brushes of rheostats should make positive contact with the exposed turns of wire. These exposed turns should be kept clean. Carbon resistors that have been overheated because of operation at power levels beyond their rated value will change resistance substantially. In such instances, they must be replaced.

Diodes: Diodes can be checked with an ohmmeter or a dc voltmeter and a dc source. When using the ohmmeter, switch to the lowest scale and connect the diode so that a very low resistance reading is obtained. If a significantly higher reading is obtained when the meter leads are reversed, the diode checks out properly. If only a low resistance can be obtained, the diode is shorted. If only high-resistance readings can be obtained, the diode is open.

Using the diode principle of unidirectional current flow, a dc voltage test can be set up.

Transistors

Transistors (Fig. 9.6) require very little maintenance, except to check lead connections. Transistor leads go directly to silicon crystal junctions. Therefore, when soldering, excessive heat must be prevented from reaching the crystal junction. A pair of pliers grasping the lead when applying heat should prevent this. In addition, when soldering, it is important to quickly apply the iron tip and immediately remove it from the lead.

Transistors are very susceptible to transient voltage spikes that can occur in the supply power line. Excessive spikes can cause undesirable transistor conduction as well as damage to the component. Double-anode zener diodes or similar voltage clippers effectively prevent these spikes from reaching the transistor.

Overcurrent devices

Important steps in fuse maintenance include a periodic check of fuse clips to ensure that ferrules are in good contact with clips (Fig. 9.7). Also look for cor-

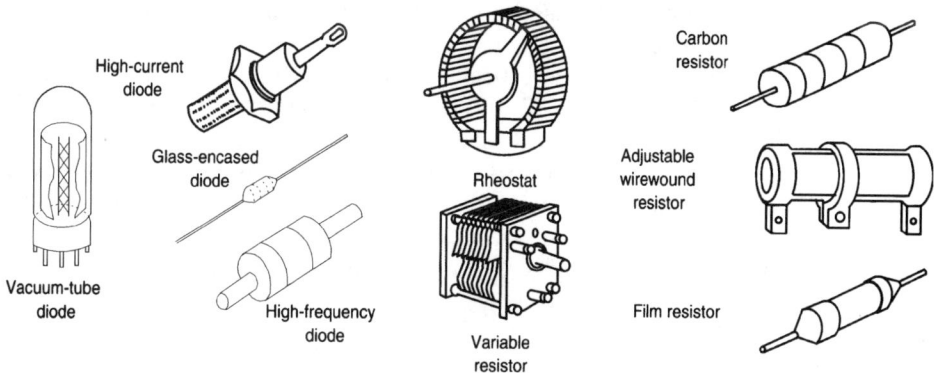

Figure 9.5 Several types of resistors, rheostats, and diodes.

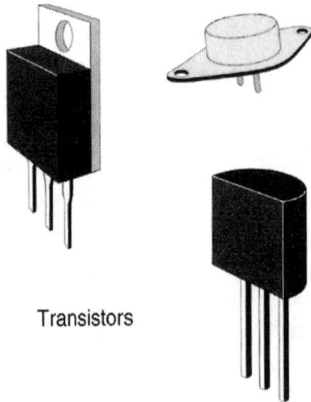

Transistors

Figure 9.6 Transistors are very susceptible to transient voltage spikes.

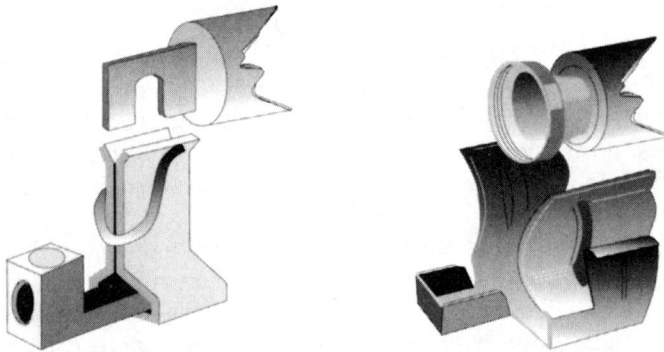

Figure 9.7 Periodic inspection of fuse clips is necessary to ensure that ferrules are in good contact with clips.

rosion or overheating at fuse clips. Hot fuses or clips usually indicate either poor contact or an overloaded fuse, probably close to blowing. A supply of fuses of proper ratings should always be kept on hand for fast replacement of blown fuses.

Molded-case circuit breakers (Fig. 9.8) require little maintenance. They should be kept free of dust and grime to ensure proper mechanical operation. Regular inspections of circuit breakers should be made, including dynamic tests. Every 3 or 4 years, operating trip settings may be checked with special load-testing instruments. These tests are often performed by security maintenance contractors or special service organizations. When these tests are performed, it is advisable to load-test overcurrent protective relays.

Batteries

Batteries should be located in a clean, dry place preferably on shelves or racks, or in a cabinet. Maintenance includes regular checking of battery electrolyte

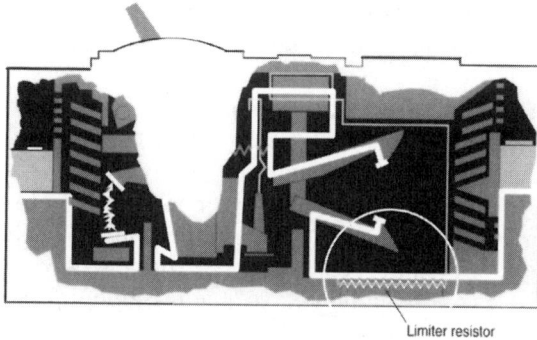

Limiter resistor

Figure 9.8 Molded-case circuit breakers require very little maintenance.

levels, specific gravity, voltage output, "ball" floats, and the battery exterior for dirt or corrosion of terminals.

When the electrolyte is low, bring the liquid up to proper level by adding distilled water (never tap water). Specific gravity and voltage should be checked at least twice a year. Batteries furnished with transparent glass or plastic cases often are equipped with colored ball floats that indicate discharge condition of the battery. When balls are floating at the top of the cell, the battery is in a fully or nearly fully charged condition. When the balls are "down," the battery is nearing complete discharge.

Be sure the battery is kept clean and dry. Terminals should be tight and free of corrosion. Terminals of lead-acid cells can be cleaned with a cloth moistened with aqueous ammonia. A coating of petroleum jelly will retard further corrosion of battery terminals.

Battery chargers should receive regular inspections to head off trouble. The need for more detailed maintenance will become apparent when batteries are not maintained at full charge. For example, lower-than-normal charge rates will cause the batteries to slowly discharge. When this happens, check input voltage to the charger, and inspect components such as selenium rectifiers for aging, and rheostats for loose or corroded slide bands.

Wiring

The key to reliable operation of circuits between panels and remote devices is proper installation. Be sure that conduits will not be subject to water or other contamination, which could cause insulation breakdown. In addition, proper wire stripping and connections at terminals are important. Improper stripping sometimes results in nicks in solid conductors, which later break from vibration or handling at terminals.

When an installation is completed, insulation resistance tests prove the initial reliability of the installation. Use a 500-V tester. The insulation resistance should be at least 1 MΩ and preferably higher. Annual insulation resistance tests should be made and records kept for comparison of readings. A continuing downward trend of readings over a period of a few years indicates failing insulation.

If frequent grounds or partial grounds are causing erratic operation, make a 500-V RMS or a 1250-V RMS breakdown test. These tests, with voltage applied between conductor and conduit, usually will break through weak spots in the insulation, revealing the faulty conductor for immediate replacement. On circuits of 50 V or less, use the 500-V test level; on 51- to 125-V circuits, apply the 1250-V test. Portable test sets for the purpose are available. Instructions should be carefully followed to avoid damaging good wiring or components and to ensure safety.

Prior to the test, be sure to disconnect all equipment. Also, be certain that the normal power supply is removed and that all precautions are taken to prevent physical contact with conductors under test.

Control cabinets

Control enclosures should be kept free of dust and grease on the face of the panel as well as inside. Buildup can cause faulty relay operation and can be a source of corrosion of metal parts. Connecting wires coming from conduits and cables should be neatly arranged at terminating points and laced into a fanned cable form with skinners opposite the terminals they serve. This saves time when looking for a wire during troubleshooting or replacement of circuits. Be sure that the control panel enclosure is of the proper type for its environment. Check NEMA specifications to be sure.

Maintenance of Field Devices

Noncoded manual stations

Maintenance of various types of noncoded manual fire-alarm stations consists primarily of periodic operational tests, replacement of "break-glass" windows or glass rods, and inspection of terminals for loose or corroded connections. In supervised fire-alarm systems, a broken connection will sound a trouble signal (Fig. 9.9).

A test program should be set up in which one station is operated each month, or in accordance with local codes. A different station should be operated each month thereafter until all stations have been tested within a period of a year or two, depending on the number of stations installed. Where stations are installed in a presignal system, be sure to test both the presignal and general alarm function. In systems that send the alarm to a municipal fire department, check transmission of the alarm to the fire department if arrangements can be made for the test. When this is not possible, the alarm system should incorporate means to check this feature without actually transmitting the alarm to the fire department. Identification of each station, location, results of tests, etc., should be kept in a maintenance log.

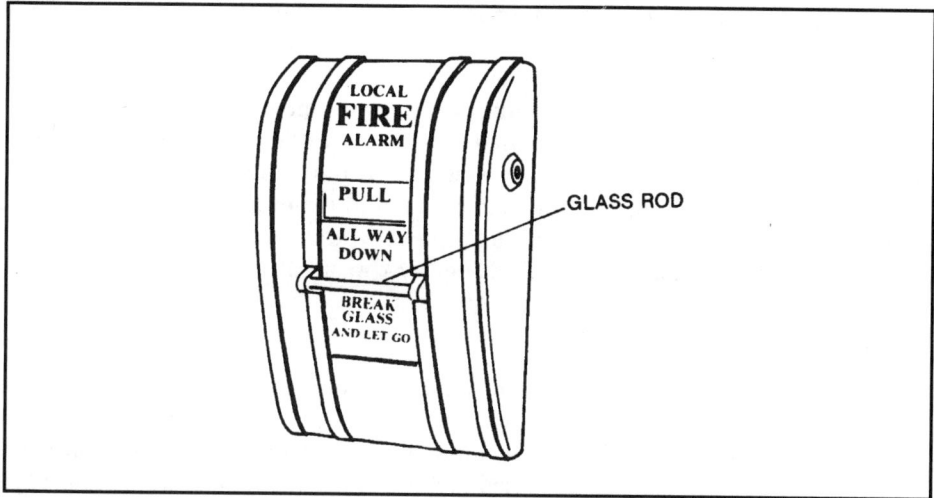

Figure 9.9   Manual fire-alarm station.

Some stations are furnished with open-type contacts, which should be checked and cleaned every year or two, depending on the environment. In some stations, contacts are enclosed in dust-tight housings, which protect from dust and dirt and eliminate contact cleaning procedures. Always keep spare glass rods, plates, and extra hammers for replacement.

Coded manual stations

Two general types of coded stations are the spring-driven type and the motor-driven type. When operating a spring-driven station, the pull handle usually winds a clock spring, which when released, unwinds and drives a gear train and code wheel. The code wheel makes or breaks contacts, which transmit code pulses to the control unit. These components are usually enclosed in transparent glass housings, which provide good protection from grime. However, some dust can penetrate the enclosure and the parts should be oiled with "watch oil" at least every 5 years. At this time, inspect terminals and conductors for tightness and corrosion and clean contacts. Be sure to check extra contacts for annunciation if the station is furnished with this feature. Usually, these contacts must be reset after operation.

Some coded stations must be wound by a key to set the spring-driven operation. On these stations, pulling the lever simply releases the spring that drives the code wheel. These stations must be rewound after operation.

Maintenance of motor-driven coded stations is similar, except that some are furnished with motors that could require attention. These motors are usually fractional-horsepower types rated in inch-ounces. They consume only 4 to 25 W of power. Those stations furnished with synchronous (hysteresis) motors will not require additional maintenance because these motors do not have brushes; normally, bearings are the sealed type. However, stations equipped with dc motors should be checked every 1 to 2 years. Inspect brushes and the commutator. Badly worn brushes will cause excessive sparking and erratic operation. Pitted commutators can be smoothed with an emery cloth or turned down in a lathe, polished, and mica undercut. If motors do not have sealed bearings, they should be oiled as recommended by the manufacturer.

A test program, log of records, and general maintenance procedures similar to that described for noncoded stations should be observed. The log can be stored on disk in computers and the client can be given a copy.

Automatic transmitters

These units are essentially manual stations, but they have the additional feature of an electromagnetic trip circuit to initiate operation. These transmitters, which can be operated either manually or electrically from a remote location, should be tested for both modes of operation in a program similar to that described previously.

Some transmitters are equipped with a local trouble buzzer, an indicating light, and a silencing switch. Others have auxiliary contacts to close a circuit to a central annunciator. Some have both features. When inspecting transmitters, be sure to check these functions. On transmitters that sound a one-round signal as a trouble indication, be certain that the unit is reset after operation or testing.

Heat detectors

One type of fixed-temperature heat detector operates when a metal with a low melting point allows electrical contacts to energize the alarm circuit. Maintenance consists of a testing program and periodic inspection of terminals and connections for looseness or corrosion. It is important that the detectors be replaced if they have been painted several times. Their function can be checked during testing.

The testing program should require that at least two of every 100 detectors be sent to a testing laboratory each year after an initial no-test period of 5 years. Testing agencies (such as UL or a similar facility) will perform the prescribed tests. This type of detector is a "one-operation" device and must be discarded after testing; however, replacement cost is low and most detectors have proven to be reliable. If either of the first two detectors are faulty, more detectors must be tested to determine if a general condition of faulty detectors exists. A record of detector location, test results, etc. should be kept in a log. A log installed in your computer and the clients can have a popup screen that signals testing and cleaning dates.

The bimetal type of fixed-temperature detector can be tested without removing from the ceiling or disconnection from the alarm circuit. These units operate on the principle of heat-expansion of dissimilar metals to energize the alarm circuit; they return to their normal position after cooling ready for another operation.

These detectors can be tested using a portable cord and lamp socket furnished with a 60- or 75-W incandescent lamp. The lamp is held within a few inches of the detector and the heat from the lamp should cause the unit to operate after a few seconds (Fig. 9.10).

Rate-of-rise detectors

Rate-of-rise detectors operate their contacts when the ceiling temperature rises faster than 12 to 15°/min. This ensures operation over a wide range of variable ambient temperatures, depending on the rating of the detector.

These rate-of-rise detectors can be of the line type or spot type. Line-type detectors consist of small diameter copper or aluminum tubing that is attached to ceilings or walls near the ceiling. Both ends of the tubing terminate at a bellows-like switch assembly, which closes alarm circuit contacts on internal pressure increase. A slow increase of pressure is bled off through a small orifice; however, a rapid rise of 15 degrees per minute or more expands the bellows, and closes the contacts to the fire-alarm circuit.

To test for proper operation, use the previously described portable incandescent lamp set held close to the tubing. Perform this test at least twice a year. Check the tubing annually for pinhole leaks with a special testing device provided by the manufacturer. These systems can be furnished with the fixed-temperature features and the addition of special fittings installed in the tubing at 15-foot intervals.

Spot-type rate-of-rise heat detectors are available. One type uses the melting solder principle of operation; the other uses the bimetal principle. Another type of detector is the *rate anticipation* heat detector. These are similar to rate-of-rise detectors; however, they have no thermal lag beyond the instant that operating temperature is reached. The fixed-temperature detectors do have this time lag.

All heat detectors discussed should receive periodic inspection for corrosion, loose connections, and testing, as previously described.

Sprinkler system actuators include: pressure-operated, water-flow switches; paddle-type switches installed in pipe risers; high and low water-level switch-

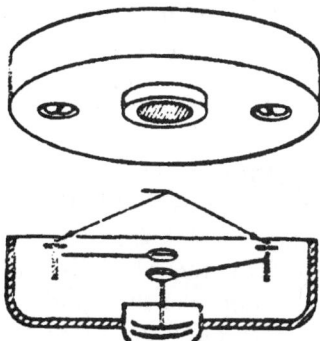

Figure 9.10 Typical fixed-temperature heat detector.

es installed in gravity water tanks; and differential pressure switches on excess pressure pump systems. In general, contacts on these devices are well protected; however, mechanical linkages are subject to corrosion and should be checked every year or two.

Smoke detectors

One type of photoelectric smoke detector consists of a photoelectric cell and an external light source. When light received by the cell drops to a predetermined value, for example when smoke blocks the light from the light source, relays will cause the alarm to sound. Because of this, it is important to keep the light source and cell lens clean. Also, the lamp in the light source should be replaced regularly before failure.

Another type of detector has an internal light source feeding into a labyrinth. Any dust accumulation in the labyrinth can be removed by blowing compressed air into the unit. This type is furnished with a 6-V, 5000-hour lamp, which should be replaced before it fails.

The balanced-bridge type of smoke detector should be cleaned two or three times a year, or after being actuated by smoke a few times. The unit should be carefully disassembled and components cleaned with ethyl alcohol.

Ionization detectors will discover fires where flame predominates and smoke generation is negligible. This is because invisible gas-combustion products become ionized when they strike the detector head, triggering the detector mechanism. These units should be kept free of dust accumulation.

Smoke detectors should be tested at least once a year. The test is accomplished simply by blowing smoke from a cigar, pipe, or other smoke toward the ceiling-mounted unit. A log should be kept to ensure that all detectors are tested on schedule.

Horns and bells

Alarm horns (Fig. 9.11) can be dc or ac. The dc alarm horns are usually of the vibrating diaphragm type. These horns are actuated by an armature which rapidly strikes a diaphragm. Connected in series with the armature actuating coil is a pair of contacts. The armature makes and breaks these contacts, alternately energizing and deenergizing the coil, creating the vibrating motion. A tubular capacitor is generally connected across the contacts to suppress arcing. To obtain desired sound-level output, an adjusting screw can be set to obtain minimum to maximum stroke of the armature that strikes the diaphragm.

The horns should be tested regularly, contacts inspected and any existing pits or burrs removed with a fine file. Contact gaps should be checked and set according to manufacturer's specifications.

Some dc horns have a small high-speed dc motor, which drives a wheel with projections that strike a diaphragm. Maintenance of this type of horn includes inspection of brushes for excessive wear and commutator for pitting or burning.

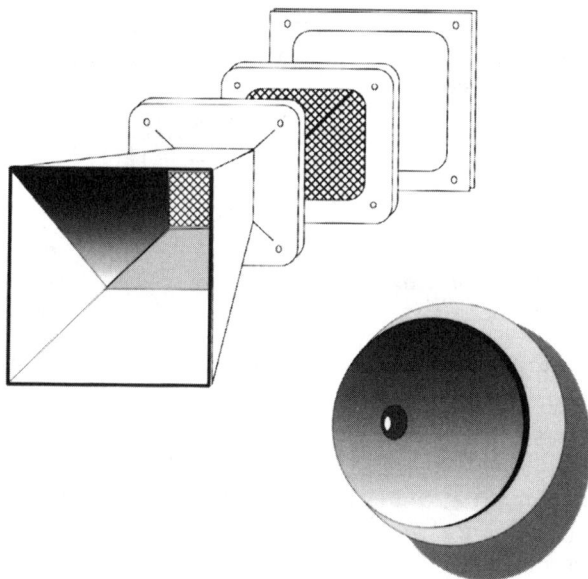

Figure 9.11   Annunciating devices.

Most ac-powered vibrating diaphragm horns require no maintenance other than a periodic test. These horns do not have contacts because the zero and peak voltage created by the 60-cycle sine wave will alternately energize and deenergize the armature.

Trumpet-type ac horns are usually furnished with contacts to obtain the right armature speed and stroke force for their characteristic resonant operation. These contacts will require attention more frequently than contacts on dc vibrating diaphragm horns.

Alarm bells, both ac and dc types, operate on the same principles as described for ac and dc horns. Construction of operating mechanisms is similar to horns, except that the armature drives a striker that strikes a gong shell. Testing, adjustment, and maintenance are similar to that recommended for horns.

It should now be obvious that preventative maintenance for any type of security/fire-alarm system is mandatory for safe and efficient operation. In fact, some jurisdictions require periodic inspections by local fire wardens and electrical inspectors.

Troubleshooting

Troubleshooting covers a wide range of problems from such small tasks as finding a ground fault in a perimeter loop (sensor circuit), to tracing out defects in a complex control circuit. In any case, troubleshooting usually requires a thorough knowledge of testing equipment and a systematic and

Figure 9.12   A ground fault or short circuit is one of the most common causes of problems in electrical systems.

methodical approach to the problem; that is, testing one part of the circuit or system after another until the trouble is located.

Those involved in the maintenance of security alarm systems should keep in mind that every electrical/electronic problem can be solved, regardless of its nature. The paragraphs that follow are designed to aid those involved in troubleshooting to better solve the more common security-system problems in a safe and logical manner.

Basic electrical problems

In general, there are only three basic electrical faults:

- A short circuit or ground fault
- An open circuit
- A change in electrical value

A ground fault is one of the most common causes of electrical problems. Such a condition can be caused by any number of faults, but in most cases the cause is faulty insulation on conductors. The defective insulation allows two conductors to touch (Fig. 9.12A) and short out. The fault can also occur between one "hot" wire and a grounded object (Fig. 9.12B).

An open circuit is an incomplete current path and is usually caused by either a blown or tripped overcurrent-protection device, a loose connection or broken conductor, or a faulty switch or control.

A change in electric value covers such items as low voltage, electrical surges, and a change in resistance.

Testing Instruments

To maintain and troubleshoot existing security systems, workers should know and apply modern testing techniques and have a good understanding of basic testing instruments.

The use of electrical/electronic testing instruments generally involves three applications:

1. Verifying proper operation of instruments and associated equipment.
2. Calibrating electronic instruments and associated equipment.
3. Troubleshooting electrical/electronic circuits and equipment.

For these applications, specific test equipment is selected to analyze circuits and to determine specific characteristics of discrete components.

The test equipment that a technician chooses for a specific task depends on the type of measurement and level of accuracy required. Additional factors that can influence selection include the following:

- Portability of the test equipment

- The amount of information that the test equipment provides

- The likelihood that the test equipment will damage the circuit or component being tested (some test equipment can generate enough voltage or current to damage an instrument or electrical circuit)

Meters

The functioning of conventional electrical measuring instruments is based upon electromechanical principles. Their mechanical components usually work on dc electricity. Mechanical frequency meters are an exception. A meter that measures ac has a built-in rectifier to change the ac to dc, and resistors to correct for the various ranges.

Today, many meters utilize solid-state digital systems for operation; they are superior because they have no moving parts. These meters will work in any position, unlike mechanical meters, many of which must remain in one position in order to be read accurately.

When using any testing instrument, always consider the operator's personal safety first. Know the voltage levels and shock hazards related to all equipment to be tested, and be certain that the instrument used has been tested and correctly calibrated; this should be done at least once a year. To prevent damage to the test instrument, select a range (on meters with different ranges) that ensures less than full-scale deflection on readouts in the cases of analog or digital instruments, respectively.

D'Arsonval meter movement

The conventional ammeter is based on the principle that the magnetic effect of electricity in a current-carrying conductor will rotate when placed in a magnetic field, the same as an electric motor. The readings on an ammeter of this kind have been calibrated by comparison with a primary standard instrument. This calibration provides assurance that the practical magnetic instruments used in everyday electrical measurements indicate the same amount of current as the standard instrument.

The basic d'Arsonval meter (Fig. 9.13) operates on direct current (dc) only because alternating current (ac) continually reverses its direction and the meter coil will do likewise. However, by installing a rectifier that can convert ac to dc, the output is dc and the d'Arsonval meter can read it.

A moving-coil meter movement operates on the electromagnetic principle. In its simplest form, the moving-coil meter uses a coil of very fine wire wound on a light metallic frame. A permanent magnet surrounds the coil. The metallic frame is mounted on pivots to allow it and the coil to rotate freely between the poles of the permanent magnet. When current flows through the coil, it becomes magnetized, and the polarity of the coil is such that it is repelled by the field of the permanent magnet.

This will cause the coil frame to rotate on its pivots; the distance it rotates is determined by the amount of current that flows through the coil. By attaching a pointer to the coil frame and adding a calibrated scale, the amount of current flowing through the meter can be measured.

Magnetic-vane meters

Another type of meter is called the *magnetic-vane meter*. It differs from the d'Arsonval meter in that the magnetic-vane meter contains a fixed coil and a movable iron vane, rather than the fixed magnet and moving coil used in the d'Arsonval meter.

Figure 9.14 shows a typical magnetic-vane meter. In this meter, two iron vanes are enclosed by a coil of wire. When current passes through the coil, the vanes are magnetized so that they repel each other regardless of whether the energizing current is direct or alternating. Usually, one vane is fixed and the other is pivoted. The movable vane is provided with a pointer over a calibrated scale.

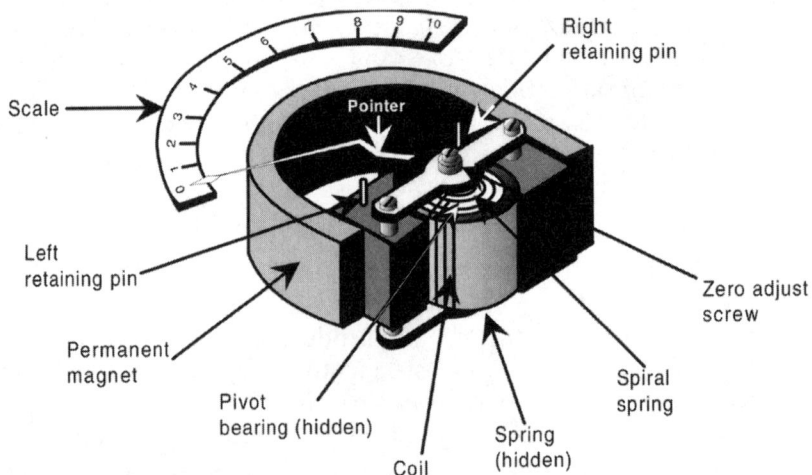

Figure 9.13 Major parts of a d'Arsonval meter.

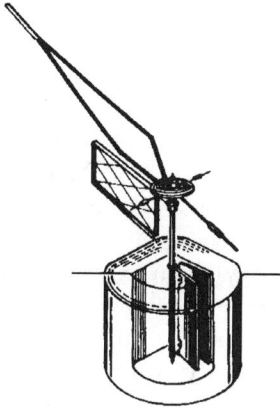

Figure 9.14 Working parts of a magnetic-vane meter.

In recent years, digital-readout meters are rapidly taking the place of many of the analog meters. The accuracy of digital meters, in many cases, far exceeds that of analog meters. Better than 0.5 percent accuracy is fairly standard for most digital meters. This improved accuracy is because of the carefully calibrated electronic circuitry contained in digital meters. Because of the readout on a digital meter, the precision of the reading accuracy by the operator is exact, not a close estimate.

Digital meters also offer an advantage in poor lighting conditions, when the face of an analog meter might be hard to see. LED (light-emitting diode) displays can easily be seen in low-light areas. They do however, consume more power and, therefore, the batteries might not last as long.

Digital meters are also smaller and are often less expensive to manufacture than the bulky analog meters of the past. Consequently, digital meters are quickly replacing the analog meters and you will probably see more digital type meters on job sites than any other types.

Ammeters

An electrical circuit is a complete conductive path through which electrons move. The movement of electrons in an electrical circuit is called *current* and is measured in *amperes* with an *ammeter.*

Most ammeters operate on a magnetic principle or the magnetic effect of an electric current producing movement of the meter pointer on the scale or the digital readout on electronic instruments. A basic dc ammeter consists of a permanent magnet that supplies a magnetic flux or field, a delicately balanced coil that rotates in the field, a pointer, and a scale. This meter design is the d'Arsonval type that was discussed previously.

Most basic d'Arsonval type ammeters will measure only small amounts of current. The typical range is in microamperes (0.000001 A) or milliamperes (0.001 A). Very few ammeters in use can measure much current without additional modifications. To increase the range of the ampere level, a shunt is used (Fig. 9.15).

The meter is connected in parallel with the shunt. Shunts can be located inside the meter housing (internal shunt) and are generally used to measure currents up to 30 A. Ampere ratings over 30 A generate too much heat for internal shunts that could damage the meter. Consequently, other types of ammeters utilize external shunts with leads connecting the shunt to the meter. External-shunt ammeters are normally used to measure currents over 30 A. The use of a shunt allows the ammeter to derive current in amperes by actually measuring the voltage drop across the shunt.

To measure current above 10 mA, a shunt with an extremely low resistance is placed in series with the load and the meter is connected across the shunt to measure the resulting voltage drop proportional to current flow. A shunt has a relatively large wattage rating to enable it to carry a relatively large current (Fig. 9.16).

As mentioned previously, alternating current can also be measured with a d'Arsonval-type meter if the alternating current is first changed to direct current by a rectifier. As the name implies, a rectifier is a device that rectifies, or converts ac to dc.

A thermocouple can also be used to adapt dc meters to measure alternating current, but because thermocouple-equipped instruments are quite sensitive to overloads, their applications are usually limited to electronic circuits where overloads seldom occur.

Caution

Never connect an ammeter in parallel with a load. Because of the low resistance in the ammeter, a short circuit can occur. Consequently, ammeters must be connected in series with the load. Furthermore, proper polarity must be observed in a dc circuit; that is, the negative lead of the meter must be connected to the negative side of the load, while the positive lead (1) must be connected to the positive side of the load. If the meter is connected with the polarities reversed, the meter coil will move in the opposite direction and can damage the meter.

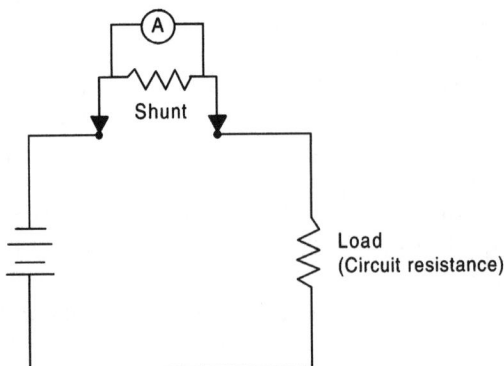

Figure 9.15 Ammeter shunt.

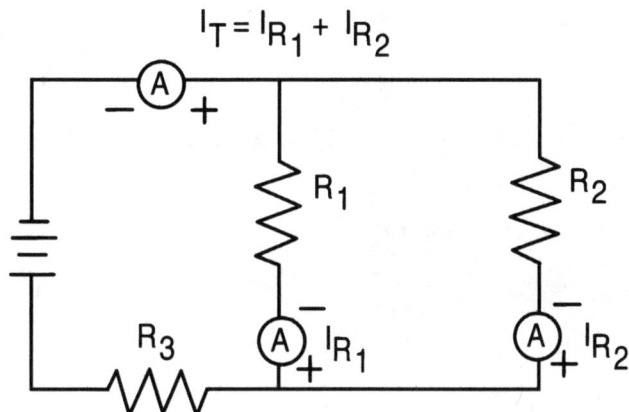

Figure 9.16 Ammeters connected at various areas of a parallel dc circuit.

Clamp-on ammeters

Figure 9.17 shows a clamp-on ammeter that can be used to measure current in an electrical circuit while the circuit carries the full load. Disconnection of the conductors is not necessary to make the test.

Although the exact operating procedures will vary with the manufacturers, most operate as follows when measuring current:

1. Be sure that the battery-attachment case (for use when in the ohmmeter mode) is removed from the instrument.

2. Release on pointer locks.

3. Turn the scale selector knob until the highest current range appears in the scale window.

4. Press the trigger button to open the jaws of the clamp before encircling one of the conductors under test with the transformer jaws.

5. Release finger pressure on the trigger slowly to allow the jaws to close around the conductor and keep an eye on the scale while doing so. If the pointer jumps abruptly to the upper range of the scale before the jaws are completely closed, the current is probably too high for the scale used. Should this happen, remove the jaws immediately from the conductor and use either a higher scale or a range-extender attachment (discussed in the next section). If the pointer deflects normally, close the jaws completely and take the reading from the scale.

Although the clamp-on ammeter is very convenient to use, the operator may be forced to work near dangerous high voltages. Therefore, special precautions must be taken to prevent accidental contact with live conductors and parts. When using this type of instrument, the current-carrying conductor should be centered inside the iron-core clamp for greatest accuracy.

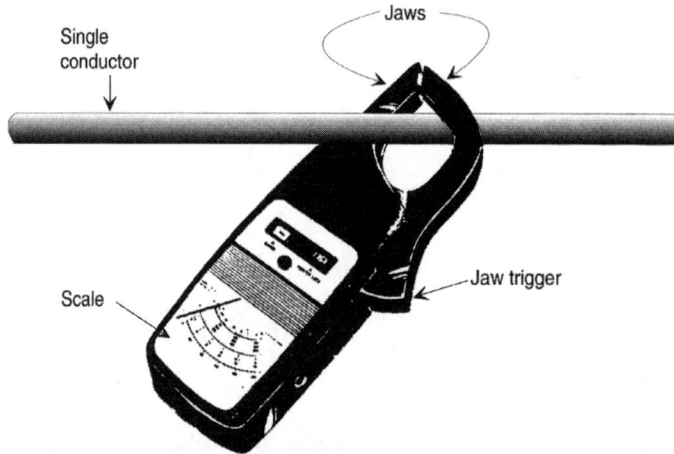

Figure 9.17 Typical clamp-on type ammeter. Both analog and digital scales are available.

Caution

Never encircle two or more conductors (Fig. 9.18A); only encircle one conductor (Fig. 9.18B).

Extending the range of ammeters

The range of an ac ammeter can be extended by using a range extender (Fig. 9.19). This device permits taking a measurement of higher current beyond the range of the regular clamp-on ammeter. The model shown extends the current range 10 times to allow an actual current reading of 1000 A on a 0- to 100-A meter scale. To illustrate its use, if the scale shows a reading of, for example, 42 A, the actual current (using the range extender) would be 420 A because $42 \times 10 = 420$.

Current multipliers

Sometimes it is desirable to use a current multiplier in conjunction with a clamp-on ammeter, such as the one shown in Fig. 9.20. This device allows current measurement on low-current equipment because the load current shown has been multiplied either two, five, or ten times; that is, if the meter scale shows a reading of, for example, 62 A and the 10× multiplier was used, the actual load current would be:

$$\frac{62}{10} = 6.2 \text{ A}$$

If a current-multiplier attachment is not available, the same effect can be had by wrapping turns of wire around the jaws of the ammeter (Fig. 9.21) In this illustration, the wires are wrapped around the jaws to multiply the cur-

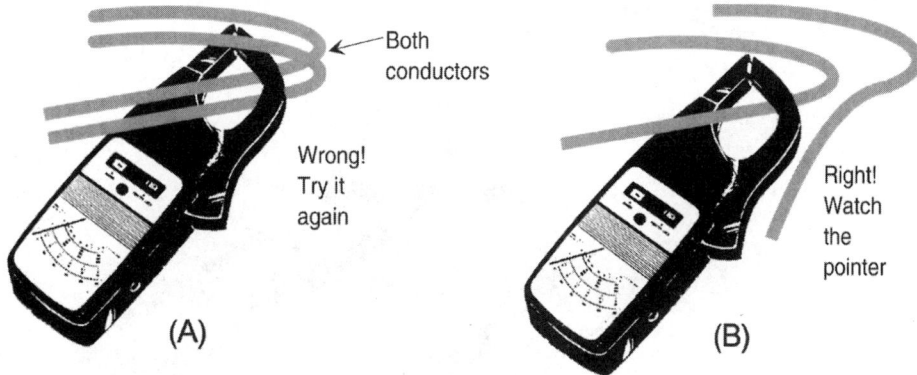

Figure 9.18 Right and wrong way to take current reading with a clamp-on ammeter.

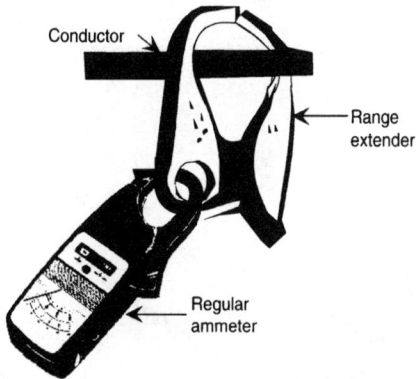

Figure 9.19 Clamp-on transformer extends ammeter range.

rent three times. If an attachment is available, and the device or apparatus to be tested has a plug-and-cord assembly, the load can be tested as shown in Fig. 9.22.

Precautions

When using clamp-on ammeters, be careful to obtain accurate readings. Some items to be considered include:

1. Be certain that the frequency of the circuit under test is within the range of the instrument. Most are calibrated at 70 hertz (Hz).

   Take care that stray magnetic fields do not affect the current reading; that is, arrange the meter leads as far away from the conductors under test as possible. Try to take current readings in a control panel at a location remote from magnetic relays that might influence the accuracy of the reading. Avoid taking current readings on conductors at a point close to a transformer.

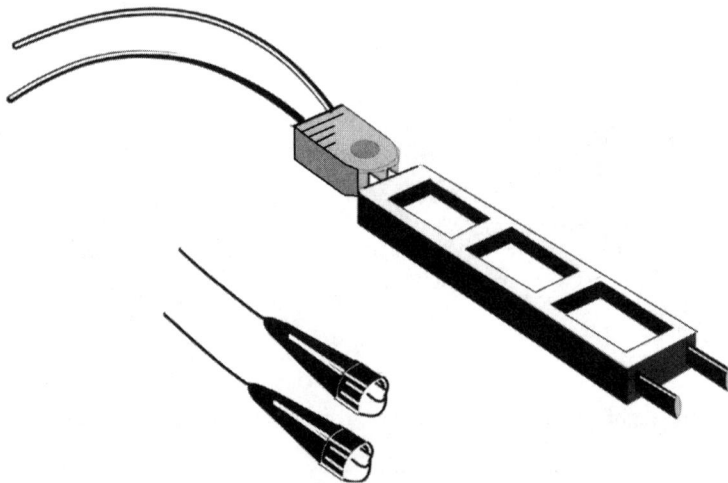

Figure 9.20   Current multiplier can be used with most clamp-on ammeters.

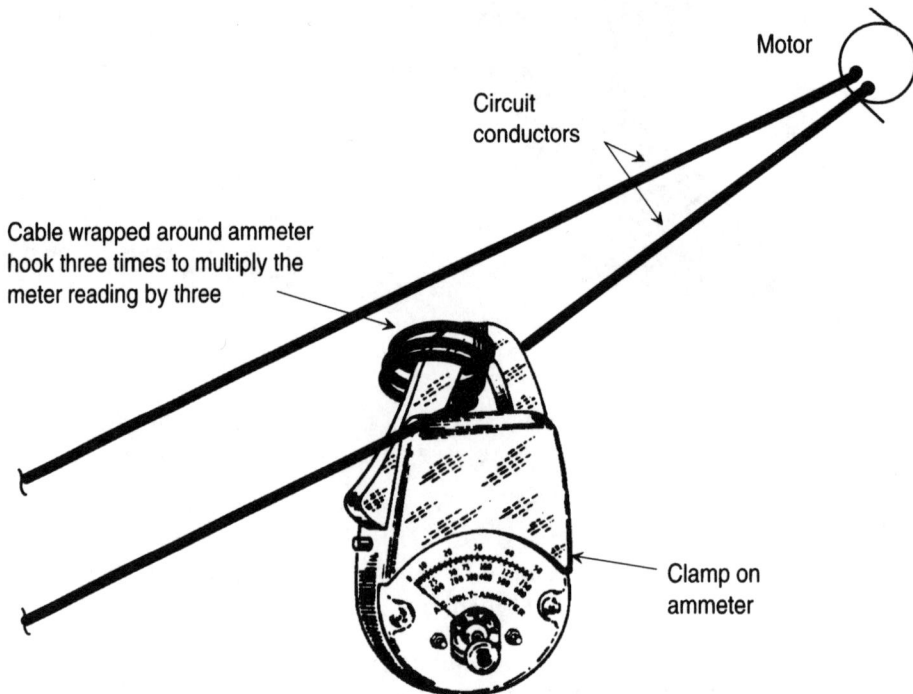

Motor

Circuit
conductors

Cable wrapped around ammeter
hook three times to multiply the
meter reading by three

Clamp on
ammeter

Figure 9.21   Wrapping turns of wire around meter jaws will increase the current reading.

Cord to motor or
portable equipment

Holes on which clamp-on
ammeter hook is inserted

Multiplier
device

Plugs into
receptacle

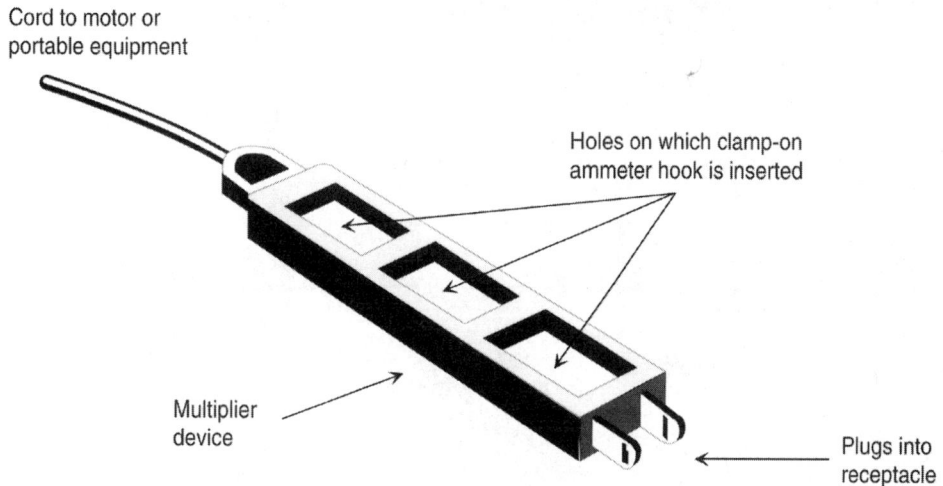

Figure 9.22   Ammeter multiplying device for use on plug-in electrical equipment.

2. When current readings are taken on high-voltage conductors, always use a hot-line extension pole specifically designed for use with a high-voltage clamp-on ammeter.

Ammeters are very useful instruments for testing and troubleshooting electrical circuits when the current in the circuit needs to be known. The ammeter, when used in conjunction with a voltmeter, can also be used to find both the resistance and volt-A (VA) in a circuit.

Recording ammeters

When it is desired to have a continuous and/or permanent record of the current in a given electrical circuit, a graphic or recording instrument can be used. Such an instrument has a meter element similar to the conventional indicating ammeters, but, in addition, it is equipped with a pen or other marking device so that a curve is drawn as current changes occur. The marking device on the recording instrument replaces the pointer on scale-indicating meters, and the marking device traces a line on a chart representing the value of the A that the instrument is measuring. See Fig. 9.23.

The charts are usually either circular or in strip form, depending on the type of instrument.

Voltmeters

The unit of electromotive force (emf) is the volt. One volt is a form of pressure that, if steadily applied to an electrical circuit having a resistance of 1 Ω, produces a current of 1 A.

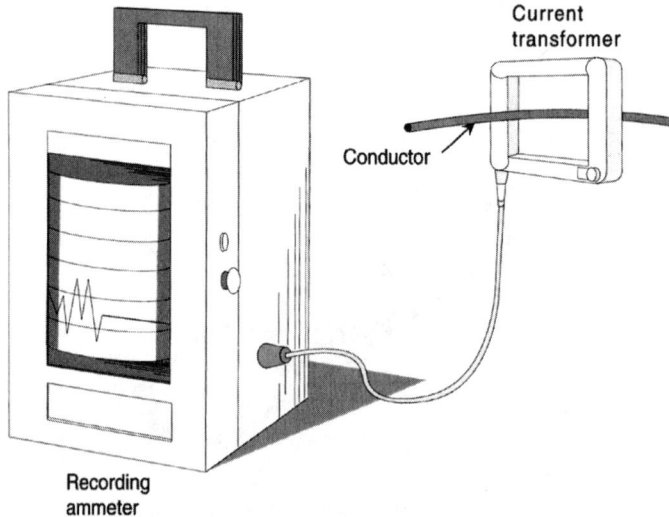

Figure 9.23   Recording ammeter.

Voltmeters are used to accurately measure the pressure or voltage in various electrical circuits. Very low voltage values are measured in milliV (a V = 1000 mV) by a millivoltmeter with low resistance.

A voltmeter should be connected across the terminals at the point where the voltage is to be measured. A voltmeter should never be connected across a circuit having a voltage higher than the rating of the instrument; that precaution must be observed particularly in the case of measurements with a millivoltmeter.

Several types of voltmeters are in common use. For example, the basic pointer-type ammeter can also be used to measure voltage. The meter coil has a fixed resistance, and, therefore, when current flows through the coil, a voltage drop will be developed across this resistance. According to Ohm's law, the voltage drop will be directly proportional to the amount of current flowing through the coil. Also, the amount of current flowing through the coil is directly proportional to the amount of voltage applied to it. Therefore, by calibrating the meter scale in units of voltage instead of current, the voltage in different parts of the circuits can be measured.

When connecting a voltmeter to a dc circuit, always observe the proper polarity. The negative lead of the meter must be connected to the negative terminal of the dc source, and the positive lead to the positive terminal. If the leads are connected to opposite terminals, the needle will move in the reverse direction. Because the voltage constantly reverses polarity in an ac circuit, there is no need to observe polarity when connecting voltmeters designed for use on ac circuits.

Many portable voltmeters are designed with two or more voltage ranges than can be read on a common scale, such as 0–150 V, 0–300 V, and 0–600 V.

The change from one meter range to another is accomplished by the selection of internal shunts, resistors, or other auxiliary components by means of a switch or plug-in terminals. When using a multirange voltmeter, it is best to select a higher range than needed to ensure that no damage will occur to the instrument. Then, if the initial reading indicates that a lower scale is needed to obtain a more accurate reading, the meter can be switched or otherwise adjusted to the next lowest range.

One of the reasons for using various ranges of voltmeters is that the greatest accuracy is obtained on the upper half of the scale. Therefore, if a single 0- to 600-V range were used, lower voltages would be harder to read and meter accuracy would be less.

Solid-state digital voltmeters are usually autoranging; that is, they do not have a selector switch for adjustment. The internal construction of the meter itself will select the proper resistance for the voltage being detected.

Voltmeter applications

Besides measuring the voltage applied to an electrical circuit, voltmeters can be used for other electrical tests, such as troubleshooting circuits, circuit tracing, and measuring low resistance. For example, a common cause of electrical problems is low equipment voltage. This problem usually occurs because of one or more of the following reasons:

- Undersized conductors
- Overloaded circuits
- Transformer taps set too low

Low-voltage test

When making a low-voltage test, first take a reading at the main switch or service entrance. If, for example, the main service is 120/240, single-phase, three-wire, the voltage reading between phases (undergrounded conductors) should be between 230 and 240 V. If the reading is much lower than this, the fault lies with the utility company supplying the power, and they should be notified to correct the problem. However, if the reading at the main service switch is between 230 and 240V, the next procedure is to check the voltage reading at various outlets throughout the system.

When a low-voltage problem is found on a circuit, leave the voltmeter terminals connected across the line and begin disconnecting all loads, one at a time, that are connected to the circuit. If the problem is corrected after several of the loads have been disconnected, the circuit is probably overloaded. Steps should be taken to reduce the load on the circuits or else increase the wire size to accommodate the load. As mentioned previously, loose connections can also cause low voltage; to check for this, the entire circuit should be deenergized, and each terminal in all disconnect switches, motor starters, and so on, should be checked for loose connections. A charred or blackened terminal screw is one sign of a loose connection.

Megohmmeter

A typical megohmmeter (megger) is composed of: a hand-driven or motor-powered ac generator and/or a transformer with voltage rectified to 100, 250, 500, and 1000 V dc; a cross-coil movement with 0- to 20,000-Ω and 0- to 1000-MΩ scales; a carrying case; and test leads. The megger is used to measure the resistance in megohms to the flow of current through and/or over the surface of electrical equipment insulation. The test results are used to detect the presence of dirt, moisture, and insulation deterioration. The instrument also typically measures resistances up to 20,000 Ω.

The test set and the sample to which it is connected are sources of high-voltage electrical energy, and all persons making or assisting in the tests must observe strict safety precautions to prevent contact with energized parts of the test equipment and associated circuits. Persons engaged in the test must stand clear of all parts of the complete high-voltage circuit unless the set is deenergized and all parts of the test circuit are grounded. As a routine safety procedure, the use of rubber gloves in making connections to the high-voltage terminals and in manipulating the controls is advised.

The instruction manuals accompanying the megger usually contain detailed instructions about preparing for tests and connecting the megger to various types of equipment and safety requirements.

Insulation resistance testing

To prepare for an insulation resistance test, first take the equipment or circuits to be tested out of service. Check between the equipment terminals and ground using the megger voltage ranges to be sure that no voltage is present. If possible, disconnect all leads to the unit being tested. When a motor or circuit is not completely isolated, be aware of all the components that will be tested when the megger is connected. Should an interconnected circuit be overlooked, the megger reading could be lower than expected.

The testing of wiring can be performed on all types of systems if two rules are kept in mind:

- Be sure all wiring is deenergized.
- Know what wiring is included in the test and make a record card of it.

Warning!
Only qualified personnel should make tests with a megger, and then only after obtaining permission from their supervisor. Megger tests can be dangerous to personnel and harmful to equipment and wiring systems if improperly handled.

When a distribution panel is present, check the entire system all the way to ground by attaching one megger lead to the dead post of the open main power switch and the other lead to a grounded conduit or grounded metal housing.

Individual circuits are tested to ground by opening distribution panel switches, fuses, or circuit breakers and testing each circuit in turn.

Multiconductor cables can be tested in several ways. For instance, measurement of insulation resistance can be made between the wire and lead sheath. Various other measurements can be made, such as wire to ground, wire to wire, wire to braid, and wire to sheath.

When testing wiring that is connected to any panelboard or equipment there could be appreciable leakage between terminals, which will show in tests as lowered insulation resistance. If previous test and record cards were made with the panels connected, continue any future test in the same manner.

Testing circuit breakers

Circuit breakers can be tested with the megger by first disconnecting the circuit breaker from the line and connecting the black megger lead to the frame or ground. Check the insulation resistance of each terminal to ground by connecting the red (positive) lead to each terminal in turn and making the measurements. Next open the breaker and measure the insulation resistance between terminals by putting one lead on one terminal and the other on the second for a two-terminal breaker; for a three-pole breaker, check between poles 1 and 2, 2 and 3, and 1 and 3, in turn.

If the breakers are of the oil type, dirty oil could cause leakage currents between the breaker contacts even when open. In dry types of breakers, dirt or grease in the mounting could cause leakage. The measurement of insulation resistance can be made from either circuit breaker terminal to ground and should also be made between terminals.

Testing safety switches and switchgear

Switches should be completely disconnected from the line and relay wiring before testing. When manual switches are being tested, measure the insulation resistance from ground to terminals and between terminals. When testing electrically operated switches, check the insulation resistance of the coil or coils and contacts. For coils, connect one megger lead to one of the coil leads and the other to ground. Next, test between the coil lead and core iron or solenoid element.

If relays are involved, measure insulation resistance with one lead connected to the relay plate or contact and the other test lead connected to the coil, core, or solenoid contact.

Although the suggested tests can appear to be excessive and most switches are durable over long periods of time, these recommendations are made because one faulty switch can actually wreck a whole plant system. Thorough testing can help to ensure against such an occurrence.

Frequency meter

Frequency is the number of cycles completed each second by a given ac voltage; usually expressed in hertz; 1 Hz = 1 cycle.

The frequency meter is used in ac power-producing devices like generators to ensure that the correct frequency is being produced. Failure to produce the correct frequency will result in heat and component damage.

Electrical thermometers

For the measurement of temperatures, three basic electrical thermometer methods are used:

- The resistance method, which makes use of the fact that the resistance of a metal varies in direct proportion to temperature. This method is normally used for temperatures up to approximately 1500°F.

- The thermocouple method, which is based on the principle that a difference in temperature in different metals generates a voltage and is used for measuring temperatures up to about 3000°F.

- The radiation-pyrometer and optical-pyrometer are generally used for temperatures above 3000°F. They combine the principle of the thermocouple with the effect of radiation of heat and light.

Infrared-sensing device

An *infrared-sensing device* is an optical device that measures the infrared heat emitted from an object.

Practical application

Most manufacturers of security/fire-alarm systems publish service manuals for their equipment, which—in a simplified and systematic manner—guide the mechanic through troubleshooting and servicing procedures for their components and equipment. Their main objective is to direct technicians to the exact cause of a malfunction and assist in repairing the system. A qualified technician, even though the person is not familiar with the system, should be able to isolate the exact cause of nearly all possible malfunctions when using such manuals. Therefore, it is recommended that all service technicians obtain a service manual (including wiring diagrams, etc.) of the particular system in question.

Although the exact content of service manuals will vary from manufacturer to manufacturer, most are divided into three main sections:

- Theory of operation
- Troubleshooting
- Repair and adjustments

The section on theory of operation should give a thorough explanation of each circuit and should be accompanied by a schematic diagram (such as the one in Fig. 9.24) that clearly identifies points that are referred to in the text. The purpose of this section is to give technicians an understanding of the proper operation of the system, which in turn, will enable them to locate the causes of those problems not found by following the regular step-by-step troubleshooting procedures.

Troubleshooting tables, such as the one in Fig. 9.25, lists the symptoms that could occur, along with the most likely causes of problems, and give page numbers of the detailed troubleshooting procedures for the various circuits. The service technician will find that consulting this table, and checking the possible causes in the order given, will prove to be the most efficient procedure to follow. The possible causes are listed either in order of most probable cause or fastest to verify cause, whichever has been found by the manufacturer's technicians to be the most efficient approach to troubleshooting.

A repair and adjustment section is usually included to provide guidance in replacing defective components, or realigning components that have gotten out of adjustment. When such instructions are given, the service technician should follow them closely to ensure successful repair. Of course, technicians performing the repair must have a working knowledge of repair procedures and precautions that pertain to printed-circuit boards. They should also be equipped with the proper tools. Lack of ability and/or use of improper tools can lead to more damage, rather than eliminate a problem.

There is really no substitute for the service manuals provided by the manufacturers, and all service technicians should obtain manuals for the equipment on which they are working.

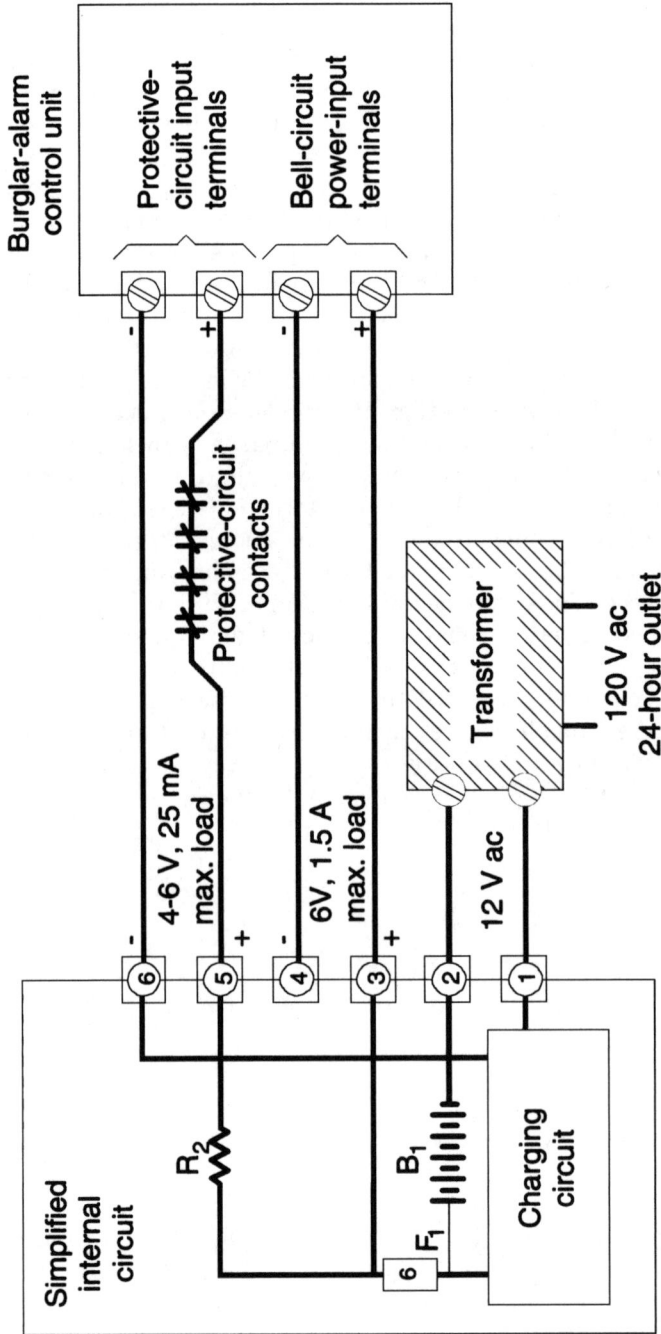

Figure 9.24 Schematic diagram accompanying the "Theory of Operation" section of a service manual.

Symptom	Possible causes	Reference
Resistors R54, R55 burnt	1. Power supply	1. Page 4B
Dialer does not trip Note: Make sure switch is in ON (center) position. Make sure potentiometer P4 is in extreme clockwise position as viewed from terminal strip end of alarm center	1. Batteries weak	1. Page 4B
	2. Power supply	2. Page 4B
	3. Trip circuits	3. Page 5B
	4. Start-delay and battery cutout circuit	4. Page 7B
	5. Channel switch and logic	5. Page 9B
Motor does not run on both channels and lamp does not light	1. Start-delay and battery cutout	1. Page 7B
	2. Motor supply	2. Page 8B
	3. Channel switch and logic	3. Page 9B
Motor does not run but lamp comes on	1. Motor supply	1. Page 8B
Motor runs but lamp does not light	1. Dialing filter and relay control	1. Page 12B
Dialer trips but does not shut off	1. Memory circuit	1. Page 6B
	2. Trip circuit	2. Page 5B
	3. Channel switch and logic	3. Page 9B
Dialer cannot switch from one channel to the other	1. Memory circuit	1. Page 6B
Dialer cannot program, erase	1. Tape head, jacks	1. Page 10B

Figure 9.25 Typical troubleshooting table.

A Deeper Understanding

10

Electric Circuits

The material covered in this chapter is the "rock" on which all electrical/electronic security-system circuits are built. The laws and equations are used on a daily basis by security technicians.

What we refer to as electricity is basically the flow of electrons—tiny atomic particles. These particles are found in all atoms. Atoms of some metals, such as copper and aluminum, have electrons that are easily pushed and guided into a stream. When a coil of metal wire is turned near a magnet, or vice versa, electricity will flow in the wire. This use of coils is applied in generating plants; water or steam is used to turn turbines, which rotate electromagnets that are surrounded by huge coils of wire. The push transmitted to the electrons by the turbine/magnet setup is measured in units called *volts*. The quantity of the flow of electricity is called *current* and it is measured in *amperes* or *amps*.

Multiply volts by amps and you get *volt-amperes* or *watts*—the power or amount of work that electricity can do. Electrical appliances and motors have certain wattage requirements depending on the task they are expected to perform. For convenience, we can use *kilowatts,* (1 kW equals 1000 W) when speaking of power production or power needs. A power plant produces kilowatts which are sold to users by the *kilowatt-hour.* For example, a 100-W lamp left on for ten hours uses 1 kWh of electricity.

Resistance, or the opposition to the flow of electricity, is another term that will be covered in this chapter. In general, a conductor enhances the flow of electricity and resistance impedes or stops the flow of electricity. Therefore, insulators are constructed of materials that offer a high resistance to the flow of electricity.

The relationship between current, voltage, resistance, and power in a basic dc circuit is common to many types of electrical circuits. Consequently, this chapter covers direct current (dc) fundamentals, including the following:

- Matter
- Energy

- Electricity
- Batteries
- Dc series circuits
- Dc parallel circuits
- Ohm's law
- Kirchoff's laws

Electrical Essentials

Matter is defined as anything that occupies space and has weight—the weight and dimensions of matter can be measured. Examples of matter include: air, water, clothing, automobiles, and even our own bodies. Matter may therefore be found in any one of three states:

- Solid
- Liquid
- Gaseous

Elements and Compounds

An *element* is a substance that cannot be reduced to a simpler substance by chemical means. Examples of elements—to name a few—include iron, gold, silver, copper, and oxygen. But there are other elements. In fact, there are over 100 known elements and all of the substances on earth are composed of one or more of these elements.

When two or more elements are chemically combined, the resulting substance is called a *compound*. A compound is a chemical combination of elements that can be separated by chemical but not by physical means. For example, water consists of hydrogen and oxygen (H_2O), and common table salt consists of sodium and chlorine. Neither of the two compounds can be separated physically.

A *mixture,* on the other hand, is a combination of elements and compounds, not chemically combined. Mixtures can, therefore, be separated by physical means. Examples of mixtures include air, which is made up of nitrogen, oxygen, carbon dioxide, and small amounts of several rare gases. Sea water is another mixture that consists of salt and water.

Molecules

A *molecule* is a chemical combination of two or more atoms and is the smallest particle in a compound that has all the characteristics of the compound itself.

Consider water, for example. Water is matter because it occupies space and has weight. Depending on the temperature, it can exist as a liquid (water), a solid (ice), or a gas (steam). Regardless of the temperature or its physical form, it will still have the same composition. However, if we start with a certain quantity of water, divide this and pour out one half, and continue this process a sufficient number of times, we will eventually end up with a quantity of water that cannot be further divided without ceasing to be water. Once this point is reached, it becomes a molecule of water. If this molecule of water is further divided in the same way as before, instead of two parts of water, there will be one part of oxygen and two parts of hydrogen (H_2O), and the resulting elements will not longer be water if they are divided.

Atoms

If it were possible to view the flow of electrons through a highly powerful microscope, at first glance the viewer might think he or she is studying astronomy rather than electricity. According to our present understanding, the atom is believed to consist of a central nucleus composed of protons and neutrons, surrounded by orbiting electrons (Fig. 10.1). The nucleus is relatively large when compared with the orbiting electrons, the same as our sun is large when compared to its orbiting planets; also the orbiting satellites are small in comparison to the satellites' planet.

In the case of an atom, the orbiting electrons are held in place by the attractive electric force between the electron and the nucleus—similar to how the earth's gravity keeps its satellite (the moon) from drifting off into space. The law of charges states that opposite charges attract and similar charges repel

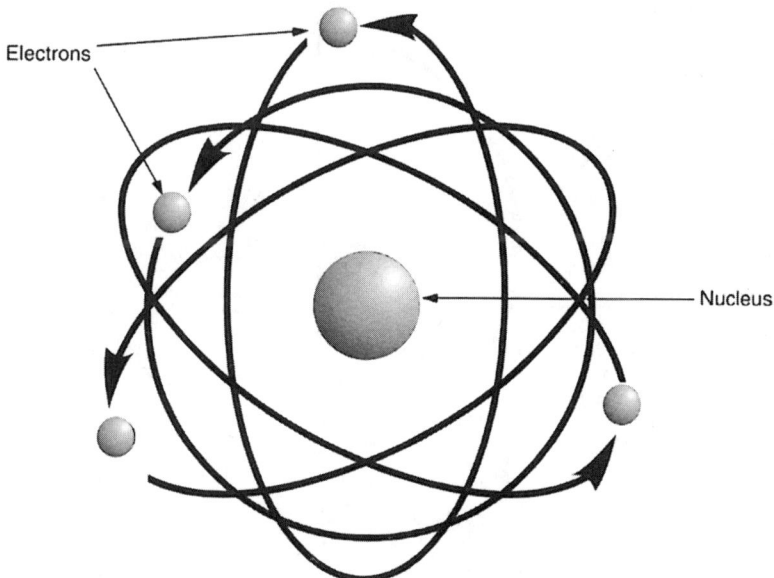

Figure 10.1 Nucleus consisting of protons and neutrons with orbiting electrons.

(Fig. 10.2). The positive charge of the protons in the nucleus, therefore, attracts the negatively charged electrons. If this force of attraction were the only one in force, the electrons would be pulled closer and closer to the nucleus and eventually be absorbed into the nucleus. However, this force of attraction is balanced by the centrifugal force that results from the motion of the electron around the nucleus as shown in Fig. 10.3. The law of centrifugal force states that a spinning object will pull away from its center point. The faster an object spins, the greater the centrifugal force becomes.

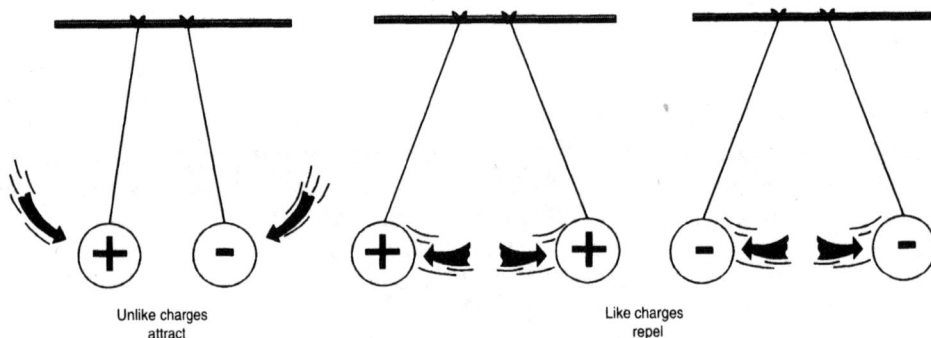

Unlike charges
attract

Like charges
repel

Figure 10.2 Reaction between charged bodies.

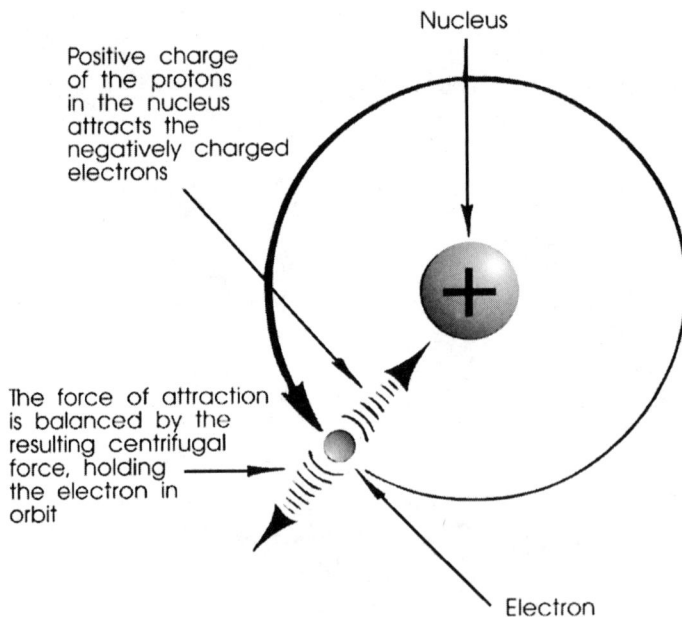

Nucleus

Positive charge
of the protons
in the nucleus
attracts the
negatively charged
electrons

The force of attraction
is balanced by the
resulting centrifugal
force, holding
the electron in
orbit

Electron

Figure 10.3 A combination of gravity and centrifugal force keeps
electrons in orbit around the nucleus.

Figure 10.4 shows an example of this principle. If an object is tied to a string, and the object is spun around, it will try to pull away from you. Compare the spinning object with a spinning electron. Compare the string—held securely by your hand—as the force of gravity. The faster the object spins, the greater the force tries to pull the object away, but the securely held string (similar to the force of gravity) keeps the object spinning at the same distance (the length of the string) from your body. If the object is spun still faster, the force becomes even greater; perhaps the string will slip slightly through your hand and the object will spin farther away from your body. Similarly, the faster an electron spins, the farther away from the nucleus it will be.

If the string holding the object being spun should slip out of your hand (similar to the absence of gravity), the object will fly off into space in a straight line. The same is true of the earth's satellite. Were it not for the earth's gravity and gravity throughout space, the moon would follow a straight line through space. Centrifugal force results from the combined effects of gravitational force and the tendency of a moving body to travel in a straight line.

Because the protons and orbital electrons of an atom are equal in number and equal and opposite in charge, they neutralize each other electrically. Consequently, each atom is normally electrically neutral—exhibiting neither a positive nor a negative charge. However, under certain conditions, an atom can become unbalanced by losing or gaining electrons. If an atom loses a negatively charged electron, the atom will exhibit a positive charge, which is then referred to as a *positive ion*. Similarly, an atom that gains an additional negatively charged electron becomes negatively charged itself and is then called a *negative ion*. In either case, an unbalanced condition is created in the atom causing the formerly neutralized atom to become charged. When one atom is

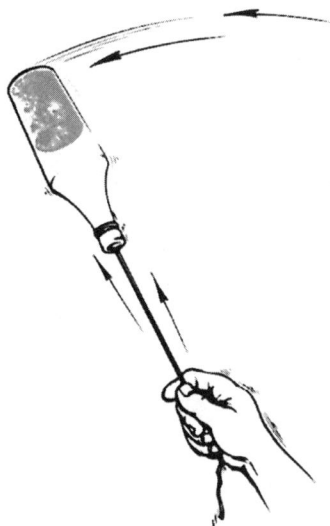

Figure 10.4 Centrifugal force causes an object to pull away.

charged and there is a dissimilar charge in another nearby atom, electrons can flow between the two. We have already learned what the flow of electrons is called. It's electricity!

The hydrogen atom is the simplest of all atoms since it consists of only one electron and one proton (Fig. 10.5).

Atoms of other materials are more complex. They consist of protons and neutrons in the nucleus with different number of electrons in multiple orbits or shells rotating around the nucleus. The copper atom, for example, has 29 protons, 35 neutrons, and a total of 29 electrons. One electron is in the outer orbit by itself (Fig. 10.6). The electrons are organized into several different orbits, or shells, surrounding the nucleus.

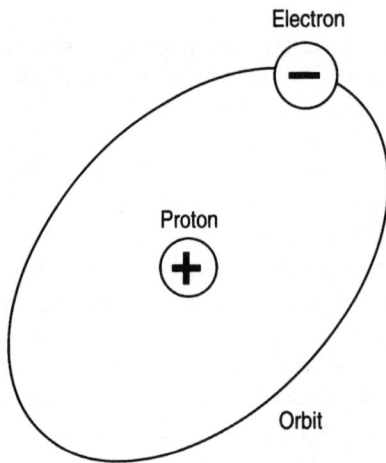

Figure 10.5 The hydrogen atom consists of only one electron in orbit around one proton.

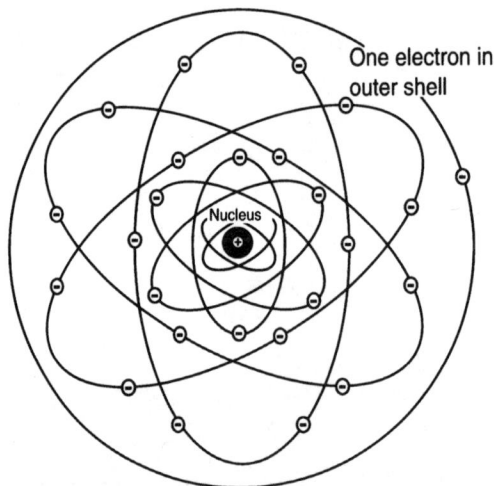

Figure 10.6 A copper atom has 29 protons, 35 neutrons, and 29 electrons.

Atoms have a set number of electrons that can be contained in one orbit or shell. The outer shell of an atom is known as the *valence shell*. Any electrons located in the outer shell of an atom are known as *valence electrons*. These valence electrons in some materials, especially metals, can be easily knocked out of their orbits. Such electrons are referred to as *free electrons,* and materials with free electrons are called *conductors*. The valence electrons are of primary concern in the study of electricity because these electrons explain much of electrical theory. A conductor, for instance, is made of a material that contains one or two valence electrons, when an atom has only one or two valence electrons, they are loosely held by the atom and are easily given up for current flow. Silver, copper, and aluminum all contain one valence electron, but silver is a better conductor than copper, and copper is a better conductor than aluminum. The reason for this is that an atom of silver is larger than an atom of copper, and an atom of copper is larger than an atom of aluminum. Because an atom of silver is larger than an atom of copper, it contains more orbits than an atom of copper. Thus, the valence electron of silver is farther away from the nucleus than an atom of copper. Because the speed an electron spins is decided by its distance from the nucleus, the valence electron of silver is spinning around the nucleus at a faster speed than the valence electron of copper. Therefore, the valence electron of silver contains more energy than the valence electron of copper. When the valence electron of silver is knocked out of orbit, it simply contains more energy than the valence electron of copper, and therefore, makes a better conductor of electricity. Copper is a better conductor of electricity than aluminum for the same reason. Figure 10.7 shows an atom of silver and an atom of copper for comparison.

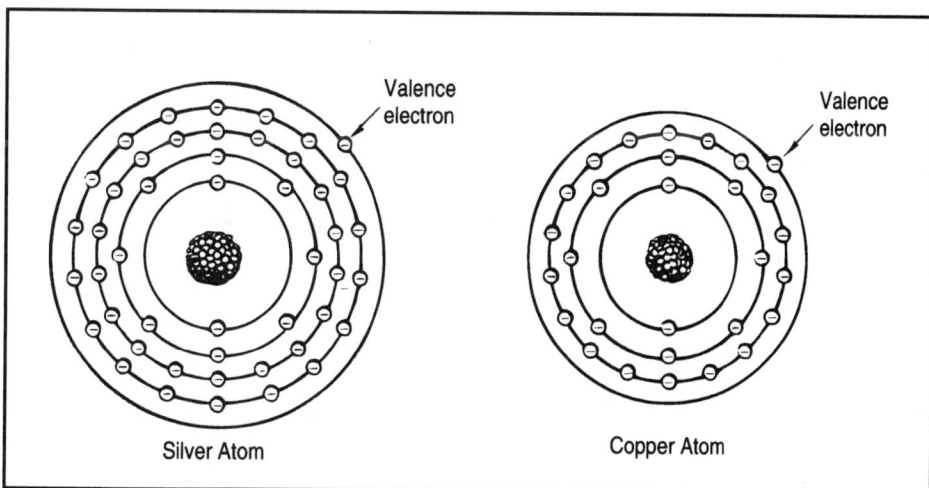

Figure 10.7   Silver atom (*left*) is larger than the copper atom (*right*).

Conductors, insulators, and semiconductors

Materials used in the electrical and electronic fields can be classified in three groups, according to their electrical properties:

- Conductors
- Insulators
- Semiconductors

Conductors are generally made from materials that have large, heavy atoms to provide an easy path for electron flow. A variety of materials are used to transmit electrical energy, but copper (because of its excellent cost-to-conductivity ratio) still remains the basic and most ideal conductor. Other electrical conductors are made from aluminum. Sometimes materials are silver plated to provide an even better flow of electrons.

Theoretically, conductors are materials that have only one or two valence electrons in their atom (Fig. 10.8). An atom that has only one valence electron makes the best electrical conductor because the electron is loosely held in orbit and is easily given up for current flow.

Insulators

Insulators are generally made from lightweight materials that have small atoms. Such materials as glass, rubber, and certain plastics have practically no free electrons. This makes it very difficult for current to flow through them.

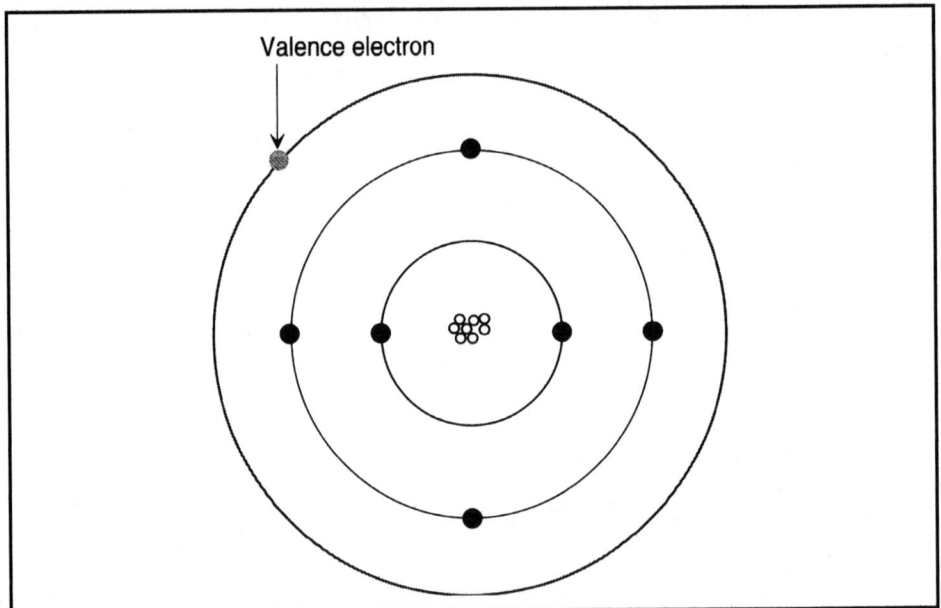

Figure 10.8 Atom of a conductor.

The atoms of an insulating material have their outer orbits filled or almost filled with valence electrons; thus, an insulator will have seven or eight valence electrons (Fig. 10.9). Because an insulator has its outer orbit filled or almost filled with valence electrons, they are tightly held in orbit and not easily given up for current flow. Insulators are used in a variety of applications ranging from the covering on conductors to the dielectric in capacitors.

Semiconductors

Semiconductors, as the name implies, are materials that are neither good conductors nor good insulators. Semiconductors are made from materials that have some free electrons—more than an insulator, but fewer than a conductor and have four valence electrons in their outer orbit (Fig. 10.10).

Electricity

Electricity, as stated previously, is the flow of electrons. Atoms tend to remain neutral because the outer orbits of electrons repel other electrons, which prevents their movement or flow between atoms. To have electricity, these atoms must be charged by knocking the one or more of its electrons out of orbit by another electron. There are several known ways to accomplish this:

- Friction—Voltage produced by rubbing certain materials together.
- Chemical action—Voltage produced by chemical reaction in a battery cell.

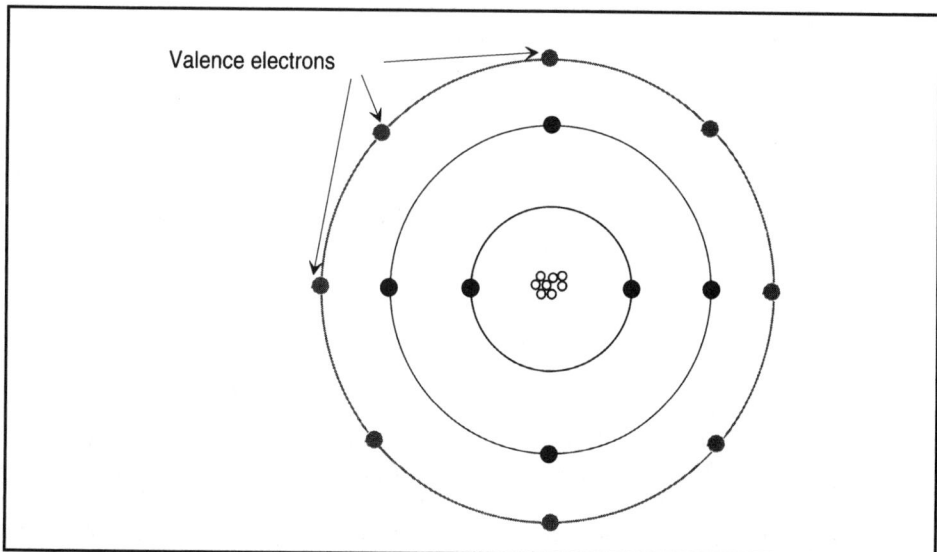

Figure 10.9   Atom of an insulator.

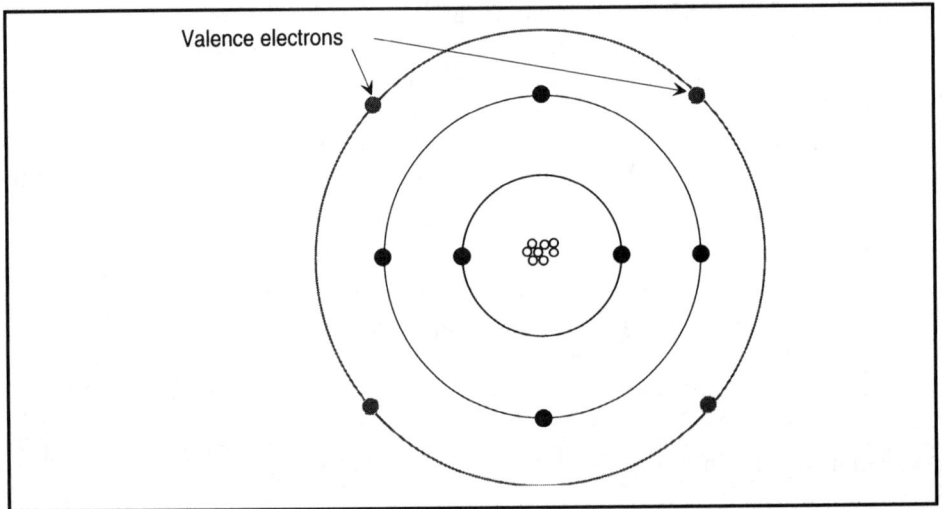

Figure 10.10   Atom of a semiconductor.

- Magnetism—Voltage produced in a conductor when the conductor moves through a magnetic field, or a magnetic field moves through the conductor in such a manner as to cut the magnetic lines of force of the field.
- Light—Voltage produced by light striking photosensitive (light-sensitive) substances.
- Heat—Voltage produced by heating the joint (junction) where two dissimilar metals are joined.
- Pressure—Voltage produced by squeezing crystals of certain substances.

Static electricity

Friction or static electricity can be produced by walking across a carpet with leather shoes on a dry winter day. After doing so, the next metal object you touch will often cause a spark and perhaps a shock. Sliding across car seat covers during the winter will do the same thing. Even pulling off a sweater can create a static charge. Static electricity can also be experienced by combing your hair on a dry, cold day that will sometimes result in sparks and a crackling sound.

Lightning is another form of static electricity. Air currents striking the face of clouds causes condensation of the moisture in them. When the wind strikes the cloud, these small particles of moisture are blown upward, carrying negative charges to the top of the cloud and leaving the bottom with positive charges. As very heavy rain or other forms of heavy condensation falling through a part of the cloud, can create a situation in which one side of the cloud becomes charged positively and the other side negatively, leaving millions of volts of difference in potential.

When clouds (under the condition described) come near enough to the ground or to another cloud with opposite charges, they will discharge to the ground or to another cloud with explosive violence. Because there is a strong tendency for lightning discharges to strike trees, structures, and other objects and travel on metals that extend in the general direction of the discharge, lightning rods and properly grounded electrical systems can prevent much of the potential damage inherent in lightning.

Static electricity, even in the small charges obtained from combing your hair can damage the "memory" of computer software, hardware, and other types of electronic equipment.

Static electricity, whatever its source, has little use in practical applications because the flow of electrons is temporary and erratic; that is, one brief transfer of electrons from one material to another takes place and then the effect is over. To make electricity useful for practical applications, a constant and steady flow of electrons must be produced.

Chemical electricity

Chemical action, contained in electric cells and batteries, is another source for generating electricity.

In general, a typical basic battery consists of a container filled with a liquid that is called the *electrolyte*. Into this liquid are placed two plates of different metals, which, in turn, are separated from each other. The purpose of the chemical (electrolyte) is to push electrons onto one plate while at the same time, taking them off from the other. This action results in an excess of electrons or a negative charge on one of the plates, called the *negative terminal*. The other plate loses electrons and becomes positively charged and provides the positive terminal (Fig. 10.11).

The action of the electrolyte in carrying electrons from one plate to the other is actually a chemical reaction between the electrolyte and the two plates. This action changes chemical energy into electrical charges on the cell plates and terminals.

Without a load connected to the battery terminals, for example, an electric lamp, electrons would be pushed or transferred onto the negative plate until there is no more room. The electrolyte would take enough electrons from the positive plate to make up for those it pushed onto the negative plate. Both plates would then be fully charged and no electrons would be moving between the plates.

However, when a load is connected to the terminals of the battery, the electrons will leave the negative plate, travel through the load, and back to the positive plate in the battery. During this process, the electrolyte will carry

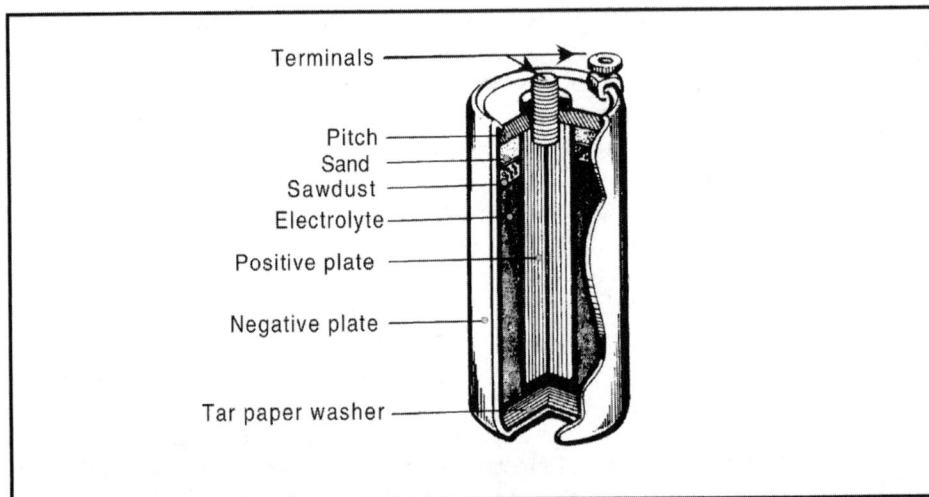

Figure 10.11 Components of a typical primary cell battery.

more electrons across from the positive plate to the negative plate. As long as electrons leave the negative plate and travel through the load to the positive plate, the battery will produce electricity.

In a primary cell, during this use of electricity, the charge on the negative plate is being used up as the electrons are being transferred. Eventually, the negative plate will be completely dissolved in the electrolyte by the chemical action, and the cell would become "dead," or unable to furnish a charge, until the negative plate is replaced. Consequently, this type of cell is called a *primary cell,* meaning that once it is completely discharged, it cannot be charged again, except by using new materials.

Plates in primary cells can be made from carbon and most other metals, while acids or salt compounds can be used for the electrolyte. Flashlight dry cell batteries are examples of primary cells.

A storage battery of secondary cells can furnish more power than a primary cell battery, and can be recharged. Most are of the lead-acid type and the electrolyte is sulfuric acid, while the positive plate is lead peroxide, and the negative plate is lead. During discharge of the cell, the acid becomes weaker and both plates change chemically to lead sulfate (Fig. 10.12).

The housing or cases of lead-acid cells are made of hard rubber or glass, which prevents corrosion and acid leaks. A space at the bottom of the cell collects the sediment formed as the cell is used. The top of the case is removable and acts as the support for the plates. Two or more cells connected together make up a storage battery, as used in cars or for emergency standby power. This type of battery stores electricity and can be recharged after discharge by reversing the current flow into the battery.

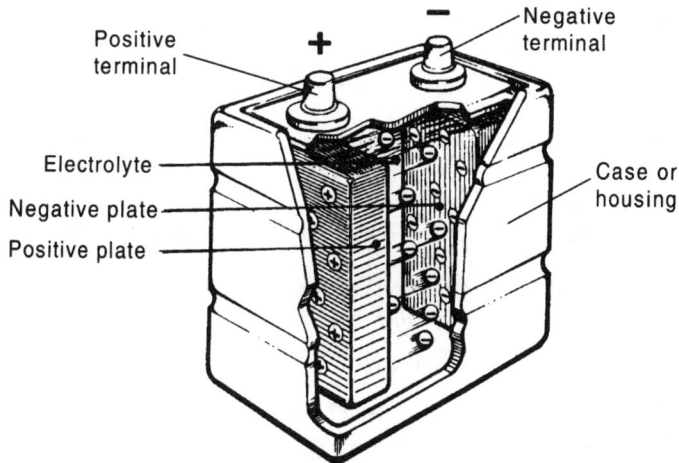

Figure 10.12   Components of a secondary cell.

The active materials of most secondary cells are not rigid enough to be mounted independently. Therefore, a special grid structure of inactive metal is normally used to hold them. For maximum chemical action, a large plate area is needed, so each positive plate is interlaced between two negative plates. Separators, made of wood or porous glass, hold each positive and negative plate apart, but let the electrolyte pass through.

Although the cells are constructed somewhat differently, the chemical action of both types is the same; that is, electrons are pushed onto the negatively charged plates by the electrolyte and are discharged through the load to the positive plates. In secondary cells, the electrolyte that becomes weaker, not the dissolving of the negative plate (as in primary cell batteries).

Magnetism

Magnetism is one of the most commonly used methods to produce voltage. Vast quantities of electric power are produced daily from mechanical sources. The mechanical power can be provided by a number of different sources, such as gasoline or diesel engines, and water or steam turbines. However, the final conversion of these source energies to electricity is accomplished by generators using the principle of electromagnetic induction.

Three fundamental conditions must exist before a voltage can be produced by magnetism.

1. There must be a conductor in which the voltage will be produced.
2. There must be a magnetic field in the conductor's vicinity.

3. There must be relative motion between the field and conductors. The conductor must be moved to cut across the magnetic lines of force, or the field must be moved so the lines of force are cut by the conductor.

Therefore, when a conductor or conductors move across a magnetic field so as to cut the lines of force, electrons within the conductor are propelled in one direction or another. Thus, an electric force (voltage) is created.

Figure 10.13 shows the three basic conditions needed to create an induced voltage.

1. A magnetic field exists between the poles of the C-shaped magnet.

2. There is a conductor (copper wire).

3. There is a relative motion. The wire is moved back and forth across the magnetic field.

A voltage will also be produced by holding the wire stationary and moving the magnetic field back and forth.

Electricity produced by light

When light strikes the source of a substance, it can dislodge electrons from their orbits around the surface atoms of the substance. This occurs because light has energy, the same as any moving force.

Some substances (mostly metallic ones) are far more sensitive to light than others. That is, more electrons will be dislodged and emitted from the surface of a highly sensitive metal, with a given amount of light, than will be emitted from a less-sensitive substance. Upon losing electrons, the photosensitive (light-sensitive) metal becomes positively charged, and an electric force is created. Voltage produced in this manner is referred to as a *photoelectric voltage*.

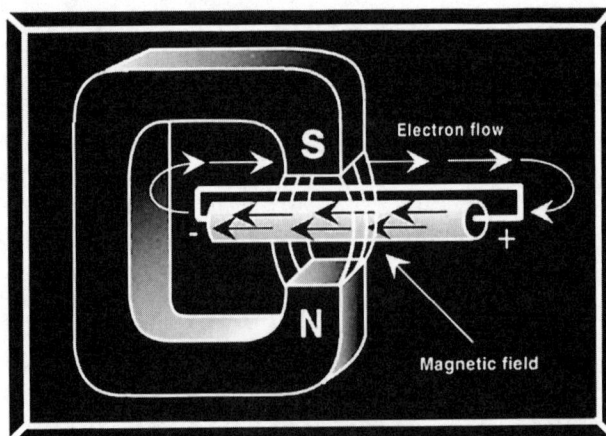

Figure 10.13 Voltage produced by magnetism.

The photosensitive materials most commonly used to produce a photoelectric voltage are various compounds of silver oxide or copper oxide. A complete device that operates on the photoelectric principle is referred to as a *photoelectric cell*. Many different sizes and types of photoelectric cells are in use, and each serves the special purpose for which it is designed. Nearly all, however, have some of the basic features of the photoelectric cells shown in Fig. 10.14.

The cell in Fig. 10.14 has a curved light-sensitive surface focused on the central anode. When light from the direction shown strikes the sensitive surface, it emits electrons toward the anode. The more intense the light, the greater the number of electrons emitted. When a wire is connected between the filament and the back, or dark side of the cell, the accumulated electrons will flow to the dark side. These electrons will eventually pass through the metal of the reflector and replace the electrons leaving the light-sensitive surface. Thus, light energy is converted to a flow of electrons, and a usable current is developed.

Another type of cell is shown in Fig. 10.15. This cell is constructed in layers; that is, a baseplate of pure copper is coated with light-sensitive copper oxide. An extremely thin semi-transparent layer of metal is placed over the copper oxide. This additional layer serves two purposes:

- It permits the penetration of light to the copper oxide.
- It collects the electrons emitted by the copper oxide.

An externally connected wire completes the electron path, the same as in the reflector-type cell (Fig. 10.14). The photocell's voltage is used (as needed) by connecting the external wires to some other device, which amplifies (enlarges) it to a usable level.

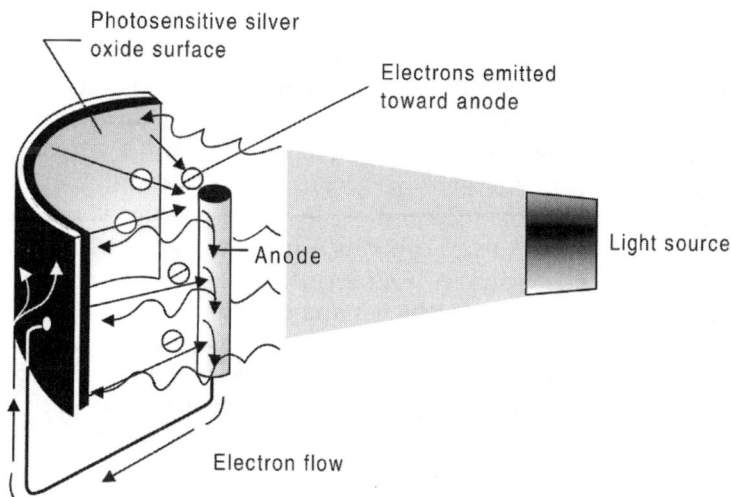

Figure 10.14 Photocell with curved surface.

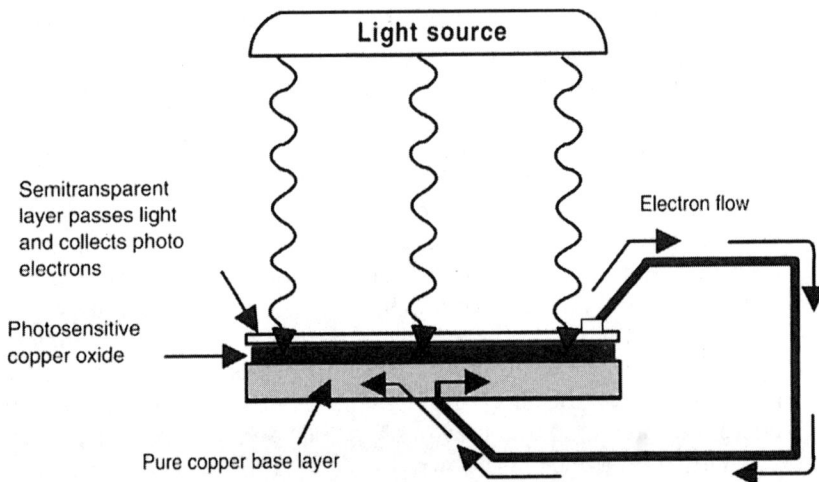

Figure 10.15   Photoelectric cell constructed in layers.

The power capacity of a photocell is very small. However, it reacts to light-intensity variations in an extremely short time. This characteristic makes the photocell very useful in detecting or accurately controlling a great number of operations. For instance, the photoelectric cell, or some form of the photoelectric principle, is used in television cameras, automatic manufacturing process controls, door openers, burglar alarms, etc.

Electricity produced by heat

When a length of metal, such as copper, is heated at one end, electrons tend to move away from the hot end toward the cooler end. This is true of most metals. However, in some metals, such as iron, the opposite is true; that is, electrons tend to move toward the heat or hot end (Fig. 10.16). The negative charges (electrons) are moving through the copper away from the heat and through the iron toward the heat. They cross from the iron to the copper through the current meter to the iron at the cold junction. This device is generally referred to as a *thermocouple*. Such a device is frequently encountered in HVAC temperature controls and controls for electrical appliances.

The thermoelectric voltage in a thermocouple depends mainly on the difference in temperature between the hot and cold junctions. Consequently, they are widely used to measure temperature, and as heat-sensing devices in automatic temperature-control equipment. Thermocouples generally can be subjected to much greater temperatures than ordinary mercury or alcohol thermometers.

Electricity produced by pressure

Pressure generating of voltage utilizes the characteristics of certain ionic crystals, for example, quartz, Rochelle salts, and tourmaline. These crystals have

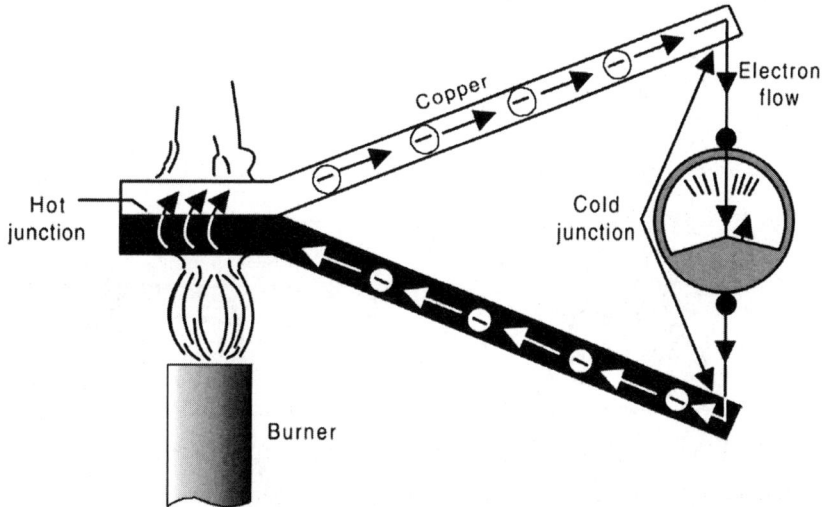

Figure 10.16 Electricity produced by heat; the device shown is called a *thermocouple.*

the remarkable ability to generate voltage whenever stresses are applied to their surfaces. Consequently, if a crystal of quartz is squeezed, charges of opposite polarity will appear on two opposite surfaces of the crystal. If the force is reversed and the crystal is stretched, charges will again appear, but will be of the opposite polarity from those produced by squeezing. If a crystal of this type is given a vibratory motion, it will produce a voltage of reversing polarity between two of its sides. Quartz or similar crystals can therefore be used to convert mechanical energy into electrical energy. This phenomenon, called the *piezoelectric effect,* is shown in Fig. 10.17. Some of the common devices that use piezoelectric crystals are microphones, phonograph cartridges, and oscillators in radio transmitters, radio receivers, and sonar equipment. This method of generating voltage is not suitable for applications having large voltage or power requirements, but is widely used in sound and communications systems, where small signal voltages can be effectively used.

Crystals of this type also possess another interesting property: the converse piezoelectric effect. That is, they have the ability to convert electrical energy into mechanical energy. A voltage impressed across the proper surface of the crystal will cause it to expand or contract its surfaces in response to the voltage applied.

Measuring Electricity

The three basic terms used to measure electricity are:

- Electromotive force—measured in volts
- Current—measured in amperes
- Resistance—measured in ohms

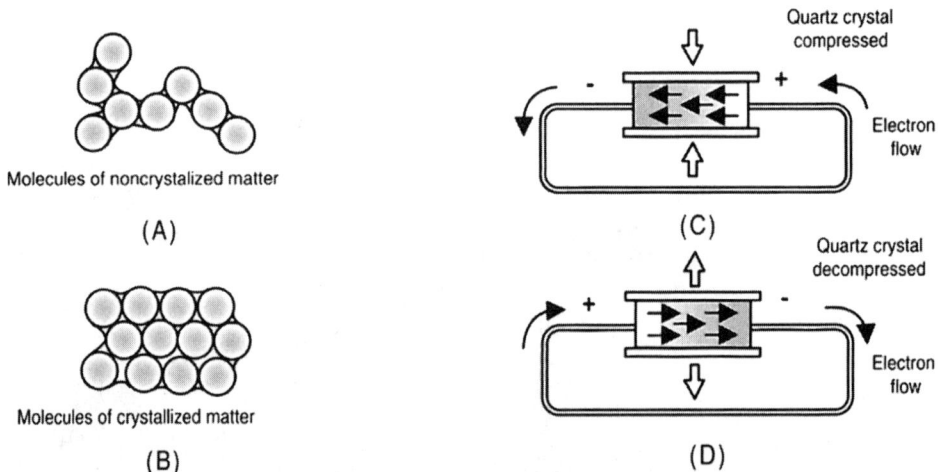

Figure 10.17   Principles of producing electricity by pressure.

In general, *electromotive force* or *voltage* is the force that causes electrons to flow. The unit of measurement of voltage is volts. In equations, voltage is represented by the letter E, which stands for electromotive force. Voltage is measured with a *voltmeter*. This and other measuring instruments are discussed in detail in Chapter 9.

Current is the rate at which electrons flow in a circuit. A current of 1 A is said to flow when 1 C of charge passes a point in 1 second. One coulomb is equal to the charge of 6.28×10^{18} electrons. Current is measured in units called *amperes* or *amps*. In equations, current is represented by the letter I, which stands for intensity of current.

Note
The rate of electron flow is not the speed at which the electrons are moving. Rather, it is the amount of electrons that flows past a given point in a given period of time.

In many cases, the ampere is too large a unit for measuring current. Therefore, either the *milliampere* (mA) or the *microampere* is used. A milliampere equals one-thousandth of an ampere, while the microampere (μA) equals one-millionth of an ampere. The device used to measure current is called an *ammeter*.

Resistance is the opposition to the flow of current in a circuit. All circuits have some resistance and the amount of resistance is measured in ohms. In equations, resistance is represented by the letter R or the Greek letter omega (Ω). A conductor has one ohm of resistance when an applied potential of one volt produces a current of one ampere.

Resistance, although an electrical property, is determined by the physical structure of a material. The resistance of a material is governed by many of the same fctors that control current flow.

Factors that affect resistance

The magnitude of resistance is determined in part by the number of free electrons available within a substance. Because a decrease in the number of free electrons will decrease the current flow, it can be said that the opposition to current flow (resistance) is greater in materials with few free electrons. Consequently, the resistance of a conductor is determined by the number of free electrons available in that material.

Depending upon their atomic structure, different materials will have different quantities of free electrons. Therefore, the various conductors used in electrical applications have different values of resistance.

Effect of cross-sectional area

Cross-sectional area greatly affects the magnitude of resistance. If the cross-sectional area of a conductor is increased, a greater quantity of electrons is available for movement through the conductor. Therefore, a larger current will flow for a given amount of applied voltage. An increase in current indicates that when the cross-sectional area of a conductor is increased, the resistance must have decreased. If the cross-sectional area of a conductor is decreased, the number of available electrons decreases and, for a given applied voltage, the current through the conductor decreases. A decrease in current flow indicates that when the cross-sectional area of a conductor is decreased, the resistance must have increased.

The diameter of many electrical conductors is often only a fraction of an inch. Therefore, the diameter of conductors is usually expressed in mils (thousandths of an inch). It is also standard practice to assign the unit circular mil to the cross-sectional area of a conductor (Fig. 10.18). The circular mil is found by squaring the diameter, when the diameter is expressed in mils. Consequently, if the diameter of a conductor is, for example, 35 mils (0.035 in), the circular mil area is equal to:

$$(35) 2 = 1225 \text{ circular mils}$$

Note
Resistance of a conductor is inversely proportional to its cross-sectional area.

Effect of conductor length

The length of a conductor is also a factor that determines the resistance of a conductor. If the length of a conductor is increased, the amount of energy given up increases. As free electrons move from atom to atom, some is given off as heat. The longer a conductor is, the more energy is lost to heat. The additional energy loss is subtracted from the energy being transferred through the conductor, resulting in a decrease in current flow for a given applied voltage. A decrease in current flow indicates an increase in resistance because the volt-

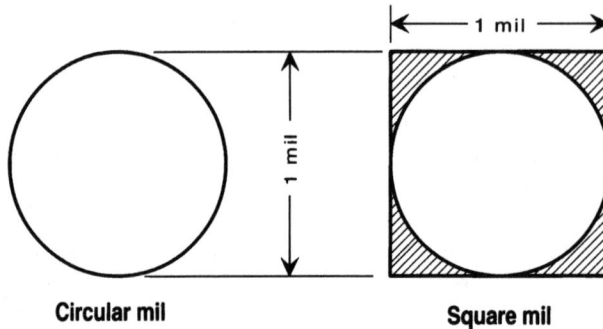

Circular mil   **Square mil**

*Note: Shading represents
difference in area between
circular and square mils.*

Figure 10.18   Comparison between a square mil and a circular mil.

age was held constant. Therefore, if the length of a conductor is increased, the resistance of that conductor increases.

Effect of temperature

Temperature changes affect the resistance of materials in different ways. In some materials, an increase in temperature causes an increase in resistance, whereas in others, an increase in temperature causes a decrease in resistance. The amount of change of resistance per unit change in temperature is known as the *temperature coefficient*. When the resistance of a material increases with an increase in temperature, the material is said to have a *positive temperature coefficient*. When the resistance of a material decreases with an increase in temperature, the material is said to have a *negative temperature coefficient*. Most conductors used in electrical applications have a positive temperature coefficient. However, carbon (a frequently used material) has a negative temperature coefficient.

Note
The resistance of a conductor is directly proportional to its length.

Some materials, such as the alloys constantan and manganin, are considered to have a zero temperature coefficient because their resistance remains relatively constant for changes in temperature.

Conductance

Conductance is the ability of a material to pass electrons, and is the opposite of resistance. The factors that affect the magnitude of resistance are exactly the same for conductance, but they affect conductance in the opposite manner. Therefore, conductance is directly proportional to area, and inversely proportional to the length of the material. The unit of conductance is the MHO, which is ohm spelled backward. Another term for mho is *siemans*.

Direct Current

The common flashlight is an example of a basic electric circuit. It contains a source of electrical energy (the dry cells in the flashlight), a load (the bulb), which changes the electrical energy into light energy, and a switch to control the energy delivered to the load.

The technician's main aid in studying circuits is the schematic diagram. In general, a schematic diagram is a "picture" of the circuit that uses symbols to represent the various circuit components and lines to connect these components. Basic symbols used in schematic diagrams are shown in Fig. 10.19. Refer to this symbol list frequently as you study the various schematic diagrams in the remaining pages of this chapter. In doing so, you will have symbols memorized by the time you have completed Chapter 10.

Figure 10.20 shows a pictorial representation of a common flashlight (*A*), along with the appropriate symbols, at the approximate locations, for the bulb, switch, and battery. A line is used to connect these components. In actuality, drafters would not draw the flashlight because we are concerned only with the electrical circuit. Consequently, schematic diagrams are used as shown in (*B*) and (*C*) below the pictorial drawing. These are actually simplified representations of the drawing in (*A*).

——— Wire	Lamp incandescent
Conductors —●— Connected	—⌒⌒— Fuse
—◆— Connected	Resistors —ᴧᴧ— Fixed
⊥ Not connected	—ᴧᴧ— Variable (potentiometer)
⏚ Ground	—ᴧᴧ— Rheostat
—⊣⊢ Cell	—∘∕∘— Switch
—⊣⊪⊢ Battery	—(V)— Voltmeter
⁺—⊣⊪⊢⁻ or	—(A)— Ammeter

Figure 10.19 Symbols commonly used in electrical schematic diagrams.

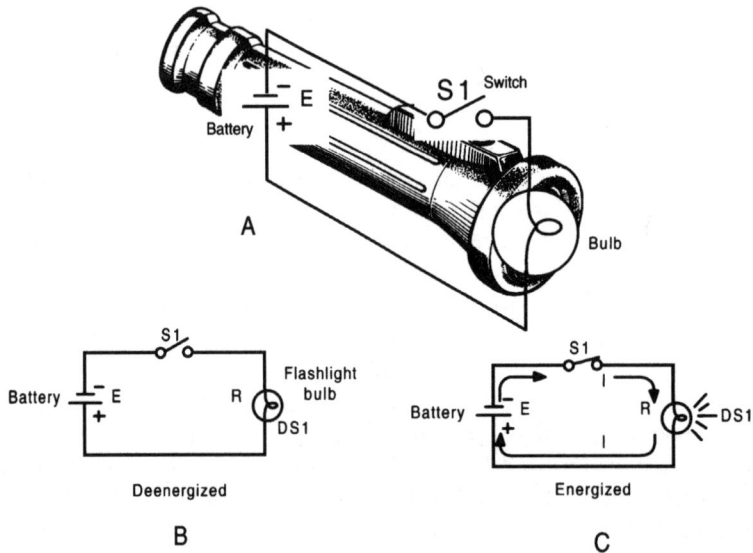

Figure 10.20   Basic electric circuit in a common flashlight.

Refer again to Fig. 10.20 and notice the schematic in (B). This drawing shows the flashlight in the OFF (deenergized) state. The switch (S1) is open. There is no complete path for current (I) through the circuit, and the bulb (DSI) does not light. In Fig. 10.20C, however, switch S1 is closed (moved to the ON position) and current flows in the direction of the arrows from the negative terminal of the battery, through the switch (S1), through the lamp (DS1), and back to the positive terminal of the battery. With the switch closed, the path for current is complete and current will continue to flow until the switch (S1) is moved to the open (OFF) position, or the battery is completely discharged.

Ohm's law

In the early part of the nineteenth century, George Ohm proved by experiment that a precise relationship exists between current, voltage, and resistance. This relationship is called *Ohm's law* and is stated as follows:

The current in a circuit is directly proportional to the applied voltage and inversely proportional to the circuit resistance.

Ohm's law can be expressed as an equation:

$$I = \frac{E}{R}$$

where: I = current in amperes
E = voltage in volts
R = resistance in ohms

As stated in Ohm's law, current is inversely proportional to resistance. This means, as the resistance in a circuit increases, the current decreases proportionately.

Refer again to Fig. 10.20C—the schematic of the flashlight in the ON state. If the battery supplies a voltage of 1.5 V and the lamp (DS1) has a resistance of 5 Ω, then the current in the circuit can be determined by substituting values in the equation previously described:

$$I = \frac{E}{R} = 1.5 \ \frac{V}{5 \ \Omega} = 0.3 \ A$$

Note
In using Ohm's law, if any two quantities are known, the third one can be determined.

If the flashlight contained two 1.5-V batteries, the voltage would double and 3 V would be applied to the circuit. Using this voltage in the equation:

$$I = \frac{3.0 \ V}{5 \ \Omega} = 0.6 \ A$$

In comparing these two sample equations, you can see that both the voltage and current have doubled in the latter example. This demonstrates that the current is directly proportional to the applied voltage.

If the value of resistance of the lamp is double, the equation will be:

$$I = \frac{E}{R} = \frac{3.0 \ V}{10 \ \Omega} = 0.3 \ A$$

The current has been reduced to one-half of the value of the previous equation, or 0.3 A. This demonstrates that the current is inversely proportional to the resistance. Doubling the value of the resistance of the load reduces circuit current value to one-half of its former value.

Application of Ohm's law

In using Ohm's law, if any two of the variables are known, the unknown can be found. For example, if current (*I*) and voltage (*E*) are known, resistance (*R*) can be determined as follows:

1. Use the basic equation:

$$I = \frac{E}{R}$$

2. Remove the divisor by multiplying both sides by *R*:

$$R \times I = \frac{E}{R} \times \frac{R}{I}$$

3. Note result of Step 2: $R \times I = E$

4. To get R alone (on one side of the equation), divide both sides by I:

$$\frac{RI}{I} = \frac{E}{I}$$

5. The basic equation, transposed for R, is:

$$R = \frac{E}{I}$$

Now let's put this equation to practical use. Refer to Fig. 10.21 and notice that the voltage (E) is 10 V, and the current (I) equals 1 A. Solve for R, using the equation just explained.

1. Insert the known values in the equation:

$$R = \frac{10 \text{ V}}{1 \text{ A}} = 10 \text{ } \Omega$$

The basic Ohm's law equation can also be used to solve for voltage (E).

1. Use the basic equation:

$$I = \frac{E}{R}$$

2. Multiply both sides by R:

$$I \times R = \frac{E}{R} \times \frac{R}{1}$$

3. Note the results of Step 2.

$$E = I \times R$$

Now let's use this equation to find the voltage in the circuit shown in Fig. 10.22, where the amperage equals 0.5 A, and the resistance equals 45 Ω.

$$E = I \times R$$
$$E = 0.5 \text{ A} \times 45 \text{ } \Omega$$
$$E = 22.5 \text{ V}$$

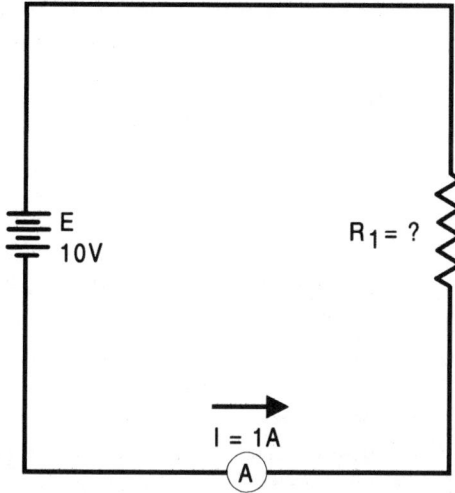

Figure 10.21 dc circuit with unknown resistance.

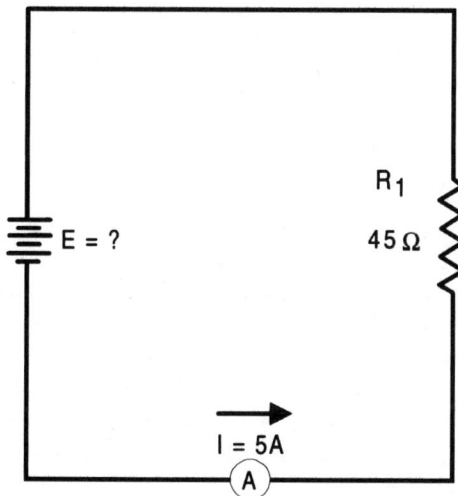

Figure 10.22 dc circuit with unknown voltage.

The Ohm's law equation and its various forms can be readily obtained with the aid of the circle in Fig. 10.23. The circle is divided into three parts with each part containing E, I, and R respectively. The letter E is above the horizontal line, while I and R are below the horizontal line. To determine the unknown quantity, first cover that quantity with a finger. The position of the uncovered letters in the circle will indicate the mathematical operation to be performed. For example, to find current (I), cover the letter I with a finger. The uncovered letters indicate that E is to be divided by R, or $I = \frac{E}{R}$. To find the equation for E, cover E with your finger. The result indicates that I is to be multiplied by R, or $E = IR$. To find the equation for R, cover R. The result indicates that E is to be divided by I, or $R = \frac{E}{I}$.

Although some handheld computers and calculators can now extend these calculations and are great time-saving devices when used on the job or in practical applications, the technician should have a basic knowledge of how these calculations are obtained. Figure 10.23 should be used to supplement your knowledge of the algebraic method, not as a substitution for the algebraic method. Algebra is a basic tool in the solution of electrical problems.

Power

Power, whether electrical or mechanical, pertains to the rate at which work is being done. Work is done whenever a force causes motion. When a mechanical force is used to lift or move a weight, work is done. When voltage causes electrons to move, work is done. The instantaneous rate at which this work is done is called the electric power rate, and is measured in watts or volt-amperes. Power, in watts, is equal to the voltage across a circuit multiplied by current through the circuit. This represents the rate at any given instant at which work is being done. The symbol *P* indicates electrical power, and the basic power equation is:

$$P = E \times I$$

where: E = voltage and I = current.

The amount of power changes when either voltage or current, or both voltage and current, change. The power equation also has variations similar to those discussed previously for determining either voltage, current, or resistance in a circuit.

Electrical components are often given a power rating. The power rating, in watts, indicates the rate at which the device converts electrical energy into another form of energy, such as light, heat, or motion.

As with other electrical quantities, prefixes may be attached to the word *watt* when expressing very large or very small amounts of power. Some common examples of these are the kilowatt (1000 W), the megawatt (1,000,000 W), and the milliwatt (1/1000 of a watt). The National Electrical Code® is now referring to watts as "volt-amperes" in most NEC Articles and Sections, so the technician should be aware that power may be expressed in either watts or volt-amperes (VA).

Four of the most important electrical quantities have been discussed thus far:

- Voltage (*E*)
- Current (*I*)
- Resistance (*R*)
- Power (*P*)

Anyone involved in the electrical industry in any capacity must understand the relationships that exist among these quantities because they are used

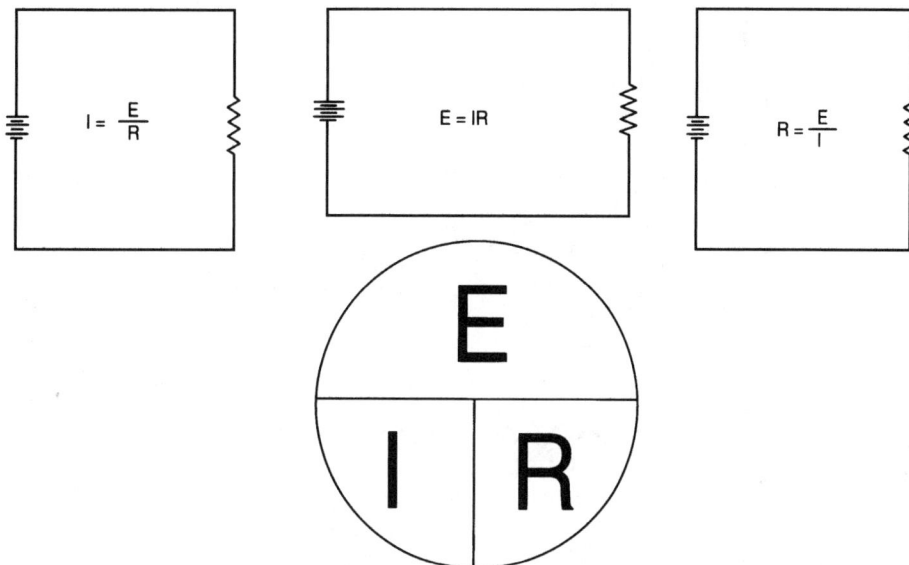

Figure 10.23 Ohm's law in diagram form.

throughout this book and will be used throughout the person's career. Figure 10.24 is a summary of 12 basic equations that you should know. The four quantities E, I, R, and P are at the center of the circle. Adjacent to each quantity are three segments. Notice that in each segment, the basic quantity is expressed in terms of two other basic quantities, and no two segments are alike.

Series Dc Circuits

A series is defined as a circuit that contains only one path for current flow. To compare the basic flashlight circuit (Fig. 10-20) with a more complex series circuit, the reader can imagine adding lamps. Remember that our basic flashlight circuit had only one lamp, whereas a more complex series circuit might contain three lamps connected in series; and the current in a three-lamp circuit must flow through each lamp to complete the electrical resistance. Consequently, in a series circuit, the total circuit resistance (R_T) is equal to the sum of the individual occurences of resistance.

$$R_T = R_1 + R_2 + R_3 + \ldots + R_n$$

The series circuit in Fig. 10.26 consists of three resistors; one rated at 10 Ω, one at 15 Ω, and one at 30 Ω. A voltage source provides 120 V. What is the total resistance?

1. Use the basic equation for finding resistance in a series circuit.

2. Substitute known values in the equation.

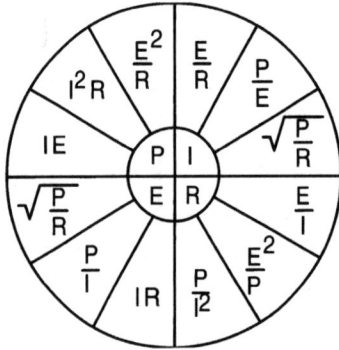

Figure 10.24 Summary of basic electrical equations.

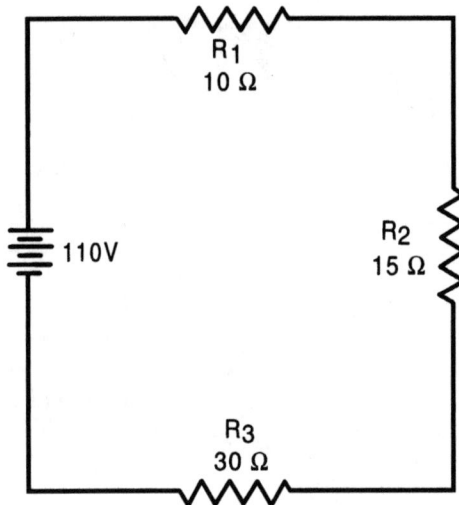

Figure 10.25 Solving for total resistance in a series circuit.

$$R_T = 10\ \Omega + 15\ \Omega + 30\ \Omega$$

$$R_T = 55\ \Omega$$

In some applications, the total resistance is known and the value of one of the circuit resistors has to be determined. The former equation for finding resistance in a series circuit can be transposed to solve for the value of the known resistance.

Note

The subscript n in this equation denotes any number of additional resistances that might be in the equation.

For example, the series circuit in Fig. 10.26 has a total resistance of 40 Ω. Two of the resistors are rated at 10 Ω each, while the rating of resistance R_3 is unknown. Here's how to find the value of the unknown resistor.

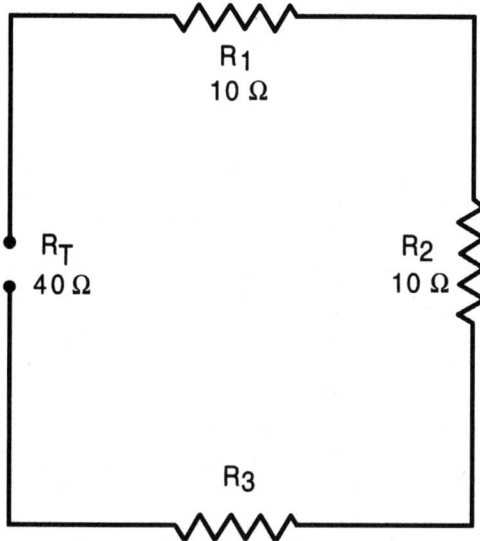

Figure 10.26 Calculating the value of one unknown resistance in a series circuit.

1. Use the basic equation.
2. Subtract $R_1 + R_2$ from both sides of the equation.
3. Continue solving for R_3 as follows:

$$R_T - R_1 - R_2 = R_3$$
$$R_3 = R_T - R_1 - R_2$$
$$R_3 = 40 \ \Omega - 10 \ \Omega - 10 \ \Omega$$
$$R_3 = 40 \ \Omega - 20 \ \Omega$$
$$R_3 = 20 \ \Omega$$

Current in a series circuit

Because there is only one path for current in a series circuit, the same current must flow through each component of the circuit. To determine the current in a series circuit, only the current through one of the components need be known. Ohm's law can be used to calculate the current in a series circuit if the voltage and resistance quantities are known.

The current flow through each component of a series circuit can be verified by inserting meters into the circuit at various points (Fig. 10.27). Upon examining these meters, each meter would be found to indicate the same value of current.

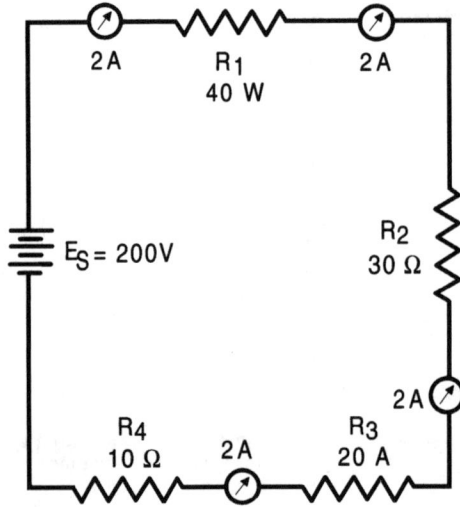

Figure 10.27 Current in a series circuit.

Voltage in a series circuit

The voltage drop across the resistor in a circuit consisting of a single resistor and a voltage source is the total voltage across the circuit and is equal to the applied voltage. The total voltage across a series circuit that consists of more than one resistor is also equal to the applied voltage, but consists of the sum of the individual resistor voltage drops.

In any series circuit, the sum of the resistor voltage drops must equal the source voltage. This statement can be proven by examining the series circuit in Fig. 10.28. In this circuit, a source voltage (E_T) of 20 V is dropped across a series circuit consisting of two 5-Ω resistors. The total resistance of the circuit (R_T) is equal to the sum of the two individual resistances, 10 Ω. Using Ohm's law, the circuit current can be calculated as follows:

$$I_T = \frac{E_T}{R_T}$$

$$I_T = 20 \, \frac{V}{10 \, \Omega}$$

$$I_T = 2 \, A$$

Because the value of the resistors is known to be 5 Ω each, and the current through the resistors is known to be 2 A, the voltage drops across the resistors can be calculated as:

$$E_1 = I_1 \times R_1$$

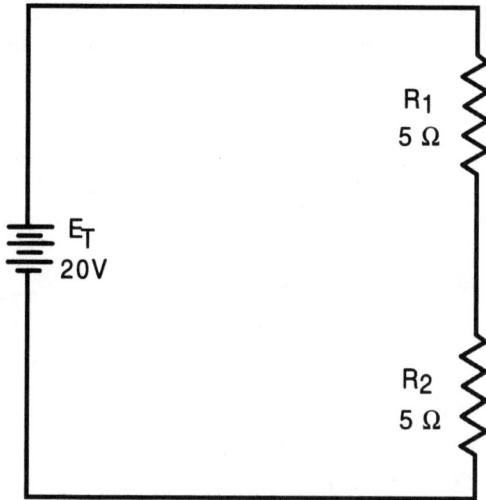

Figure 10.28 Calculating individual voltage drops in a series circuit.

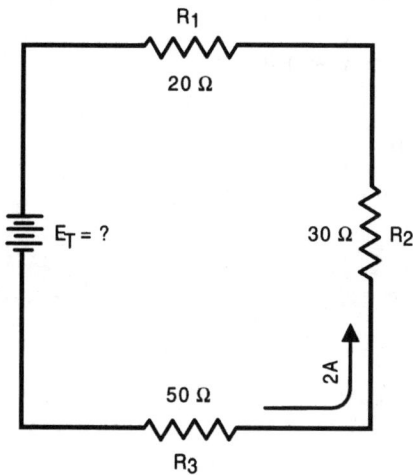

Figure 10.29 Solving for applied voltage in a series circuit.

$$E_1 = 2\,\text{A} \times 5\,\Omega$$

$$E_1 = 10\,\text{V}$$

Because R_2 is the same ohmic value as R_1, and carries the same current, the voltage drop across R_2 also equals 10 V. Then, adding the voltage drops for R_1 and R_2, we obtain (10 + 10 =) 20 V, which is equal to the applied voltage in the circuit. From the previous explanation, we see that the total voltage in a dc series circuit may be obtained by using the following equation:

$$E_T = E_1 + E_2 + E_3 + \ldots + E_n$$

To use the above equation in a practical application, let's assume that a circuit has three resistors connected in series of 20, 30, and 50 Ω, respectively. An ammeter shows that 2 A of current flows through the circuit. What is the total applied voltage?

1. Draw a circuit diagram as shown in Fig. 10.29.

2. Solve for the voltage drop across resistor R_1:

$$E_1 = R_1 \times I$$

$$E_1 = 20 \text{ Ω} \times 2 \text{ A} = 40 \text{ V}$$

3. Solve for the voltage drop across resistor R_2:

$$E_2 = R_2 \times I$$

$$E_2 = 30 \text{ Ω} \times 2 \text{ A} = 60 \text{ V}$$

4. Solve for the voltage drop across resistor R_3:

$$E_3 = R_3 \times I$$

$$E_3 = 50 \text{ Ω} \times 2 \text{ A} = 100 \text{ V}$$

5. Add the three voltage drops:

$$E_T = 40 \text{ V} + 60 \text{ V} + 100 \text{ V}$$

$$E_T = 200 \text{ V}$$

Power in a series circuit

Each of the resistors in a series circuit consumes power, which is dissipated in the form of heat. Because this power must come from the source, the total power must be equal to the power consumed by the circuit resistances. In a series circuit, the total power is equal to the sum of the power dissipated by the individual resistors. The equation to find the total power in a series circuit follows:

$$P_T = P_1 + P_2 + P_3 + \dots + P_n$$

Next, let's determine the total power in watts for the series circuit in Fig. 10.30. Note that this circuit has an applied potential of 120 V and three resistors are connected in series, each rated at 5 Ω, 10 Ω, and 15 Ω respectively.

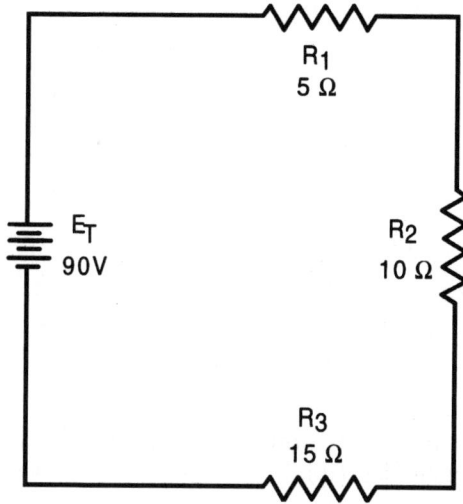

Figure 10.30 Solving for total power in a series circuit.

1. Find the total resistance in the circuit.

$$R_T = R_1 + R_2 + R_3$$

$$R_T = 5\ \Omega + 10\ \Omega + 15\ \Omega$$

$$R_T = 30\ \Omega$$

2. Determine the circuit current.

$$I = \frac{E_T}{R_T}$$

$$I = \frac{120\ \text{V}}{30\ \Omega}$$

$$I = 4\ \text{A}$$

3. Use the power equation to calculate the power for each resistor.

$$P_{1(2)(3)} = I^2 \times R_{1(2)(3)}$$

4. Calculate the power for resistor R_1.

$$P_1 = (4\ \text{A})\,2 \times 5\ \Omega$$

$$P_1 = 80\ \text{W}$$

5. Calculate the power for resistances R_2 and R_3 using the same steps as given in Step 4.

$$P_2 = 160 \text{ W}$$

$$P_3 = 240 \text{ W}$$

6. Obtain the total power by adding all power values.

$$P_T = 80 \text{ W} + 160 \text{ W} + 240 \text{ W}$$

$$P_T = 480 \text{ W}$$

When the total source voltage and the total source amperage are known, the total wattage can be found by multiplying the volts times the amps (volt-amperes). Let's double check the previous example with this method.

$$P_{\text{Source}} = E_{\text{Source}} \times I_{\text{Source}}$$

$$P_{\text{Source}} = 120 \text{ V} \times 4 \text{ A}$$

$$P_{\text{Source}} = 480 \text{ VA (W)}$$

Kirchhoff's Voltage Law

In 1847, G. R. Kirchhoff extended the use of Ohm's law by developing a simple concept concerning the voltages contained in a series circuit loop. Kirchhoff's law states:

The algebraic sum of the voltage drops in any closed path in a circuit and the electromotive forces in that path is equal to zero.

Kirchhoff's voltage law can be written as an equation as follows:

$$E_a + E_b + E_c + \ldots + E_n = 0$$

where E_a, E_b, etc. are the voltage drops or emg's around any closed circuit loop. To set up the equation for an actual circuit, the following procedure is used:

1. Assume a direction of current through the circuit.
2. Using the assumed direction of current, assign polarities to all resistors through which the current flows.
3. Place the correct polarities on any sources included in the circuit.
4. Starting at any point in the circuit, trace around the circuit, writing down the amount and polarity of the voltage across each component in succession. The polarity used is the sign after the assumed current has

passed through the component. Stop when the point at which the trace was started is reached.

5. Place these voltages, with their polarities, into the equation and solve for the desired quantity.

To place these procedures in use, assume that three resistors are connected in series with a 50-V source. What is the voltage across the third resistor if the voltage drops across the first two resistors are 25 V and 15 V respectively?

1. Draw a diagram such as the one shown in Fig. 10.31.

2. Draw an arrow indicating the assumed direction of current flow. Again, see Fig. 10.31.

3. Using the current direction arrow as made in Step 2, mark the polarity (− or +) at each end of each resistor and also on the terminals of the source. These markings are also shown in Fig. 10.31.

4. Starting at point A, trace around the circuit in the direction of current flow, recording the voltage and polarity of each component.

5. Starting at point A and using the components from the circuit, we have:

$$(+ E_x) + (+ E_2) + (+ E_1) + (-E_a) = 0$$

6. Substitute known values in the equation from the circuit.

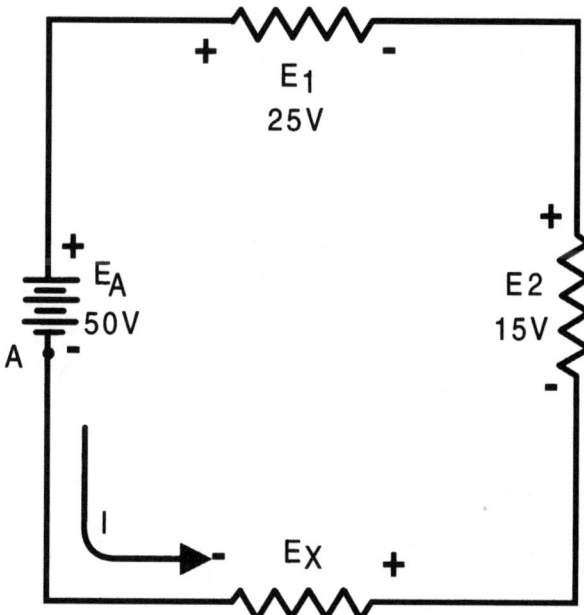

Figure 10.31   Determining unknown voltage in a series circuit.

$$E_X + 15\text{ V} + 25\text{ V} - 50\text{ V} = 0$$

$$E_X - 10\text{ V} = 0$$

$$E_X = 10\text{ V}$$

The unknown voltage (E_X) is found to be 10 V.

Solving for unknown current

Using the same procedure, problems can be solved in which the current is the unknown quantity.

For example, assume that a series circuit has a source voltage of 60 V and contains three resistors of 5 Ω, 10 Ω, and 15 Ω. Find the circuit current.

1. Draw and label the circuit as shown in Fig. 10.32.

2. Start at any point and write out the loop equation.

$$E_2 + E_1 + E_A + E_3 = 0$$

3. Because $E = IR$, substitute known values in the equation.

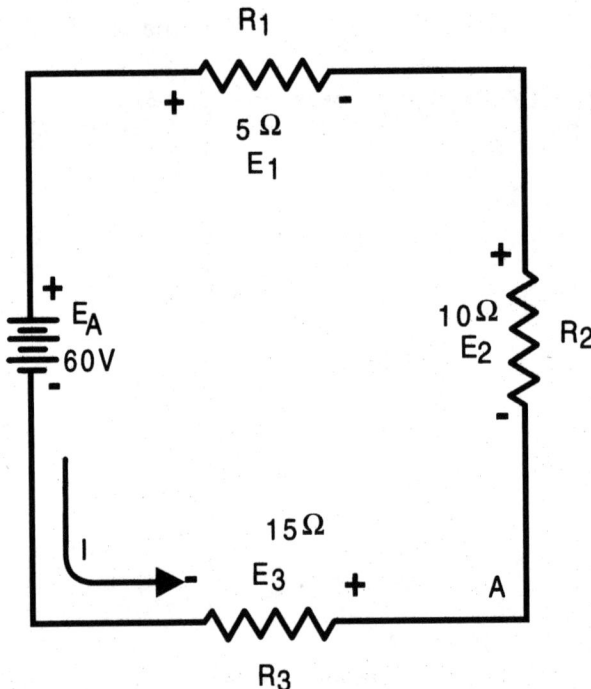

Figure 10.32 Correct direction of assumed current.

$$(I \times R_2) + (I \times R_1) + E_A + (I \times R_3) = 0$$

$$(I \times 10\ \Omega) + (I \times 5\ \Omega) + (-60\ V) + (I \times 15\ \Omega) = 0$$

4. Combine similar terms.

$$(I \times 30\ \Omega) + (-60\ V) = 0$$

$$I = \frac{60\ V}{30\ \Omega}$$

$$I = 2\ A$$

Because the current obtained in the preceding calculation is a positive 2 A, the assumed direction of current was correct. However, if the calculation had been a negative value, the assumed direction of current flow would be incorrect. Even if the wrong current direction is assumed, the amount of current in the calculation will be the same. The polarity, however, is negative if the wrong current direction is chosen. In this case, all that is required is to reverse the direction of the assumed current flow. However, should it be necessary to use this negative current value in further calculations on the circuit using Kirchhoff's law, the negative polarity should be retained in the calculations.

Parallel Dc Circuits

A *parallel circuit* is defined as one having more than one current path connected to a common voltage source. Parallel circuits, therefore, must contain two or more resistances that are not connected in series. An example of a basic parallel circuit is shown in Fig. 10.33.

Start at the voltage source (ES) and trace counterclockwise around the circuit. Two complete and separate paths can be identified in which current can flow.

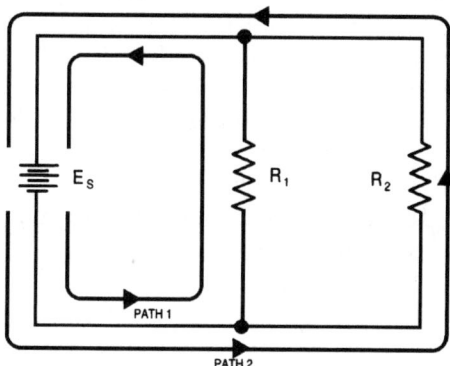

Figure 10.33 A basic parallel circuit.

One path is traced from the source, through resistance R_1, and back to the source. The other path is from the source, through resistance R_2, and back to the source.

Voltage in a parallel circuit

The source voltage in a series circuit divides proportionately across each resistor in the circuit. However, in a parallel circuit, the same voltage is present in each branch. In Fig. 10.33, this voltage is equal to the applied voltage (E_S) and can be expressed in the following equation:

$$E_S = E_{R1} = E_{R2}$$

Taking voltage measurements across the resistors of a parallel circuit is illustrated in Fig. 10.34. Each voltmeter indicates the same amount of voltage. Also notice that the voltage across each resistor in the circuit is the same as the applied voltage.

For example, assume that the current through a resistor of a parallel circuit is known to be 4.5 mA and the value of the resistor is 30,000 Ω (30 kΩ). What is the source voltage?

The circuit in question is shown in Fig. 10.35 on the next page and the source voltage can be found by using the basic Ohm's law equation:

$$E = IR$$

Substituting the known values in the equation, we have the following:

$$E_{R2} = 0.0045 \text{ A} \times 30,000 \text{ }\Omega$$

$$E_{R2} = 135 \text{ V}$$

Because the source voltage is equal to the voltage of a branch, 135 V is the source voltage and is also the voltage applied to each branch of this circuit.

Current in a parallel circuit

Ohm's law states that the current in a circuit is inversely proportional to the circuit resistance. This fact is true in both series and parallel circuits.

There is a single path for current in a series circuit. The amount of current is determined by the total resistance of the circuit and the applied voltage. In a parallel circuit, the source current divides among the available paths.

Part A of Fig. 10.36 shows a basic series circuit. Here, the total current must pass through the single resistor (R_1). Notice that the applied voltage equals 50 V and the resistance of R_1 is 10 Ω. The amount of current can be determined by using Ohm's law and is calculated as follows:

$$I = \frac{E}{R}$$

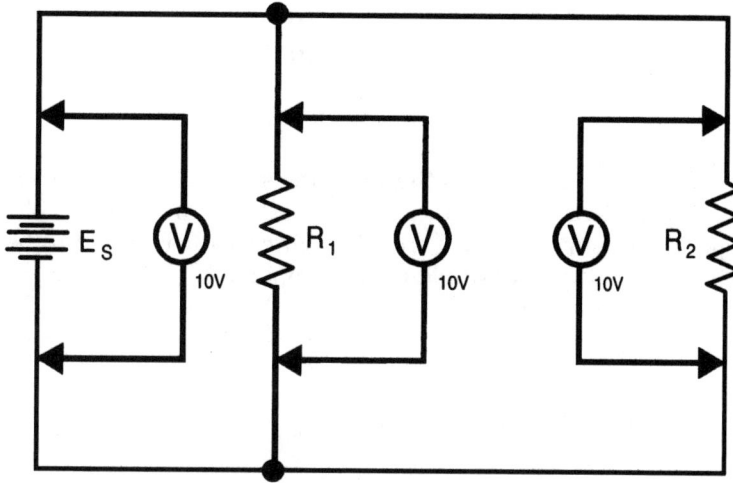

Figure 10.34   Voltage comparison in a parallel circuit.

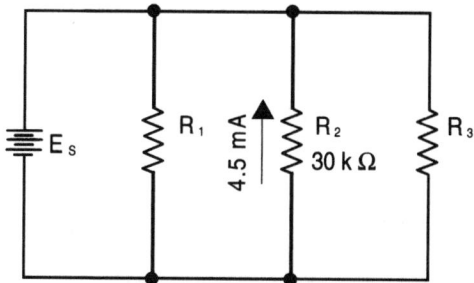

Figure   10.35 Finding   source voltage in a parallel circuit.

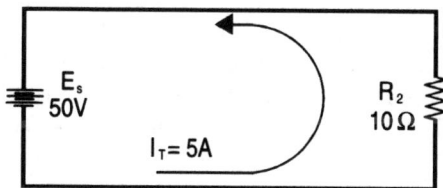

(A)

Series Circuit

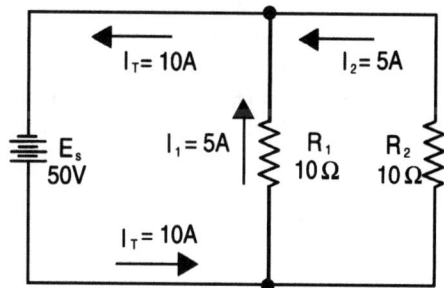

(B)

Parallel Circuit

Figure 10.36   Analysis of current in a series and parallel circuit.

$$I_T = \frac{50 \text{ V}}{10 \text{ }\Omega}$$

$$I_T = 5 \text{ A}$$

Part B of Fig. 10.36 shows the same resistor (R_1) with a second resistor (R_2) of equal value connected in parallel across the voltage source. When Ohm's law is applied, the current flow through each resistor is found to be the same as the current through the single resistor in part A.

$$I = \frac{E}{R}$$

$$ES = ER_1 = ER_2$$

$$I_{R1} = \frac{50 \text{ V}}{10 \text{ }\Omega}$$

$$I_{R1} = 5 \text{ A}$$

$$I_{R2} = \frac{50 \text{ V}}{10 \text{ }\Omega}$$

$$I_{R2} = 5 \text{ A}$$

It is apparent that if 5 A of current are passing through each of the two resistors, the total current must be 10 A drawn from the source.

The total current of 10 A leaves the negative terminal of the battery and flows to point a. Because point a is a connecting point for the two resistors, it is called a *junction*. At junction a, the total current divides into two currents of 5 A each. These two currents flow through their respective resistors and rejoin at junction b. The total current then flows from junction b

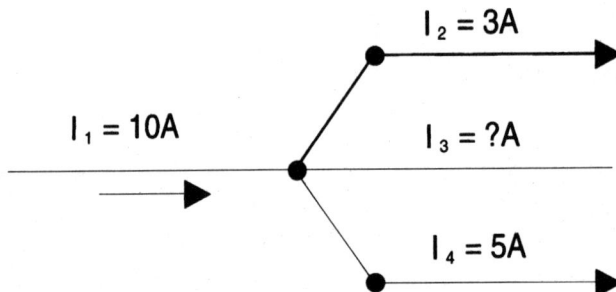

$I_2 = 3A$

$I_1 = 10A$

$I_3 = ?A$

$I_4 = 5A$

Figure 10.37   Circuit with four current values.

back to the positive terminal of the source. The source supplies a total current of 10 A and each of the two equal resistors carries one-half of the total current.

Each individual current path in the circuit in Fig. 10.36B is referred to as a *branch*. Each branch carries a current that is a portion of the total current. Two or more branches form a *network*.

From the previous explanation, the characteristics of current in a parallel circuit can be expressed in terms of the following general equation:

$$I_T = I_1 + I_2 + \ldots + I_n$$

Kirchhoff's current law

The division of current in a parallel network follows a definite pattern. This pattern is described by *Kirchhoff's current law* which states:

> *The algebraic sum of the currents entering and leaving any junction of conductors is equal to zero.*

This law can be stated mathematically as:

$$I_a + I_b + \ldots + I_n = 0$$

where: I_a, I_b, etc. are the currents entering and leaving the junction. Currents entering the junction are considered to be positive and currents leaving the junction are considered to be negative. When solving a problem using Kirchhoff's current law, the currents must be placed into the equation with the proper polarity signs attached.

Now let's use Kirchhoff's current law to solve for the value of I_3 in Fig. 10.37. The known values are first substituted in Kirchhoff's current law equation.

$$I_1 + I_2 + I_3 + I_4 = 0$$

$$10\text{ A} + (-3\text{ A}) + I_3 + (-5\text{ A}) = 0$$

$$I_3 + 2\text{ A} = 0$$

$$I_3 = -2\text{ A}$$

Current I_3 has a value of 2 A, and the negative sign shows it to be a current leaving the junction.

Resistance in parallel circuits

Figure 10.38 shows two resistors connected in parallel across a 5-volt battery. Each has a resistance value of 10 Ω. A complete circuit consisting of two parallel paths is formed and current flows as shown.

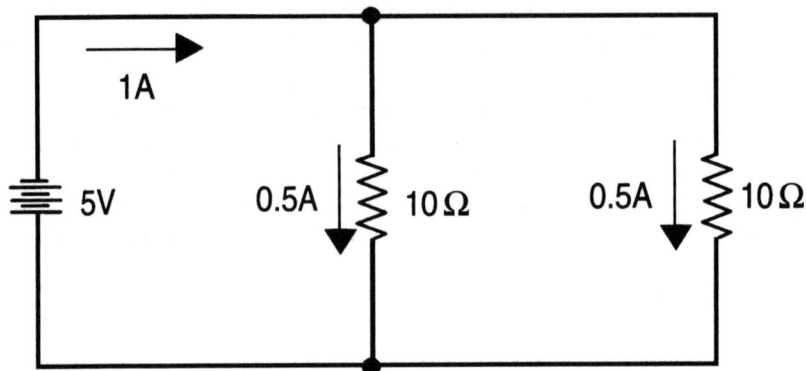

Figure 10.38   Two equal resistors connected in parallel.

Computing the individual currents show 0.5 A flows through each resistor. The total current flowing from the battery to the junction of the resistors, and returning from the resistors to the battery, is equal to 1 A.

The total resistance of the circuit can be calculated by using the values of total voltage (E_T) and total current (I_T).

$$R = \frac{E}{I}$$

$$R_T = \frac{5\ V}{1\ A}$$

$$R_T = 5\ \Omega$$

This computation shows the total resistance to be 5 Ω; one-half the value of either of the two resistors.

The total resistance of a parallel circuit is smaller than any of the individual resistors.

In other words, the total resistance in a parallel circuit is not the sum of the individual resistor values, as was the case in a series circuit.

There are several methods used to determine the total or equivalent resistance of parallel circuits. The best method to use for a given circuit depends on the number and value of the resistors. Where all resistors in the circuit are of the same value, the following simple equation can be used:

$$R_T = \frac{R}{N}$$

where: R_T = total parallel resistance

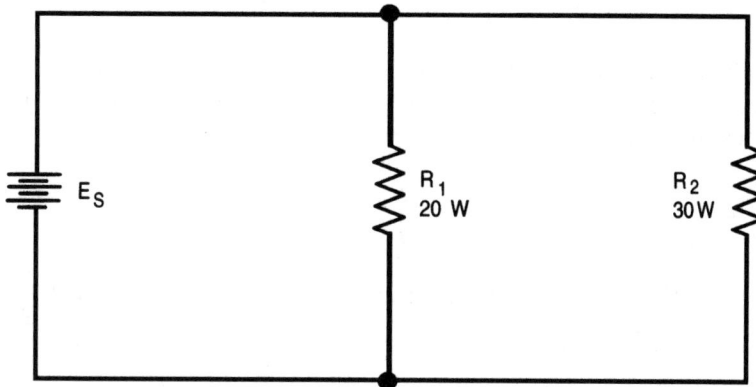

Figure 10.39 Parallel circuit with two unequal resistors.

R = ohmic value of one resistor
N = number of resistors

The equation is valid for any number of parallel resistors of equal value.

The total resistance of parallel circuits can be found if the individual resistance values are known along with the source voltage. The following equation is the most common:

$$\frac{1}{R_T} = \frac{1}{R_1} + \frac{1}{R_2} + \frac{1}{R_3} + \cdots \frac{1}{R_n}$$

When using the preceding equation to determine the resistance in a parallel circuit, convert the fractions to a common denominator. For example, find the total resistance in a parallel circuit with two resistors rated at 3 Ω and 6 Ω respectively.

$$\frac{1}{R_T} = \frac{1}{3\,\Omega} + \frac{1}{6\,\Omega}$$

$$\frac{1}{R_T} = \frac{2}{6\,\Omega} + \frac{1}{6\,\Omega}$$

$$\frac{1}{R_T} = \frac{3}{6}\,\Omega$$

$$\frac{1}{R_T} = \frac{1}{2}\,\Omega$$

Because both sides are reciprocals (divided into one), disregard the reciprocal function.

$$R_T = 2\ \Omega$$

When only two resistors, each of either the same or different values, are in a parallel circuit, the following equation may be used to find the total resistance in the circuit.

$$R_T = \frac{R_1 \times R_2}{R_1 + R_2}$$

Using the above equation, what is the total resistance in the circuit shown in Fig. 10.39

$$R_T = \frac{20\ \Omega \times 30\ \Omega}{20\ \Omega + 30\ \Omega}$$

$$R_T = \frac{600}{50\ \Omega}$$

$$R_T = 12\ \Omega$$

Equivalent circuits

In dealing with electrical circuits, it is sometimes necessary to reduce a complex circuit into a simpler form. Any complex circuit consisting of resistances can be redrawn (reduced) to a basic equivalent circuit containing the voltage source and a single resistor representing total resistance. This process is called reduction to an *equivalent circuit*.

Figure 10.40 shows a parallel circuit with three resistors of equal value and the redrawn equivalent circuit. The parallel circuit shown in part A shows the original circuit. To create the equivalent circuit, first calculate the total resistance in the circuit.

$$R_T = \frac{R}{N}$$

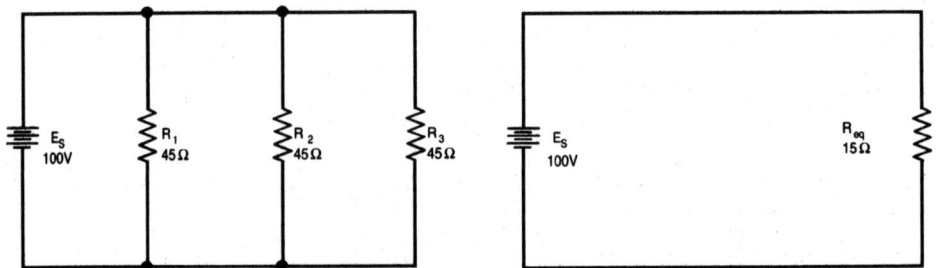

Figure 10.40   Parallel circuit (*left*) with equivalent circuit (*right*).

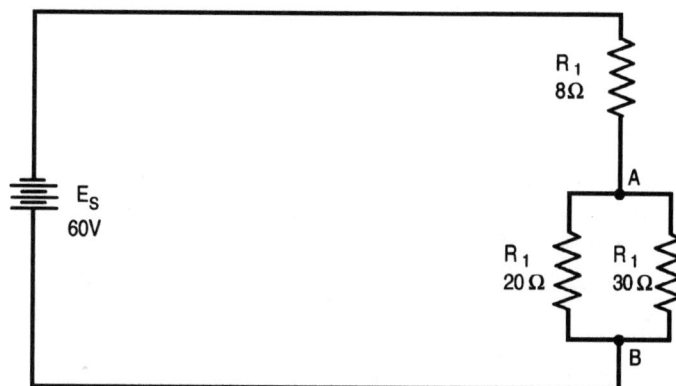

A

B

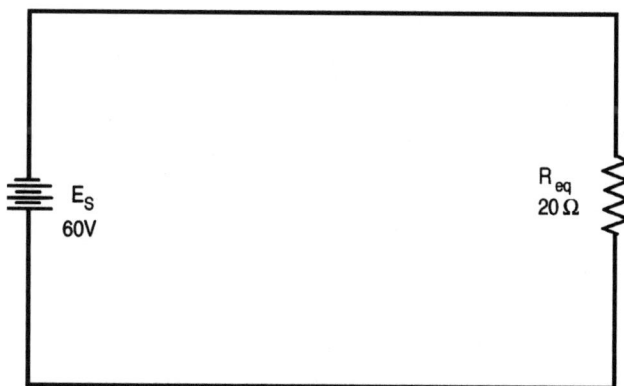

C

Figure 10.41   Steps in reducing a combination circuit to its simplest form.

$$R_T = \frac{45\ \Omega}{3}$$

$$R_T = 15\ \Omega$$

Once the equivalent resistance is known, a new circuit is drawn consisting of a single resistor (to represent the equivalent resistance) and the voltage source, as shown in part B.

Series-Parallel Dc Circuits

In the preceding sections, series and parallel dc circuits have been considered separately. Electricians often encounter circuits consisting of both series and parallel elements. A circuit of this type is referred to as a *combination circuit*. Determining quantities and elements in a combination circuit is simply a matter of applying the laws and rules discussed up to this point.

The basic technique used for solving dc combination-circuit problems is the use of equivalent circuits. To simplify a complex circuit to a simple circuit containing only one load, equivalent circuits are substituted (on paper) for the complex circuit they represent—the technique briefly discussed in the preceding section of this chapter.

To demonstrate the method used to solve combination circuit problems, refer to the circuit in Fig. 10.41A. Examination of this circuit shows that the only quantity that can be computed with the given information is the equivalent resistance of R_2 and R_3. Because only two resistors are contained in this part of the circuit and these resistors are connected in parallel, the product over the sum equation can be used to obtain the total resistance for this portion of the circuit.

$$R_T = \frac{R_2 \times R_3}{R_2 + R_3}$$

$$R_T = \frac{20\ \Omega \times 30\ \Omega}{20\ \Omega + 30\ \Omega}$$

$$R_T = \frac{600}{50\ \Omega}$$

$$R_T = 12\ \Omega$$

Now that the equivalent resistance for R_2 and R_3 has been calculated, the circuit can be redrawn as a series circuit (Fig. 10.41B). The total resistance for the entire circuit can now be calculated as:

$$R_\mathrm{T} = 8\ \Omega + 12\ \Omega$$

$$R_\mathrm{T} = 20\ \Omega$$

The original circuit can be redrawn with a single resistor that represents the equivalent resistance of the entire circuit as shown in Fig. 10.41C. After doing this, the total current in the circuit may be found.

$$I_\mathrm{T} = \frac{60\ \text{V}}{20\ \Omega}$$

$$I_\mathrm{T} = 3\ \text{A}$$

Summary

- A basic electric circuit consists of a source of electrical energy connected to a load. The load uses the energy and changes it to a useful form.

- A schematic diagram is a "picture" of a circuit that uses symbols to represent components. The space required to depict an electrical or electronic circuit is greatly reduced by the use of a schematic diagram.

- Ohm's law can be transposed to find one of the values in a circuit if the other two values are known.

- The same current flows through each part of a series circuit.

- The total resistance of a series circuit is equal to the sum of the individual resistances.

- The total voltage across a series circuit is equal to the sum of the individual voltage drops.

- The voltage drop across a resistor in a series circuit is proportional to the ohmic value of the resistor.

- The total power in a series circuit is equal to the sum of the individual power used by each circuit component.

11

The Code

The design and installation of security/fire-alarm systems employ a wide variety of techniques, often involving special types of equipment and materials designed for specific applications. Many security systems operate on low-voltage circuits, but are installed by the use of techniques similar to those conventional electrical circuits for light and power. All installations, when used in buildings, must conform to applicable National Electrical Code® (NEC) requirements, local ordinances, and instructions provided by equipment and component manufacturers. This chapter explains the key terms and basic layout of the NEC. A brief review of the individual NEC sections that apply to security systems is also covered.

Introduction to the NEC

Since the first central-station electric generating plant was developed in New York City in 1882 by Thomas A. Edison, the electrical construction industry has grown at an astonishing rate to become one of the largest industries in the United States. The first generating plant created public demand for the use of electric lighting and power in existing buildings, as well as in new construction.

These first electrical wiring installations were usually laid out by workers employed and trained by the power companies, and the majority of these installations were "designed" by the mechanics on the job, often as the work progressed. Building contractors then began hiring mechanics of their own to install electrical wiring systems, but because of the special skills and knowledge required, these same builders soon began leaving the wiring installations to independent mechanics. With time, the independents began to specialize in this work as electrical contractors.

As electrical construction continued to become a more and more important part of the construction industry, architects began to add layouts of the desired electrical system to their architectural drawings. This layout usually indicated the lighting outlets, base "plugs," and light switches by means of certain symbols. A line was sometimes drawn from a lighting outlet to a wall switch to indicate how the various lamps were to be controlled, but this was usually the extent of the electrical design. The details of wiring, number of circuits, and device location decisions, were still left to the mechanics (electricians) installing the system. As electrical applications became more extensive and complex, electrical contractors began to hire draftspeople to prepare working drawings. The thorough outlet layout on the architectural drawings provides a basis for preparing estimates, and to give instructions to electricians in the field.

From that point on, electrical construction continued to become a more important part of general building construction, and soon the architects began to prepare more extensive layouts of the electrical systems, until finally separate drawings were included along with the architectural drawings. As the volume of layout work increased and electrical systems became still more extensive and complex, a greater engineering knowledge of power and illumination requirements became necessary. Persons with the proper knowledge and training began to devote their time exclusively to designing and layout out electrical installations as consulting engineers. These consulting engineers conveyed their designs by means of working drawings that used symbols, lines, notations, and written specifications. Thus, the electrical designer became a very important cog in the wheel of electrical construction. Yet, the best electrical designs, specifying the best materials and equipment, are useless without trained workers to properly install the systems. Consequently, the trained electrical technician is indispensable in the building construction industry… and a thorough knowledge of the NEC is one of the first requirements in becoming a trained electrical technician. In fact, the NEC is probably the most widely used and generally accepted code in the world. It is used as an electrical installation, safety, and reference guide in the United States, and in many other parts of the world as well.

Purpose and history of the NEC

Owing to the potential fire and explosion hazards caused by the improper handling and installation of electrical wiring, certain rules in the selection of materials, quality of workmanship, and precautions for safety must be followed. To standardize and simplify these rules and provide a reliable guide for electrical construction, the National Electrical Code (NEC) was developed. The NEC (Fig. 11.1), originally prepared in 1897, is frequently revised to meet changing conditions, improved equipment and materials, and new fire hazards. It is a result of the best efforts of electrical engineers, manufacturers of electrical equipment, insurance underwriters, fire fighters, and other experts throughout the country.

The NEC is now published by the National Fire Protection Association (NFPA), Batterymarch Park, Quincy, Massachusetts 02269. It contains speci-

Figure 11.1 The NEC has become the bible of the electrical construction industry. Keep your old versions of the code in case there are future disputes about your installations.

fic rules and regulations intended to help in the practical safeguarding of persons and property from hazards arising from the use of electricity.

Although the NEC itself states, "This Code is not intended as a design specification nor an instruction manual for untrained persons," it does provide a sound basis for the study of electrical installation procedures—under the proper guidance. The probable reason for the NEC's self-analysis is that the code also states, "This Code contains provisions considered necessary for safety. Compliance therewith and proper maintenance will result in an installation essentially free from hazard, but not necessarily efficient, convenient, or adequate for good service or future expansion of electrical use."

The NEC, however, has become the bible of the electrical construction industry, and anyone involved in electrical work, in any capacity, should obtain an up-to-date copy, keep it handy at all times, and refer to it frequently.

Whether you are installing a new security system or repairing an existing one, all electrical work must comply with the current National Electrical Code (NEC) and all local ordinances. Like most laws, the NEC is easier to work with once you understand the language and know where to look for the information you need.

This chapter is not a substitute for the NEC. You need a copy of the most recent edition and it should be kept handy at all times. The more you know about the Code, the more you are likely to refer to it.

NEC Terminology

The two basic types of rules in the NEC are mandatory rules and advisory rules. Here is how to recognize the two types of rules and how they relate to all types of electrical systems.

- Mandatory rules—All mandatory rules have the word *shall* in them. The word *shall* means must. If a rule is mandatory, you must comply with it.

- Advisory rules—All advisory rules have the word *should* in them. The word *should,* in this case, means recommended, but not required. If a rule is advisory, compliance is discretionary. If you want to comply with it, do so. But the guidelines are not mandatory.

Be alert to local amendments to the NEC. Local ordinances might amend the language of the NEC, changing it from *should* to *shall.* This means that you must do in that county or city what might only be recommended in some other area. The office that issues building permits will either sell you a copy of the code that's enforced in that area or tell you where the code is sold. In rare instances, the electrical inspector, having jurisdiction over the area, might issue these regulations verbally.

You will encounter a few other "landmarks" while looking through the NEC. These are summarized in Fig. 11.2, and a brief explanation of each follows:

Explanatory material: Explanatory material in the form of Fine Print Notes is designated (FPN). Where these appear, the FPNs normally apply to the NEC Section or paragraph immediately preceding the FPN.

Change bar: A change bar in the margins indicates that a change in the NEC has been made since the last edition. When becoming familiar with each new edition of the NEC, always review these changes. Also, several illustrated publications on the market point out changes in the NEC with detailed explanations of each. Such publications make excellent reference material.

Bullets: A filled-in circle, called a *bullet,* indicates that something has been deleted from the last edition of the NEC. Although not absolutely necessary, many electricians like to compare the previous NEC edition to the most recent one when these bullets are encountered, just to see what has been omitted from the latest edition. The most probable reasons for the deletions are errors in the previous edition, or obsolete items.

Extracted text: Material identified by the superscript letter x includes text extracted from other NFPA documents, as identified in Appendix A of the NEC.

As you open the NEC book, you will notice several different types of text used. Here is an explanation of each.

1. *Black Letters:* Basic definitions and explanations of the NEC.

Mandatory rules are characterized by
the use of the word:

SHALL

A recommendation or that which is
advised but not required is
characterized by the use of the word:

SHOULD

Explanatory material in the form of
Fine Print Notes is designated:

(FPN)

A change bar in the margins
indicates that a change in the
NEC has been made since the
last edition.

A bullet indictates that something
has been deleted from the last
edition of the NEC.

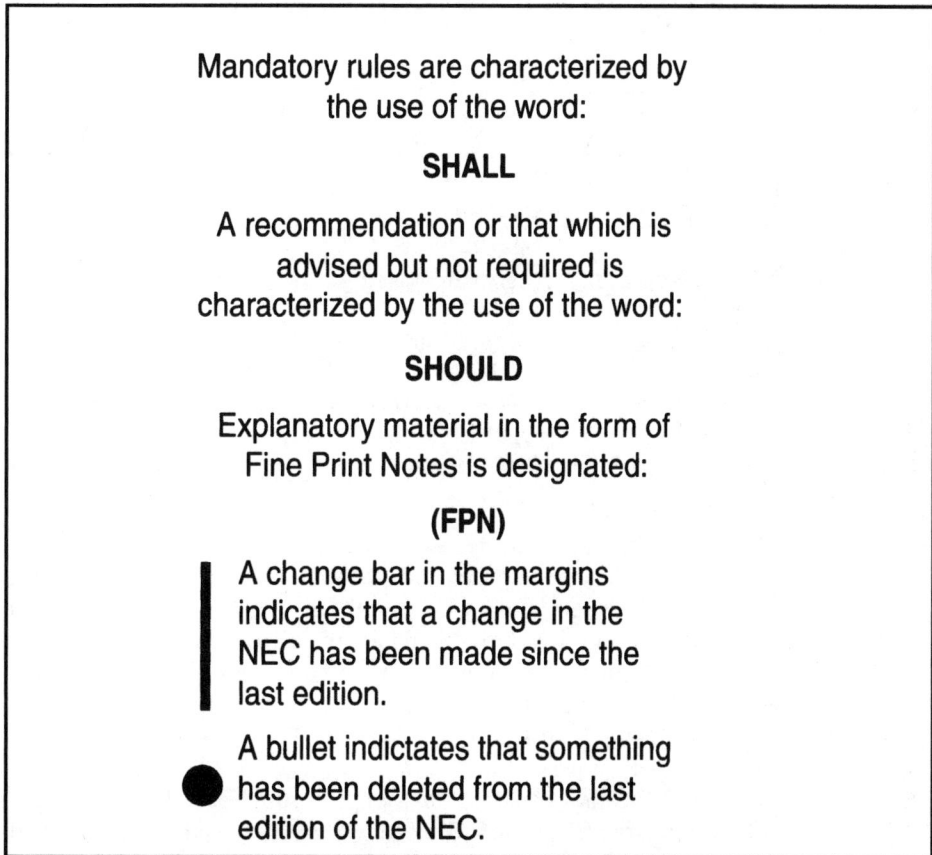

Figure 11.2 NEC terminology.

2. *Bold Black Letters:* Headings for each NEC application.

3. *Exceptions:* These explain the situations when a specific rule does not apply. Exceptions are written in italics under the section or paragraph to which they apply.

4. *Tables:* Tables are often included when there is more than one possible application of a requirement (Fig. 11.3).

5. *Diagrams:* A few diagrams are scattered throughout the NEC to illustrate certain NEC applications (Fig. 11.4).

Learning the NEC Layout

The NEC is divided into the Introduction (Article 90) and nine chapters. Chapters 1, 2, 3, and 4 apply generally; Chapters 5, 6, and 7 apply to special occupancies, special equipment, or other special conditions. These latter chap-

Size of Largest Service-Entrance Conductor or Equivalent Area for Parallel Conductors		Size of Grounding Electrode Conductor	
Copper	Aluminum or Copper-Clad Aluminum	Copper	Aluminum or Copper-Clad Aluminum
2 or smaller	1/0 or smaller	8	6
1 or 2	2/0 or 3/0	6	4
2/0 or 3/0	4/0 or 250 kcmil	4	2
Over 3/0 through 350 kcmil	Over 250 kcmil through 500 kcmil	2	1/0
Over 350 kcmil through 600 kcmil	Over 500 kcmil through 900 kcmil	1/0	3/0
Over 600 kcmil through 1100 kcmil	Over 900 kcmil through 1750 kcmil	2/0	4/0
Over 1100 kcmil	Over 1750 kcmil	3/0	250 kcmil

Figure 11.3 Typical NEC table.

ters supplement or modify the general rules. Chapters 1 through 4 apply, except as amended by Chapters 5, 6, and 7 for the particular conditions.

While looking through the book, if you encounter a word or term that is unfamiliar, look in Chapter 1, Article 100—Definitions. Chances are, the term will be found here. If not, look in the Index for the word and the NEC page number. Many terms are included in Article 100, but others are scattered throughout the book.

For definitions of terms not found in the NEC, obtain a copy of *Illustrated Dictionary for Electrical Workers,* available from Delmar Publishers, Inc., Albany, New York.

Chapter 8 of the NEC covers communications systems and is independent of the other chapters, except where they are specifically referenced therein.

Chapter 9 consists of tables and examples.

Also, the NEC Contents are at the beginning of the book and a comprehensive index is at the back of the book. You will find frequent use for both of these helpful tools when searching for various installation requirements.

Each chapter is divided into one or more Articles. For example, Chapter 1 contains Articles 100 and 110. These articles are subdivided into sections. For example, Article 110 of Chapter 1 begins with *Section 110-2. Approval.* A bullet in the margin indicates that Section 110-1 has been deleted from the last NEC edition. Some sections might contain only one sentence or a paragraph, but others are further subdivided into lettered or numbered paragraphs, such as (a), (1), (2), etc.

Begin your study of the NEC with Articles 90, 100, and 110. These three articles have the basic information that will make the rest of the book easier to understand. Article 100 defines terms you will need to understand the code. Article 110 gives the general requirements for electrical installations. Read these three articles over several times until you are thoroughly familiar with

Art. 430 Part H
Sec. 430-101
thru 430-113

Controller

Disconnecting Means

Disconnects motor and controllers from circuit.
1. Continuous rating of 115% or more of motor FLC. Also see Art. 430, Part B, Par. 1 below.
2. Disconnecting means shall be a motor circuit switch rated in horsepower or a circuit breaker, with exceptions. (See 430-109)
3. Must be in sight from controller, and not more than 50 feet distant.

Art. 430 Part D
Sec. 430-51
thru 430-58

Branch Circuit Short-Circuit Protection

Protects branch circuit from short circuits or grounds.
1. Must carry starting current of motor.
2. Rating must not exceed values in Table 430-152 unless not sufficient to carry starting current of motor.
3. Values for branch circuit protective devices shall in no case exceed exceptions listed in 430-52.

Art. 430 Part G
Sec. 430-81
thru 430-90

Controller

Used to start and stop motors.
1. Must have current rating of 100% or more of motor FLC.
2. Must be able to interrupt LRC.
3. Must be horsepower rated, not less than motor, with exceptions. (See 430-83)

Art. 430 Part C
Sec. 430-31
thru 430-44

Controller

Motor Running Protection

Protects motor and controller against excessive heat due to motor overload.
1. Must trip at following percent or less of motor FLC, for continuous motors rated more than one horsepower.
   a) 125% FLC for motors with a marked service factor not less than 1.15 or a marked temperature rise not over 40 C.
   b) 115% FLC for all others. (See part C of NEC for other types of protection)
2. Three thermal units required for any three phase ac motor.
3. Must allow motor to start.
4. Select size from FLC on motor nameplate.

Art. 430 Part B
Sec. 430-21
thru 430-28

Specifies the sizes of conductors capable of carrying the motor current without overheating.
1. To determine the ampacity of conductors, switches, branch circuit overcurrent devices, etc., the full-load current values given in Tables 430-147 thru 430-150 shall be used instead of actual current rating marked on motor nameplate. (See 430-6)
2. Branch circuit conductors supplying a single motor shall have ampacity not less than 125% of motor FLC, with exceptions. (See 430-22)

Motor

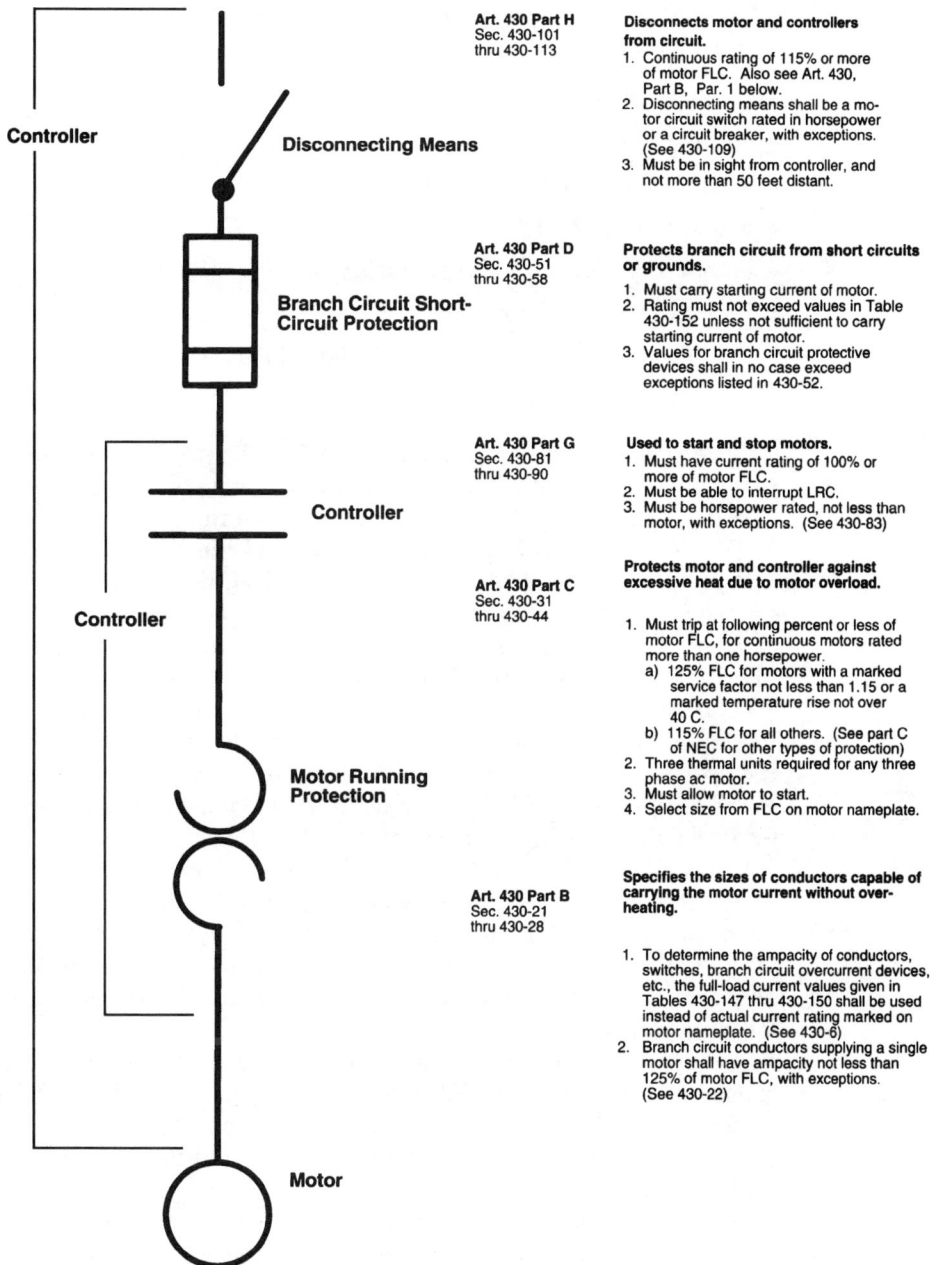

Figure 11.4 Typical NEC diagram; several are scattered throughout the NEC for clarity.

all the information. It's time well spent. For example, Article 90 contains the following sections:

- Purpose (90-1)
- Scope (90-2)
- Code Arrangement (90-3)
- Enforcement (90-4)
- Mandatory Rules and Explanatory Material (90-5)
- Formal Interpretations (90-6)
- Examination of Equipment for Safety (90-7)
- Wiring Planning (90-8)

Once you are familiar with Articles 90, 100, and 110, you can move on to the rest of the NEC. There are several key sections you will use often in servicing electrical systems. Let's discuss each of these important sections.

Wiring Design and Protection: Chapter 2 of the NEC discusses wiring design and protection, the information electrical technicians need most often. It covers the use and identification of grounded conductors, branch circuits, feeders, calculations, services, overcurrent protection and grounding. This is essential information for any type of electrical system, regardless of the type.

Chapter 2 is also a "how-to" chapter. It explains how to provide proper spacing for conductor supports, how to provide temporary wiring and how to size the proper grounding conductor or electrode. If you run into a problem related to the design/installation of a conventional electrical system, you can probably find a solution for it in this chapter.

Wiring Methods and Materials: Chapter 3 has the rules on wiring methods and materials. The materials and procedures to use on a particular system depend on the type of building construction, the type of occupancy, the location of the wiring in the building, the type of atmosphere in the building or in the area surrounding the building, mechanical factors and the relative costs of different wiring methods (Fig. 11.5).

The provisions of this article apply to all wiring installations except remote control switching (Article 725), low-energy power circuits (Article 725), signal systems (Article 725), communication systems and conductors (Article 800) when these items form an integral part of equipment, such as motors and motor controllers.

Four basic wiring methods are used in most modern electrical systems. Nearly all wiring methods are a variation of one or more of these four basic methods:

- Sheathed cables of two or more conductors, such as nonmetallic-sheathed cable and armored cable (Articles 330 through 339)
- Raceway wiring systems, such as rigid steel conduit and electrical metallic tubing (Articles 342 to 358)

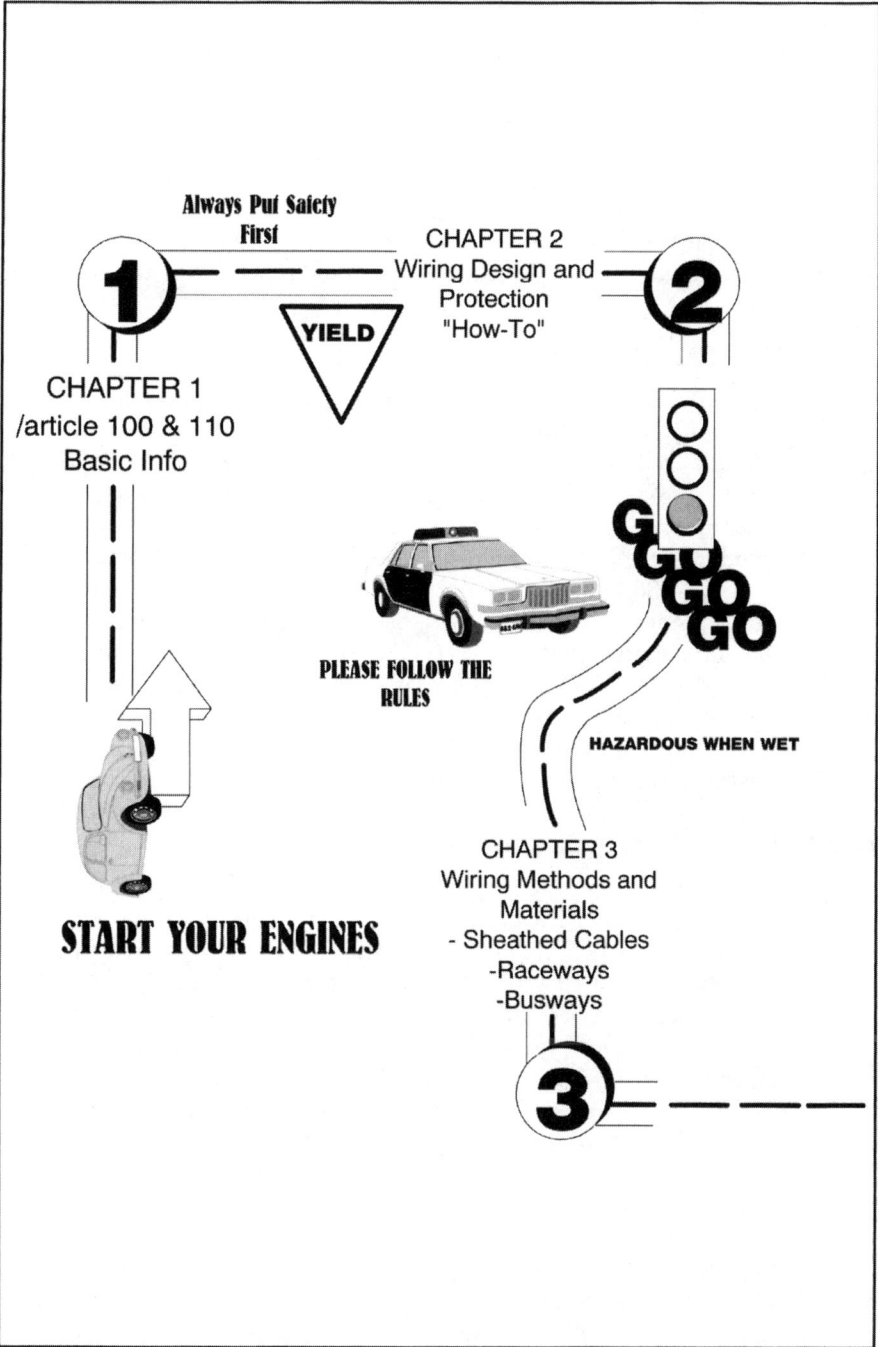

Figure 11.5 Become thoroughly familiar with Articles 90, 100, and 110 before continuing.

■ Busways (Article 364)

■ Cabletray (Article 318)

Article 310 in Chapter 3 gives a complete description of all types of electrical conductors. Electrical conductors come in a wide range of sizes and forms. Be sure to check the working drawings and specifications to see what sizes and types of conductors are required for a specific job. If conductor type and size are not specified, choose the most appropriate type and size that meets standard NEC requirements. If there is still doubt, check with the electric system designer.

Articles 318 through 384 give rules for raceways, boxes, cabinets, and raceway fittings. Outlet boxes vary in size and shape, depending on their use, the size of the raceway, the number of conductors entering the box, the type of building construction and atmospheric conditions of the areas. Chapter 3 should answer most questions on the selection and use of these items.

The NEC does not describe all types and sizes of outlet boxes in detail. But manufacturers of outlet boxes have excellent catalogs illustrating all of their products. Collect these catalogs. They are essential to your work.

Article 380 covers the switches, push buttons, pilot lamps, receptacles, and convenience outlets you will use to control electrical circuits or to connect portable equipment to electric circuits. Again, get the manufacturers' catalogs on these items. They will provide you with detailed descriptions of each of the wiring devices.

Article 384 covers switchboards and panelboards, including their location, installation methods, clearances, grounding, and overcurrent protection.

Equipment for general use

Chapter 4 of the NEC begins with the use and installation of flexible cords and cables, including the trade name, type letter, wire size, number of conductors, conductor insulation, outer covering and a description of their uses. The chapter also includes fixture wires, again giving the trade name, type letter, and other important details.

Article 410 on lighting fixtures is especially important. It gives installation procedures for fixtures in specific locations. For example, it covers fixtures near combustible material and fixtures in closets. The NEC does not describe how many fixtures will be needed in a given area to provide a certain amount of illumination.

Article 430 covers electric motors, including mounting the motor and making electrical connections to it. Motor controls and overload protection are also covered.

Articles 440 through 460 cover air conditioning and heating equipment, transformers and capacitors.

Article 480 gives most requirements related to battery-operated electrical systems. Storage batteries are seldom thought of as part of a conventional electrical system, but they often provide standby emergency lighting service.

They can also supply power to security systems that are separate from the main ac electrical system (Fig. 11.6).

Special occupancies

Chapter 5 of the NEC covers special occupancy areas. These are areas where the sparks generated by electrical equipment could cause an explosion or fire. The hazard is typically created by the presence of volatile liquids or fine particulates in the area. Commercial garages, aircraft hangers, and service stations are typical special occupancy locations.

Articles 500 to 501 cover the different types of special occupancy areas where an explosion is possible. The atmospheric groups were established to make it easy to test and approve equipment for various types of uses.

Articles 501-4, 502-4, and 503-3 cover the installation of explosion-proof wiring. An explosion-proof system is designed to prevent the ignition of a surrounding explosive atmosphere when arcing occurs within the electrical system.

The three main classes of special-occupancy location are:

- Class I (Article 501): Areas containing flammable gases or vapors in the air. Class-I areas include paint spray booths and dyeing plants, where hazardous liquids are used and gas generator rooms.

- Class II (Article 502): Areas where combustible dust is present, such as grain-handling and storage plants, dust and stock collector areas, and sugar pulverizing plants. These are areas where, under normal operating conditions, there may be enough combustible dust in the air to produce explosive or ignitable mixtures.

- Class III (Article 503): Areas that are hazardous because of the presence of easily ignitable fibers or flyings in the air, although not in large enough quantity to produce ignitable mixtures. Class-III locations include cotton mills, rayon mills, and clothing manufacturing plants.

Article 511 and 514 cover garages and similar locations where volatile or flammable liquids are used. Although these areas are not always considered critically hazardous locations, there may be enough danger to require special precautions in the electrical installation. In these areas, the NEC requires that volatile gases be confined to an area not more than 4 feet above the floor. So, in most cases, conventional raceway systems are permitted above this level. If the area is judged critically hazardous, explosion-proof wiring (including seal-offs) might be required (Fig. 11.7).

Article 520 regulates theaters and similar occupancies in which fire and panic can cause hazards to life and property. Drive-in theaters do not present the same hazards as enclosed auditoriums. But the projection rooms and adjacent areas must be properly ventilated and wired for the protection of operating personnel and others using the area.

CHAPTER 5
Special Occupancies
511-514 Commercial
Garages, Hangars, etc.
Explosion Hazards
520 Theaters

Class I   Flammable Gases, Vapors
Class II Combustible Dust
Class III Ignitable Air-borne Fibers

KEEP ON TRUCKIN

CHAPTER 6
Special Equipment
Seldom Used in
Residential Design
600 Electrified Signs
620 Elevators
630 Electrical Welding
640 Sound-
Recording Equipment

Rough-in drawings
of equipment

CHAPTER 4
Flexible Cords
and Cables
1410 Lighting Fixtures
1430 Motors
1440-1446 :
Air-Conditioning
Heating Equipment
Transformers
Capacitors
4800 Batteries

GO TO START
If Applicable
Study This
Chapter Well in
Advance of
Starting Work

CAREFUL
OF
SHOCK
HAZARDS

CHAPTER 7
Emergency
Electrical Systems

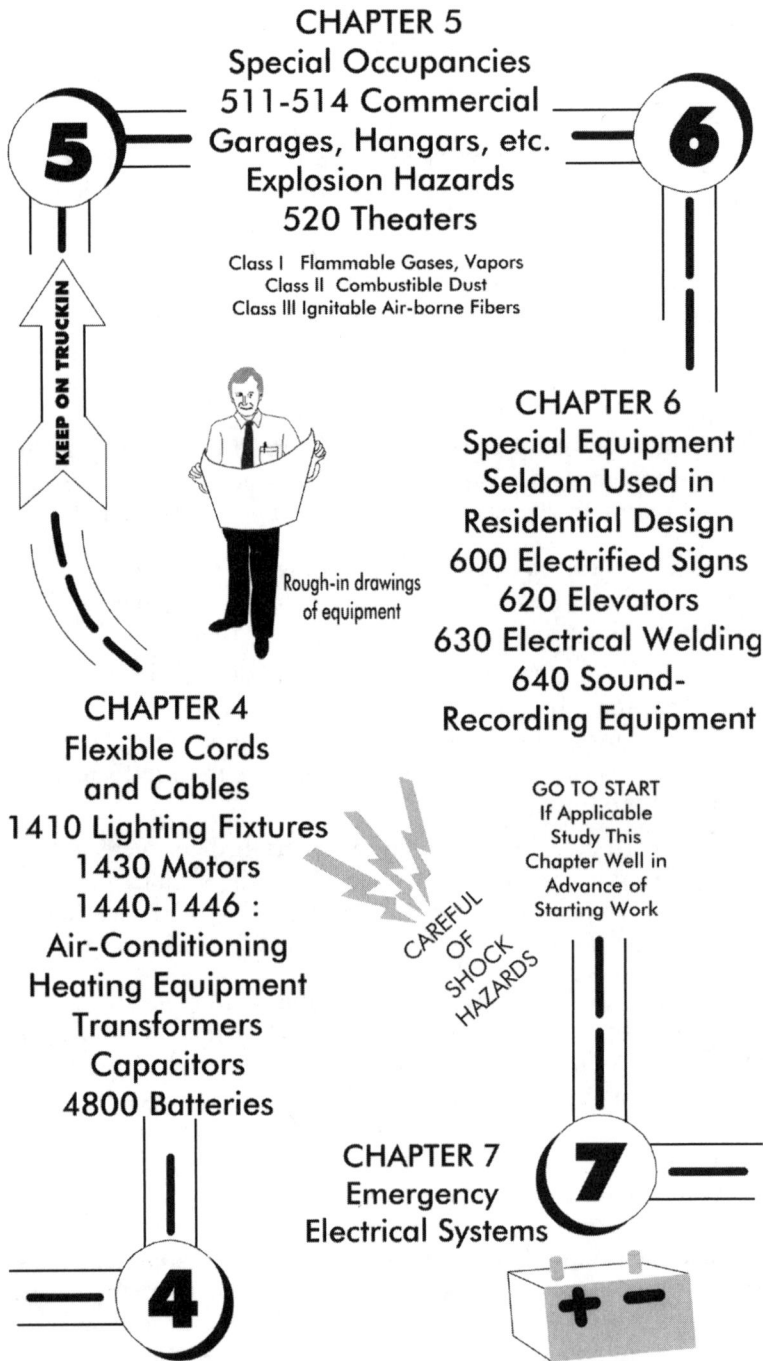

Figure 11.6   NEC Chapters 1 through 4 apply except as amended by Chapters 5, 6, and 7 for some conditions.

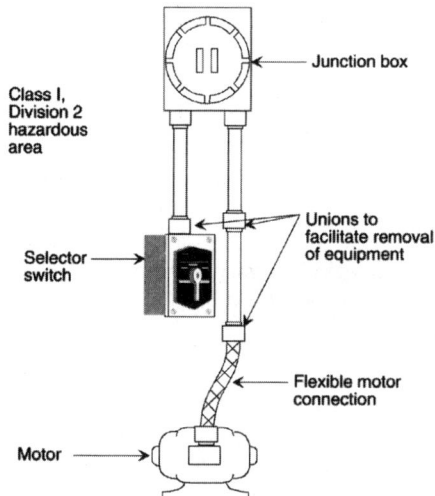

Figure 11.7 Explosion-proof fittings must be used for electrical wiring in most hazardous areas.

Chapter 5 also covers residential storage garages, aircraft hangars, service stations, bulk storage plants, health-care facilities, mobile homes and parks, and recreation vehicles and parks.

When security technicians are installing systems in hazardous locations, extreme caution must be used. You might be working with only 12 or 24 V, but a spark caused by, for example, an improper connection can set off a violent explosion. You might have already witnessed a low-voltage explosion in the common automotive battery. Although only 12 V dc is present, if a spark occurs near the battery and battery gases are leaking through the battery housing, chances are the battery will explode with a report similar to a shotgun firing.

When installing security systems in Class-I, Division-1 locations, explosion-proof fittings are required and most electrical wiring must be enclosed in rigid steel conduit (pipe).

Special equipment

Residential electrical workers will seldom need to refer to the Articles in Chapter 6 of the NEC, but the items in Chapter 6 are frequently encountered by commercial and industrial electrical workers.

Article 600 covers electric signs and outline lighting. Article 610 applies to cranes and hoists. Article 620 covers the majority of the electrical work involved in the installation and operation of elevators, dumbwaiters, escalators, and moving walks. The manufacturer is responsible for most of this work. The electrician usually just furnishes a feeder terminating in a disconnect means in the bottom of the elevator shaft. The electrician might also be responsible for a lighting circuit to a junction box midway in the elevator shaft for connecting the elevator cage lighting cable and exhaust fans. Articles in Chapter 6 of the NEC give most of the requirements for these installations.

Article 630 regulates electric welding equipment. It is normally treated as a type of industrial power equipment requiring a special power outlet. But there are special conditions that apply to the circuits supplying welding equipment. These are outlined in detail in Chapter 6 of the NEC.

Article 640 covers wiring for sound-recording and similar equipment. This type of equipment normally requires low-voltage wiring. Special outlet boxes or cabinets are usually provided with the equipment. But some items might be mounted in, or on, standard outlet boxes. Some sound-recording electrical systems require direct current, supplied from rectifying equipment, batteries or motor generators. Low-voltage alternating current comes from relatively small transformers connected on the primary side to a 120-V circuit within the building.

Other items covered in Chapter 6 of the NEC include: X-ray equipment (Article 660), induction and dielectric heat-generating equipment (Article 665), and machine tools (Article 670).

If you ever have work that involves Chapter 6, study the chapter before work begins. The preparation can save a lot of installation time. Here is another way to cut down on labor hours and prevent installation errors. Get a set of rough-in drawings of the equipment being installed. It is easy to install the wrong outlet box or to install the right box in the wrong place. Having a set of rough-in drawings can prevent those simple, but costly, errors.

Special Conditions

In most commercial buildings, the NEC and local ordinances require a means of lighting public rooms, halls, stairways and entrances. There must be enough light to allow the occupants to exit from the building if the normal building lighting is interrupted. Exit doors must be clearly indicated by illuminated exit signs.

Chapter 7 of the NEC covers the installation of emergency lighting systems. These circuits should be arranged so that they can automatically transfer to an alternate source of current, usually storage batteries or gasoline-driven generators, for their energy supply. As an alternative in some types of occupancies, you can connect them to the supply side of the main service, so disconnecting the main service switch would not disconnect the emergency circuits. See Article 700. NEC Chapter 7 also covers a variety of other equipment, systems and conditions that are not easily categorized elsewhere in the NEC.

Chapter 8 is a special category for wiring associated with electronic communications systems, including telephone and telegraph, radio and TV, fire and burglar alarms, and community antenna systems. This is one NEC chapter that the security technician must understand thoroughly. See Fig. 11.8 on the next page.

Using the NEC

Once you become familiar with the code through repeated usage, you will usually know where to look for a particular topic. While this chapter provides you

CHAPTER 9
TABLES AND EXAMPLES

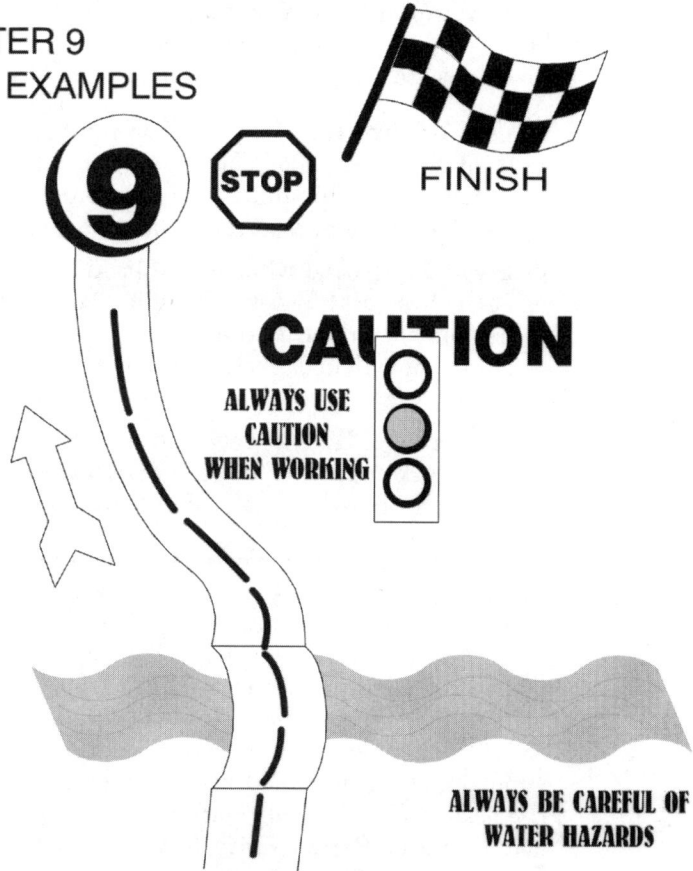

STOP

FINISH

CAUTION

ALWAYS USE
CAUTION
WHEN WORKING

ALWAYS BE CAREFUL OF
WATER HAZARDS

CHAPTER 8
Communications Systems
-Communications Circuits
-Radio & Television
-Community Antenna Television

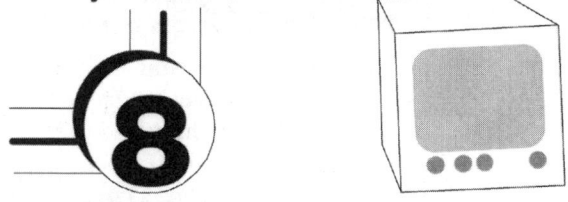

Figure 11.8   The main text of the NEC ends with Chapter 9—Tables and Examples.

with an initial familiarization of the NEC layout, much additional use of the book will be needed for you to feel comfortable with the book's content. Here's how to locate information on a specific subject.

1. Look through the Contents. You might spot the topic in a heading or sub-heading. If not, look for a broader, more general subject heading under which the specific topic might appear. Also look for related or similar topics. The Contents will refer you to a specific page number.

2. If you do not find what you're looking for in the Contents, go to the Index at the back of the book. This alphabetic listing is divided into different topics. You should locate the subject here. The index, however, will refer you to either an article or section number (not a page number) where the topic is listed.

3. If you cannot find the required subject in the Index, try to think of alternate names. For example, instead of *wire*, look under conductors; instead of *outlet box*, look under *boxes, outlet*, etc.

The NEC is not an easy book to read and understand at first. In fact, seasoned electrical workers and technicians sometimes find it confusing. Basically, it is a reference book written in a legal, contract-type language and its content does assume prior knowledge of most subjects listed. Consequently, you will sometimes find the NEC frustrating to use because terms aren't always defined, or some unknown prerequisite knowledge is required. To minimize this problem, it is recommended that you obtain one of the several NEC supplemental guides that are designed to explain and clarify the NEC. One of the best is *The National Electrical Code Handbook*, available from the NFPA, Batterymarch Park, Quincy, MA 02269 or from your local book store.

Practical application

Assume that you are installing a 120-V outlet box to provide the power supply for a surveillance camera in a commercial office. The owner wants the outlet box surface-mounted and located behind a curtain in back of their sliding glass patio doors. To determine if this is a NEC violation or not, follow these steps:

1. Turn to the Contents of the NEC book, which begins on page 70 V.

2. Find the chapter that would contain information about the general application you are working on. For this example, Chapter 4 "Equipment for General Use" should cover track lighting.

3. Now look for the article that fits the specific category you are working on. In this case, Article 410 covers lighting fixtures, lampholders, lamps, and receptacles.

4. Next locate the NEC Section within the NEC Article 410 that deals with the specific application. For this example, refer to Part R "Lighting Track."

5. Turn to the page listed. The 1993 NEC gives page 350.

6. Read NEC Section 410-100, Definition to become familiar with track lighting. Continue down the page with NEC Section 410-101 and read the information contained therein. Notice that paragraph (c) under NEC Section 410-101 states the following:

> (c) Locations Not Permitted. Lighting track shall not be installed (1) where subject to physical damage; (2) in wet or damp locations; (3) where subject to corrosive vapors; (4) in storage battery rooms; (5) in hazardous (classified) locations; (6) where concealed; (7) where extended through walls or partitions; (8) less than 5 feet above the finished floor except where protected from physical damage or track operating at less than 30 volts RMS open-circuit voltage.

7. Read NEC Section 410-101, paragraph (c) carefully. Do you see any conditions that would violate any NEC requirements if the track lighting is installed in the area specified? In checking these items, you will probably note condition (6), "where concealed." Since the track lighting is to be installed behind a curtain, this sounds like an NEC violation. But let's check further.

8. Let's get an interpretation of the NEC's definition of *concealed.* Therefore, turn to Article 100 "definitions" and find the main term *concealed.* It reads as follows:

> Concealed: Rendered inaccessible by the structure or finish of the building....

9. After reading the NEC's definition of *concealed,* although the track lighting may be out of sight (if the curtain is drawn), it will still be readily accessible for maintenance. Consequently, the track lighting is really not concealed according to the NEC definition.

When using the NEC to determine correct electrical-installation requirements, remember that you will nearly always have to refer to more than one section. Sometimes the NEC itself refers the reader to other articles and sections. In some cases, the user will have to be familiar enough with the NEC to know what other NEC sections pertain to the installation at hand. It's a confusing situation to say the least, but time and experience in using the NEC will make using it much easier to navigate.

Now let's take another example to further acquaint you with navigating the NEC.

Suppose you are installing Type SE (service-entrance) cable on the side of a home. You know that this cable must be secured, but you aren't sure of the spacing between cable clamps. To find out this information, use the following procedure:

1. Look in the NEC Table of Contents and follow down the list until you find an appropriate category.

2. Article 230 under Chapter 3 will probably catch your eye first, so turn to the page where Article 230 begins in the NEC.

3. Glance down the section numbers, 230-1, Scope, 230-2, Number of Services, etc. until you come to Section 230-51, "Mounting Supports." Upon reading this section, you will find in paragraph (a) "Service-Entrance Cables" that "Service-entrance cable shall be supported by straps or other approved means within 12 inches (305 mm) of every service head, gooseneck, or connection to a raceway or enclosure and at intervals not exceeding 30 inches (762 mm)."

After reading this section, you will know that a cable strap is required within 12 inches of the service head and within 12 inches of the meter base. Furthermore, the cable must be secured in between these two termination points at intervals not exceeding 30 inches.

Definitions

Many definitions of terms dealing with the code may be found in NEC Article 100. However, other definitions are scattered throughout the book under their appropriate category. For example, the term *lighting track,* as discussed previously, is not listed in Article 100. The term is listed under Section 410-100 and reads as follows:

Lighting track is a manufactured assembly designed to support and energize lighting fixtures that are capable of being readily repositioned on the track. Its length may be altered by the addition or subtraction of sections of track.

Regardless of where the definition is located—in Article 100 or under the appropriate NEC section elsewhere in the book—the best way to learn and remember these definitions is to form a mental picture of each item or device as you read the definition. For example, turn to page 70-5 of the 1993 NEC and under Article 100 "Definitions," scan down the page until you come to the term *Attachment Plug (Plug Cap) (Cap).* After reading the definition, you will probably have already formed a mental picture of attachment plugs. See Figure 11.9 for some of the more common attachment plugs.

Once again, scan through the definitions until the term *Appliance* is found. Read the definition and then try to form a mental picture of what appliances look like. Some of the more common appliances appear in Figure 11.10. They should be familiar to everyone.

Each and every term listed in the NEC should be understood. Know what the item looks like and how it is used on the job. If a term is unfamiliar, try other reference books such as manufacturers' catalogs for an illustration of the item. Then research the subject further to determine its purpose in electrical systems. Once you are familiar with all the common terms and definitions found in the NEC, navigating through the NEC (and understanding what you read) will be much easier.

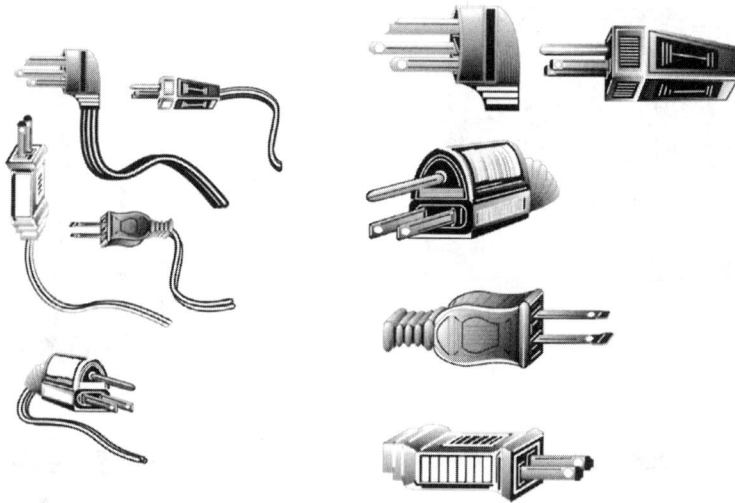

Figure 11.9 Attachment plugs in common use.

Testing Laboratories

There are many definitions included in Article 100. You should become familiar with the definitions. Because a copy of the latest NEC is compulsory for any type of electrical wiring, there is no need to duplicate them here. However, here are two definitions that you should become especially familiar with:

- *Labeled* Equipment or materials to which has been attached a label, symbol, or other identifying mark of an organization acceptable to the authority having jurisdiction and concerned with product evaluation, which maintains periodic inspection of production of labeled equipment or materials, and by whose labeling the manufacturer indicates compliance with appropriate standards or performance in a specified manner.

- *Listed* Equipment or materials included in a list published by an organization acceptable to the authority having jurisdiction and concerned with product evaluation, that maintains periodic inspection of production of listed equipment or materials, and whose listing states either that the equipment or material meets appropriate designated standards or has been tested and found suitable for use in a specified manner. Besides installation rules, you will also have to be concerned with the type and quality of materials that are used in electrical wiring systems. Nationally recognized testing laboratories (Underwriters' Laboratories, Inc. is one) offer product safety certification. They establish and operate product safety certification programs to make sure that items produced under the service are safeguarded against reasonable foreseeable risks. Some of these organizations maintain a worldwide network of field representatives who make unannounced visits to manufacturing facilities to countercheck products bearing their "seal of approval" (Fig. 11.11).

Window
air conditioner

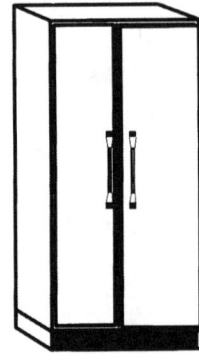

Freezer

Washer

Dryer

Electric range

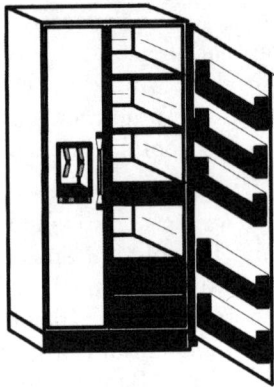

Refrigerator

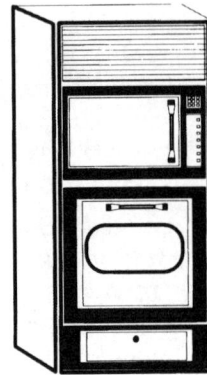

Wall-mounted
microwave and
oven

Figure 11.10   Typical appliances.

Figure 11.11   UL label.

However, proper selection, overall functional performance and reliability of a product are factors that are not within the basic scope of UL activities.

To fully understand the NEC, it is important to understand the organizations that govern it.

Nationally Recognized Testing Laboratory (NRTL)

Nationally Recognized Testing Laboratories are product safety certification laboratories. They establish and operate product safety certification programs to ensure that items produced under the service are safeguarded against reasonable foreseeable risks. NRTL maintains a worldwide network of field representatives who make unannounced visits to factories to countercheck products bearing the safety mark.

National Electrical Manufacturers Association (NEMA)

The National Electrical Manufacturers Association was founded in 1926. It consists of companies that manufacture equipment used for generation, transmission, distribution, control, and utilization of electric power. The objectives of NEMA are to maintain and improve the quality and reliability of products; to ensure safety standards in the manufacture and use of products; to develop product standards covering such matters as naming, ratings, performance, testing, and dimensions. NEMA participates in developing the NEC and the National Electrical Safety Code and advocates their acceptance by state and local authorities.

National Fire Protection Association (NFPA)

The NFPA was founded in 1896. Its membership is drawn from the fire service, business and industry, health care, educational and other institutions, and individuals in the fields of insurance, government, architecture, and engineering. The duties of the NFPA include:

- Developing, publishing, and distributing standards prepared by approximately 175 technical committees. These standards are intended to minimize the possibility and effects of fire and explosion.
- Conducting fire safety education programs for the general public.
- Providing information on fire protection, prevention, and suppression.
- Compiling annual statistics on causes and occupancies of fires, large-loss fires (over 1 million dollars), fire deaths, and firefighter casualties.
- Providing field service by specialists on electricity, flammable liquids and gases, and marine fire problems.
- Conducting research projects that apply statistical methods and operations research to develop computer modes and data-management systems.

The role of testing laboratories

Testing laboratories are an integral part of the development of the code. The NFPA, NEMA, and NRTL all provide testing laboratories to conduct research into electrical equipment and its safety. These laboratories perform extensive testing of new products to be sure they are built to code standards for electrical and fire safety. These organizations receive statistics and reports from agencies all over the United States concerning electrical shocks and fires and their causes. When the groups see trends developing that relate to equipment and dangerous situations or circumstances, the equipment will be specifically targeted for research.

Codes and Ordinances Affecting Security Systems

Electrical systems falling under the heading of "signaling" include such categories as security and fire-alarm systems—employing a wide variety of techniques, often involving special types of equipment and materials designed for specific applications. Many of these systems operate on low-voltage circuits, but are installed similarly to conventional electrical circuits for light and power. In all cases, however, when designing or installing systems for use in buildings, the installations must conform to applicable NEC requirements.

Several NEC Sections apply directly to signaling circuits. Some of these sections are:

- Alarm systems for health-care facilities, NEC Section 517-32(c) and 518-42(c).
- Burglar alarms, NEC Sections 230-82 Exception 5, and 230-94, Exception 4.
- Fire alarms, NEC Sections 230-82, Exception 5, and 230-94, Exception 4.
- Fire protection, NEC Article 760.
- Remote-control signaling, NEC Article 725.

While reviewing these NEC sections, remember that all security systems have three functions in common:

- Detection
- Control
- Annunciation (or alarm) signaling

Many systems incorporate switches or relays that operate because of entry, movement, pressure, infrared-beam interruption, etc. The control senses operation of the detector with a relay and produces an output that may operate a bell, siren, silent alarm—such as a telephone dialer to law enforcement agencies—and similar devices. The controls frequently contain ON/OFF switches, test meters, time delays, power supplies, standby batteries, and terminals for connecting the system together. The control output usually provides power to operate signaling devices or switch contacts for silent alarms.

One of the simplest and most common electric signals is the residential doorchime. These simple systems contain a low-voltage power source, one or more pushbuttons, wire, and a set of chimes.

The wiring diagram in Fig. 11.12 illustrates a typical two-note chime controlled at two locations. One button, at the main entrance, will sound two notes when pushed; the other button, at the rear door, will sound only one note when touched.

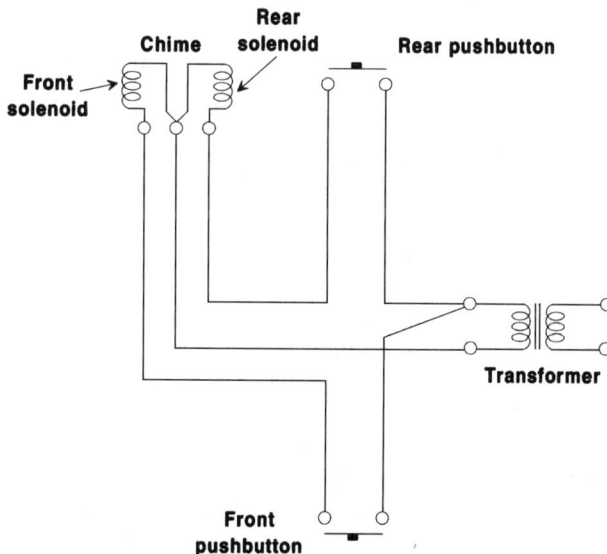

Figure 11.12 Typical two-note chime signaling circuit.

Signal circuit components

Wire sizes for the majority of low-voltage systems range from #22 to #18 AWG. However, where larger-than-normal currents are required or when the distance between the outlets is long, it might be necessary to use wire sizes larger than specified to prevent excessive voltage drop.

At least some parts of almost all security systems require line-voltage (120/240 V) connections. The control panel, for example, is almost always supplied with line voltage. These circuits are installed exactly like conventional electrical circuits for light and power, and all NEC regulations must be followed.

There are also many security components besides the main control panel that require 120-V ac power. A few are:

Direct-Wired, Single-Station Smoke Detectors: These units are 120-volt ac powered with two wires for connection to the electrical system. They usually include a power indicator lamp with a push-to-test button, but normally do not have other extra features. They are designed to be used where only one detector is desired or where units are not to be interconnected.

Direct-Wired Multiple Stations: These devices are 120-V ac powered and designed to be interconnected so that when one unit senses a fire and alarms, all units will alarm. Two types of interconnect methods are used for the stations. One method utilizes a three-wire (hot, neutral, and interconnect) interconnection at 120-V ac. This system must meet NEC requirements for 120-V ac wiring and all units in this system must be installed on the same 120-V ac circuit.

The second method incorporates a transformer in the unit to allow interconnection at a low voltage. This will allow the interconnect wiring to be low-voltage cable (signaling cable). NEC Article 725 specifies this voltage to be not more than 30 volts. Also, this type of interconnect allows the units in the system to be installed on different 120-V ac circuits, if desired. Only two wires are required for interconnection. Although the 120-V ac interconnect units are somewhat less expensive than the low-voltage interconnect units, the overall installation costs with the low-voltage units will be less.

Direct-Wired Multiple Station with Auxiliary Contacts: These 120-volt ac powered interconnect units also include a set of auxiliary contacts for connection to remote devices. The auxiliary contacts can be used to operate remote horns, lights, or shut off exhaust or ventilating fans. Such units have an additional pair of leads for the auxiliary contacts.

Battery-Powered Single Station: Most battery-powered units are single station devices powered by a 9-volt battery. UL requires that batteries last at least one year under normal use and that such units produce an audible signal indication of a low-battery condition for seven days. Use of batteries other than ones designated by the manufacturer can cause the device to fail to operate, or operate improperly. This is because of the circuitry of the unit being designed to match the performance of specified batteries.

UL Listings: Smoke and fire detectors are usually tested for compliance with UL Standard No. 217. However, periodic changes are made in this standard and manufacturers are required to resubmit their units in order to maintain their listing.

Installation: A specific mounting position for smoke and fire detectors is necessary to permit the proper entrance of aerosols to activate the detector. Most units are designed for wall or ceiling mounting and recommended positions are indicated in the instruction manual or sheet supplied with the unit. Mounting in a position not specified can make the unit less sensitive or the response unpredictable.

The locations and quantity of units required in an installation can vary, depending on the authority involved in the installation. However, almost all require one detector outside sleeping areas and many authorities are requiring interconnect units in multiple-floor dwellings.

It is further recommended, if not required, that ionization type units not be located in the kitchen or in any area where they will be affected by products of combustion.

The following are some of the authorities that regulate the number of placement of smoke and fire detectors:

- Building Officials and Code Administrators (BOCA)
- Uniform Building Code (UBC)
- Federal Housing Administration (FHA – HUD)
- National Fire Protection Association (NFPA)

In addition, many state and local ordinances now also have requirements for smoke and fire detectors.

According to a study by the NFPA, residential fires kill more than half of all the people who die in fires. It has also been established that fatal fires usually occur when the family is asleep. Studies have also shown that 85 percent of all cases of death could have been prevented if smoke and fire detectors had been installed in accordance with recommended standards.

The following illustrations, Figs. 11.13 to 11.20, summarize the NEC requirements for NEC Article 725. Please consult the actual code book for details.

Summary

The National Electrical Code specifies the minimum provisions necessary for protecting people and property from hazards arising from the use of electricity and electrical equipment. Anyone involved in any phase of the electrical industry must be aware of how to use and apply the code on the job. Using the NEC will help you install and maintain the electrical security equipment and systems that you come into contact with safely.

The NEC is composed of the following components:

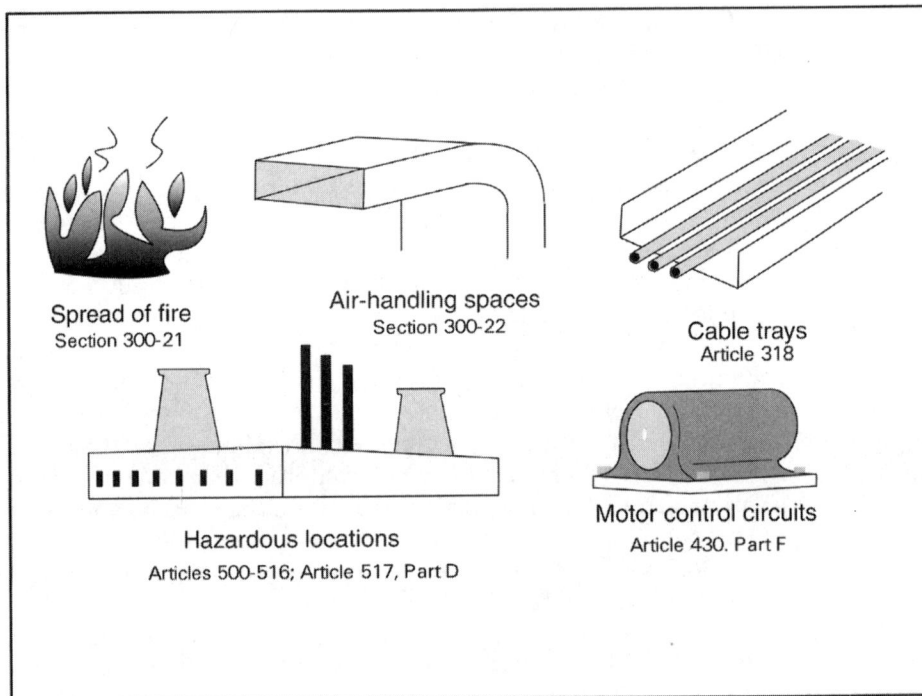

Figure 11.13　Applicable locations and other NEC articles that should be consulted.

Appendix: Appendix A includes material extracted from other NFPA documents. Appendix B is not part of the requirements of the NEC and contains additional material for informational purposes only. Appendix A and Appendix B are located at the end of the code book.

Article: Beginning with Article 90—Introduction, and ending with Article 820—Community Antenna Television and Radio Distribution Systems, the NEC Articles are the main topics in the code book.

Chapter: The NEC includes nine chapters. Chapter 1 "General," Chapter 2 "Wiring and Protection," Chapter 3 "Wiring Methods and Materials," Chapter 4 "Equipment for General Use," Chapter 5 "Special Occupancies," Chapter 6 "Special Equipment," Chapter 7 "Special Conditions," Chapter 8 "Communications Systems," and Chapter 9 "Tables and Examples." The chapters form the broad structure of the NEC.

Contents: Located among the first pages of the code book, the contents section provides a complete outline of the chapters, articles, parts, tables, and examples. The contents section, used with the index, provides excellent direction for locating answers to electrical problems and questions.

Diagrams and Figures: Diagrams and figures appear in the NEC to illustrate the relationship of articles and parts of the NEC. For example, Diagram 230-1,

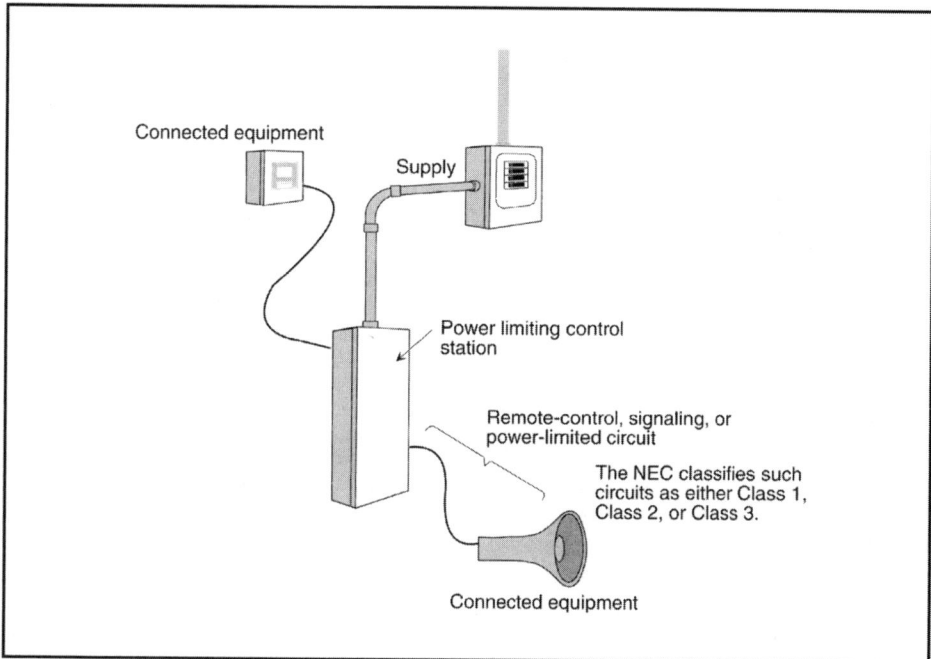

Figure 11.14 Summary of NEC Section 725-3—Classifications.

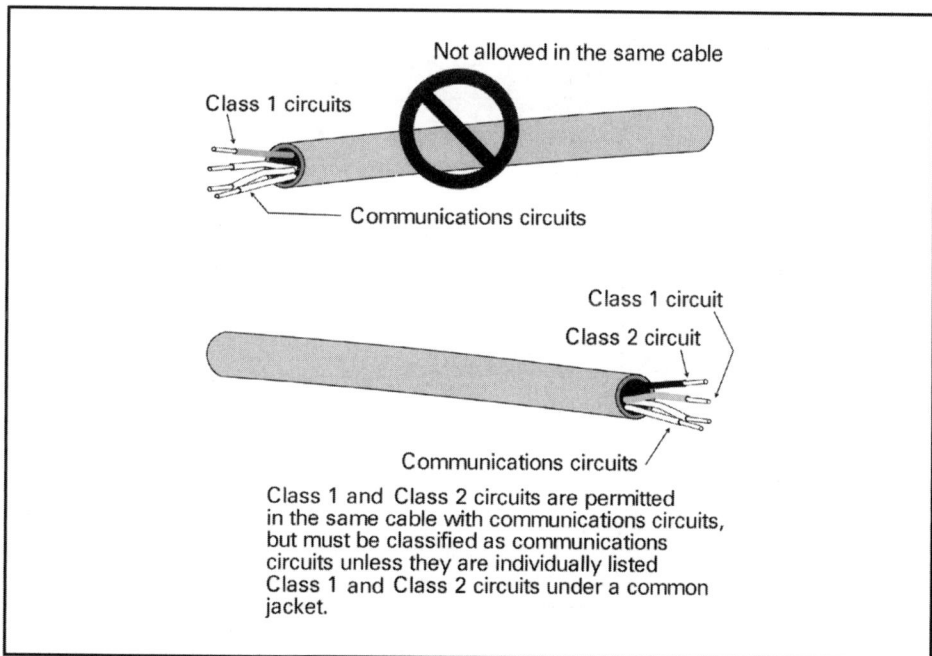

Figure 11.15 Summary of NEC Section 725-5—Communications Cables.

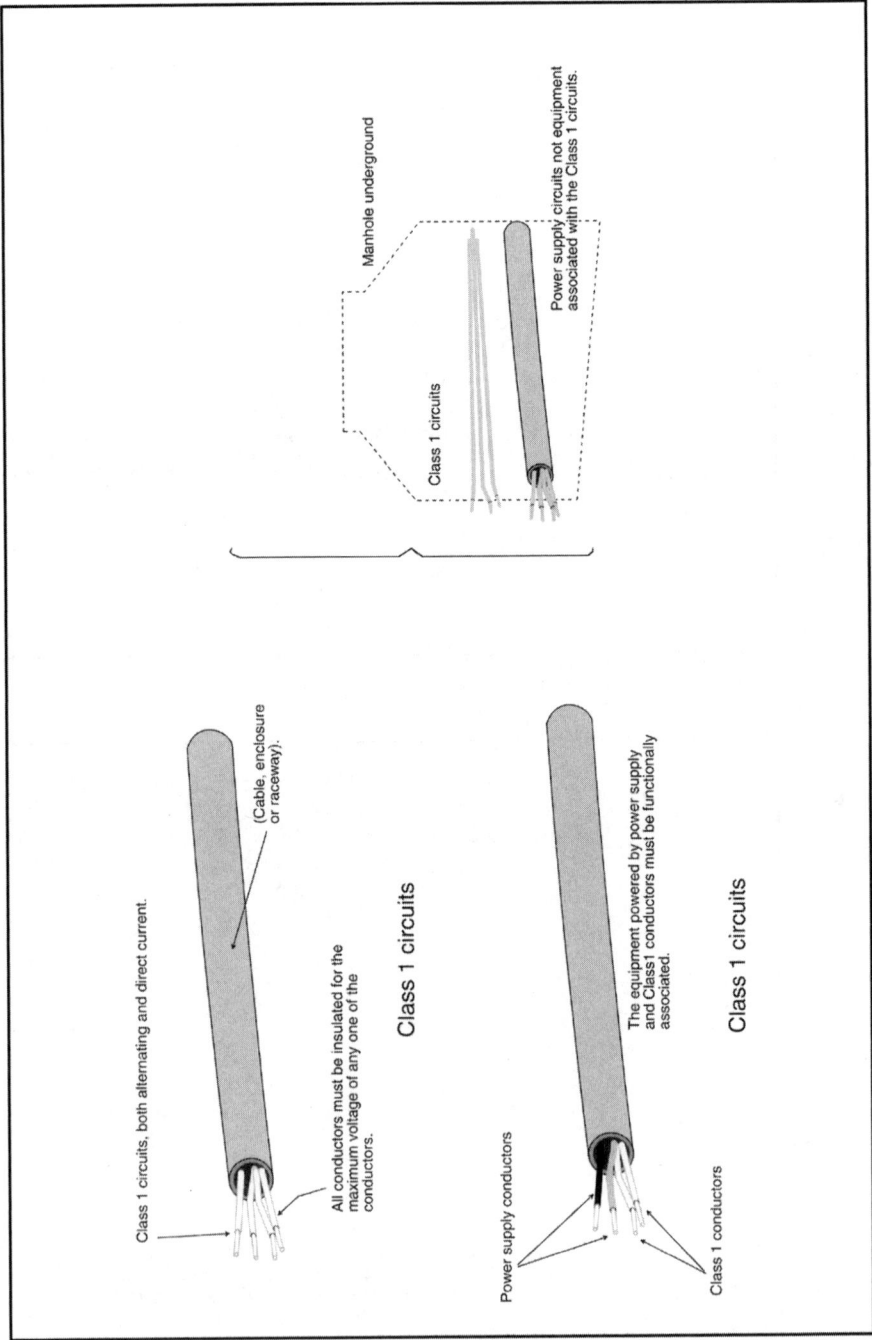

Class 1 circuits, both alternating and direct current.

(Cable, enclosure or raceway).

All conductors must be insulated for the maximum voltage of any one of the conductors.

Class 1 circuits

The equipment powered by power supply and Class1 conductors must be functionally associated.

Power supply conductors

Class 1 conductors

Class 1 circuits

Manhole underground

Class 1 circuits

Power supply circuits not equipment associated with the Class 1 circuits.

Figure 11.16   Summary of NEC Section 725-15.

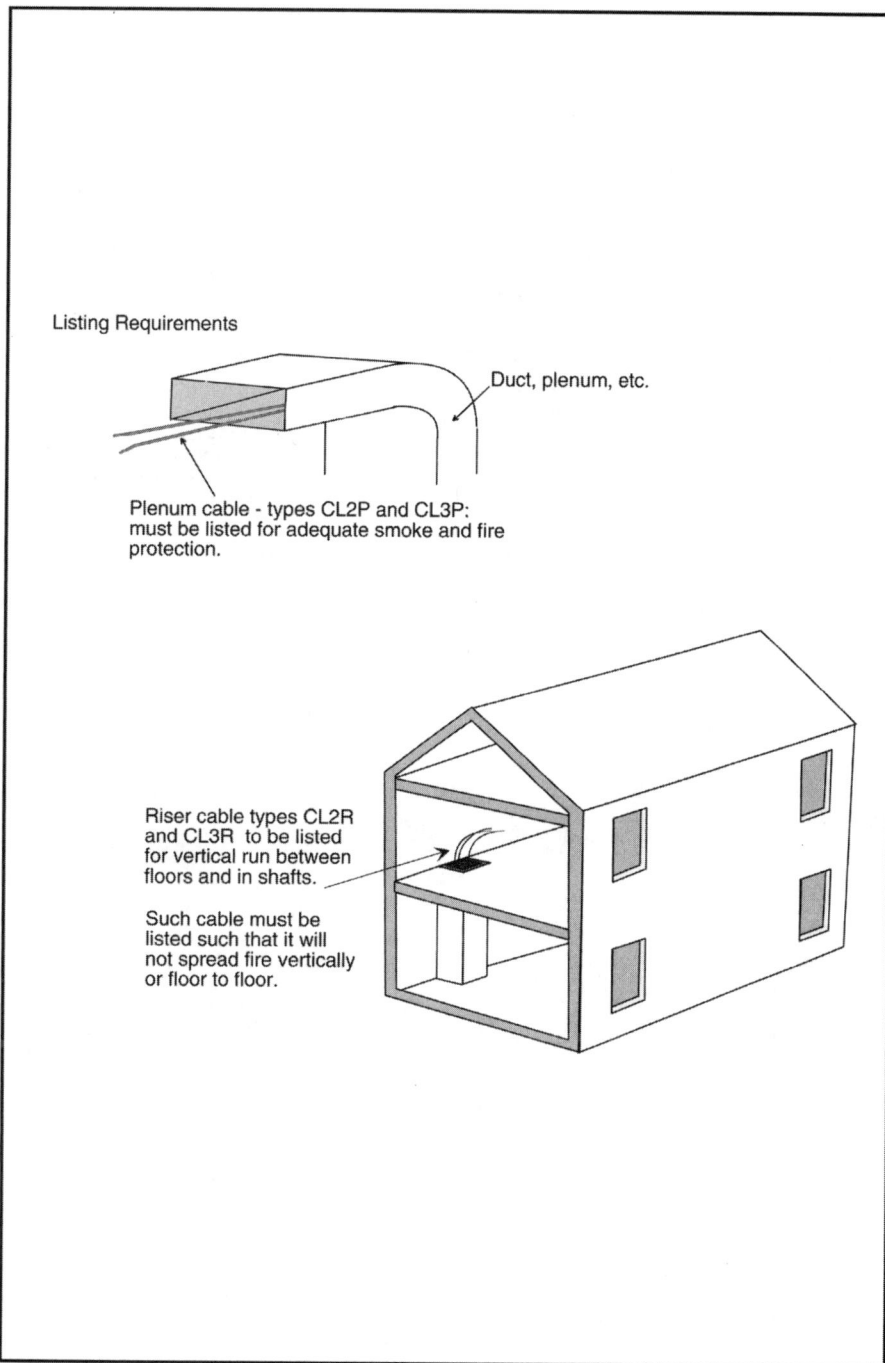

Listing Requirements

Duct, plenum, etc.

Plenum cable - types CL2P and CL3P: must be listed for adequate smoke and fire protection.

Riser cable types CL2R and CL3R to be listed for vertical run between floors and in shafts.

Such cable must be listed such that it will not spread fire vertically or floor to floor.

Figure 11.17   Summary of NEC Section 725-51.

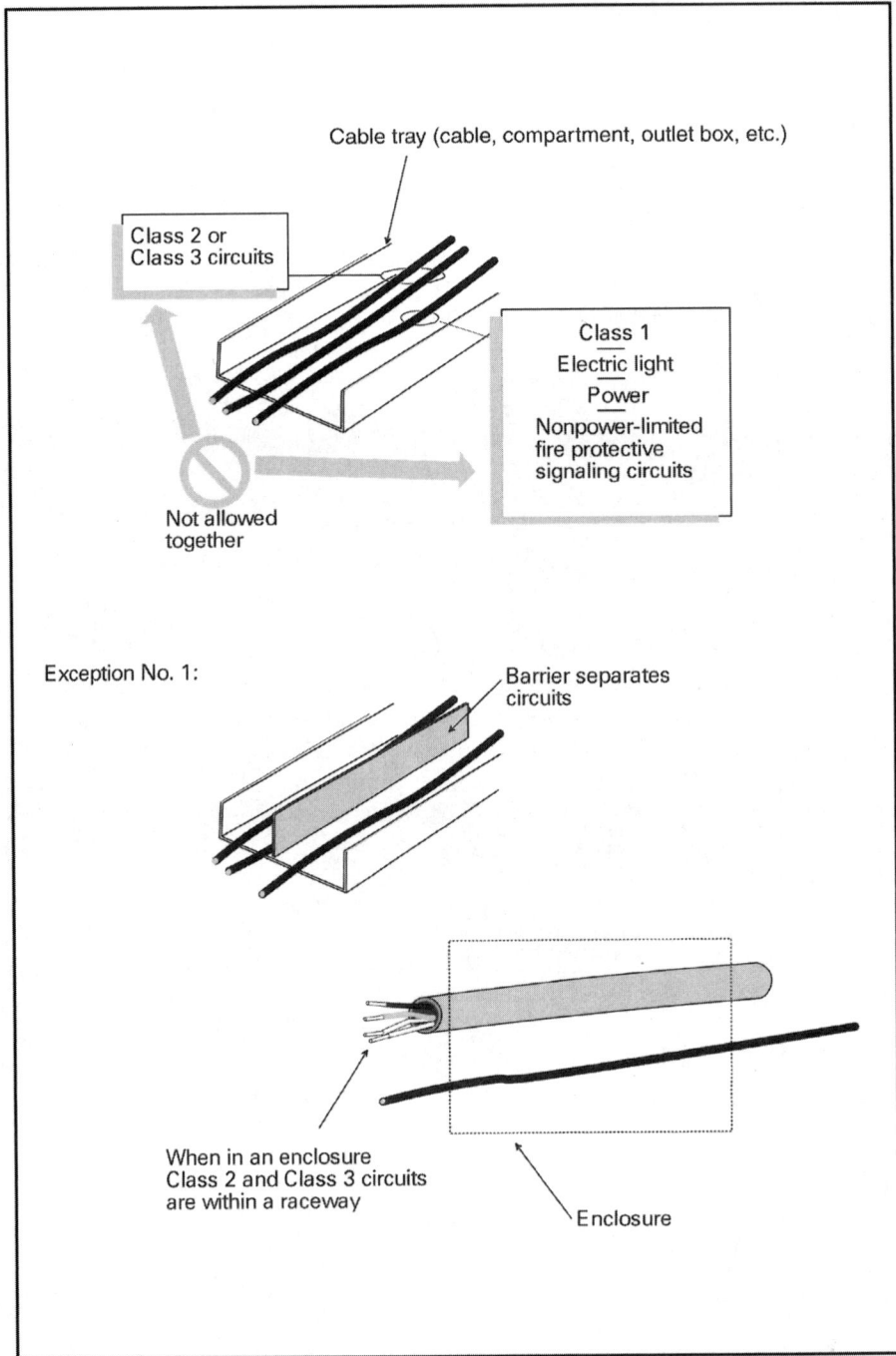

Cable tray (cable, compartment, outlet box, etc.)

Class 2 or
Class 3 circuits

Class 1

Electric light

Power

Nonpower-limited
fire protective
signaling circuits

Not allowed
together

Exception No. 1:

Barrier separates
circuits

When in an enclosure
Class 2 and Class 3 circuits
are within a raceway

Enclosure

Figure 11.18 Summary of NEC Section 725-52.

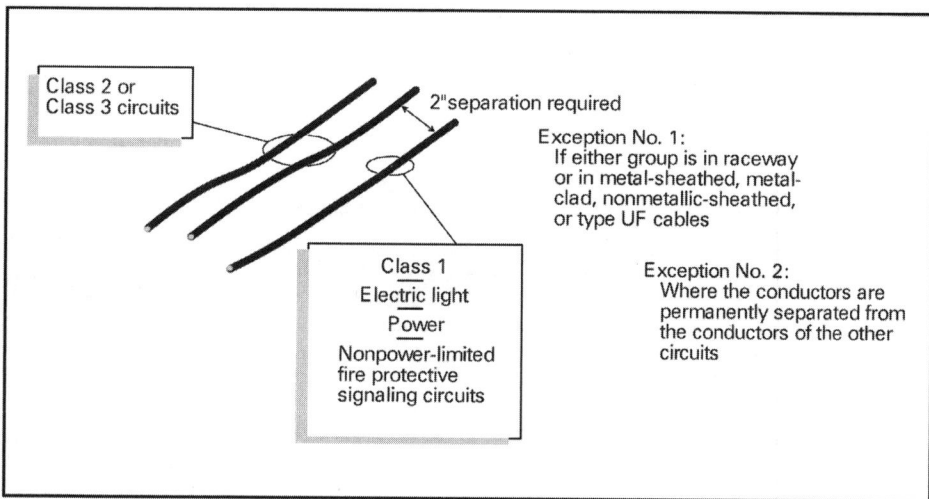

Figure 11.19 Summary of NEC Section 725-52 (*Cont.*).

Services, shows the relationship of articles and parts relating to the installation of electric services.

Examples: Examples in methods to perform for various types of buildings, feeders, and branch circuits.

Exceptions: Exceptions follow code sections and allow alternative methods, to be used under specific conditions, to the rule stated in the section.

FPN Fine Print Note: A Fine Print Note is defined in NEC Section 110-1; that is, explanatory material is in the form of Fine Print Notes (FPN).

Notes: Notes typically follow tables and are used to provide additional information to the tables or clarification of tables.

Part: Certain articles in the NEC are divided into parts. Article 220— Branch Circuit and Feeder is divided into Part A, B, C, and D.

Section: Parts and articles are divided into sections. A reference to a section will look like the following:

300-19, Supporting Conductors in Vertical Raceways or 300-19(a) Spacing Intervals—Maximum.

NEC sections provide more detailed information within NEC articles.

Table: Tables are located within chapters to provide more detailed information explaining code content. For example, Table 310-16 lists ampacities for insulated conductors for copper, aluminum, and copper-clad aluminum conductors with insulation types, sizes, temperature ratings, and ampacity correction factors. Such tables will prove invaluable for all security-system installations.

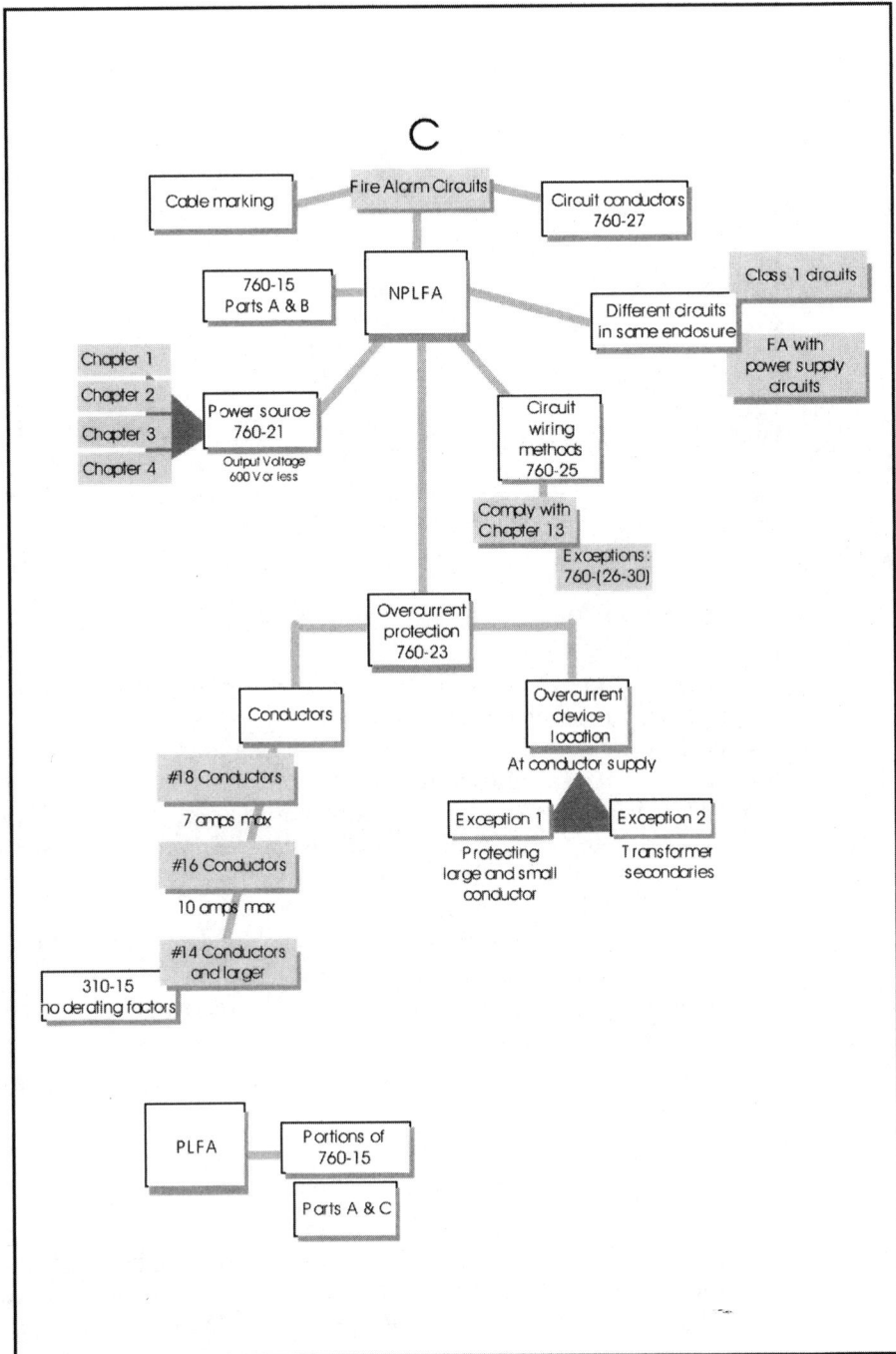

Figure 11.20 Summary of NEC fire-alarm regulations.

12

Print Reading

Anyone involved in the installation or maintenance of low-voltage systems will often encounter several types of drawings. This chapter introduces the layout of drawings and the symbols used on them. It further describes the various types of drawings and how to read them.

Architectural Drawings

In all large construction projects and in many of the smaller ones, an architect is commissioned to prepare complete working drawings and specifications for the building. These drawings usually include:

- A plot plan indicating the location of the building on the property.
- Floor plans showing the walls and partitions for each floor or level.
- Elevations of the exterior faces of the building.
- Vertical cross sections to indicate clearly the various floor levels and details of the footings, foundation, walls, floors, ceilings, and roof construction.
- Large-scale detail drawings that magnify complex parts of the structure.

The architect can use consulting engineers to prepare structural, electrical, and mechanical drawings—the latter encompassing pipe-fitting, instrumentation, plumbing, heating, ventilating, and air conditioning drawings.

Plot plan

This site plan is drawn as if the site were viewed from an airplane. It reveals the property boundaries, the existing contour lines, the new contour lines (after grading), the location of the building on the property, new and existing roadways, all utility lines, and other pertinent details. Descriptive notes can also be found on the plot plan listing names of adjacent property owners, the land surveyor, and the date of the survey. A legend or symbol list is also included so that anyone who must work with site plans can readily read the information (Fig. 12.1).

Floor plans

The plan view of any object is a drawing showing the outline and all details as a person would see it when looking directly down on the object. It shows only two dimensions: length and width. The floor plan of a building is drawn as if a slice was taken through the building (about window height) and then the top portion of the structure was removed to reveal the bottom outline (Fig. 12.2).

Let's say that we first wanted a plan view of a home's basement. The part of the house above the middle of the basement windows is imagined to be cut away. By looking down on the uncovered portion, exterior walls, partition walls, and other layout details can be illustrated. Likewise, imagine the section of the building above the middle of the first floor windows cut away and a drawing made looking straight down at the walls. This would be called the *first floor plan*. A cut through the second floor windows would be the *second floor plan,* etc.

Elevations

A plan view may represent flat surfaces, curved surfaces, or rising surfaces (like ramps and stairs), but for clarification it is usually necessary to refer to elevations and sections of the building. The elevation is an outline of an object that shows heights and can show the length or width of a particular side, but not depth. Figure 12.3 shows the front and rear elevation drawings for a build-

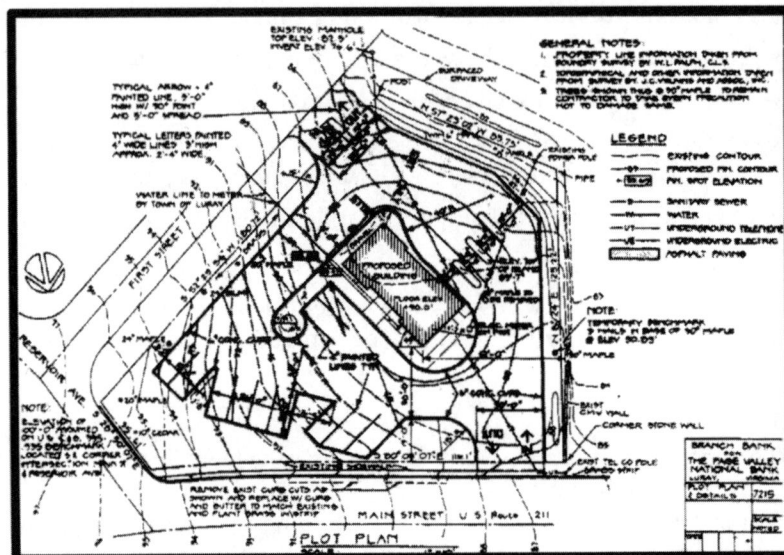

Figure 12.1 A typical plot plan.

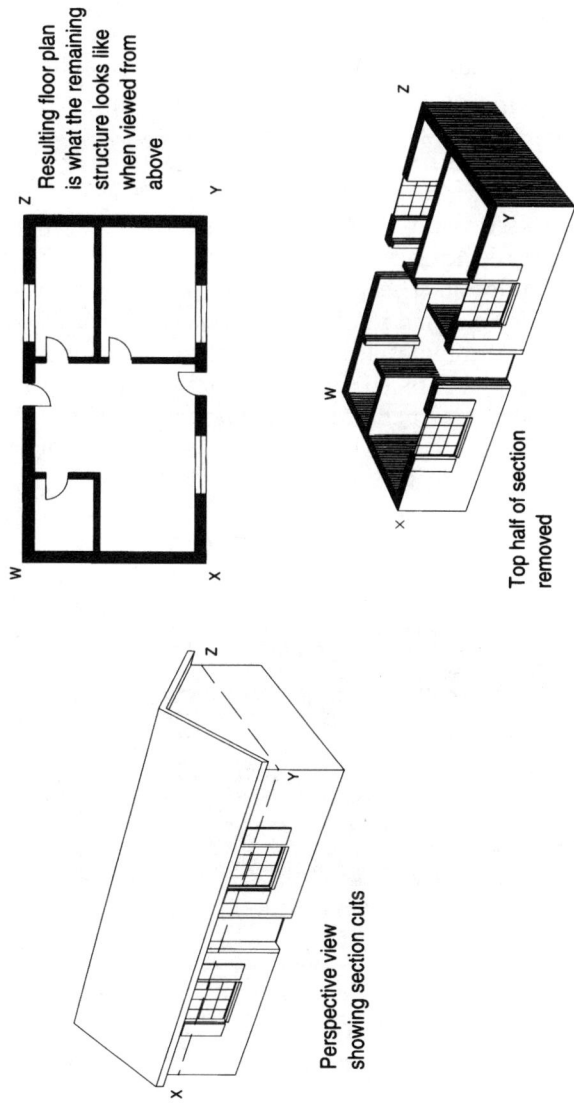

Resulting floor plan is what the remaining structure looks like when viewed from above

Top half of section removed

Perspective view showing section cuts

Figure 12.2   Principles of floor-plan layout.

ing, while Fig. 12.4 shows the side elevations of the same building. Notice that these elevation drawings show the heights of windows, doors, porches, the pitch of roofs, etc.—building components which cannot be shown conveniently on floor plans.

Elevation drawings are quite useful to security-system technicians for determining the number and sizes of all windows and doors. The types of windows and doors are also visible—giving the necessary information to select the kind of door and window contacts to use on the system.

Sections

A section or sectional view of a building is a view facing a section where a part of a structure is imagined to be sliced through, allowing the viewer to see inside the building. The point on the plan or elevation showing where the imaginary cut has been made is indicated by the section line, which is usually a very heavy double dot-and-dash line. The section line shows the location of the area imagined to be sliced through on the plan or elevation. It is, therefore, necessary to know which of the cutaway parts is represented in the sectional drawing when an object is represented as if it was cut in two. Arrow points are thus placed at the ends of the sectional lines.

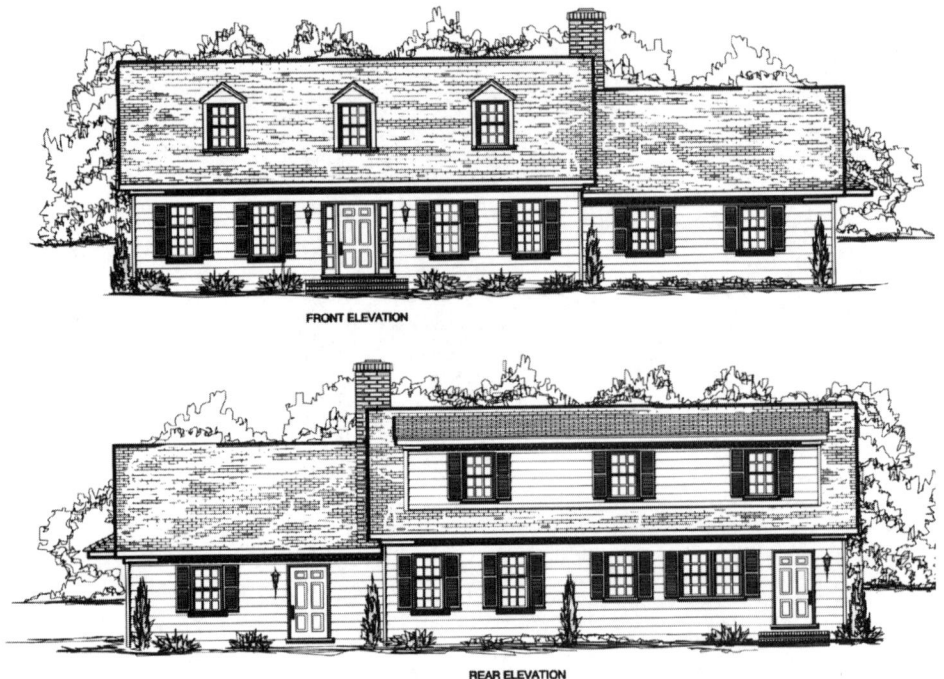

FRONT ELEVATION

REAR ELEVATION

Figure 12.3   Front and rear elevations of a building.

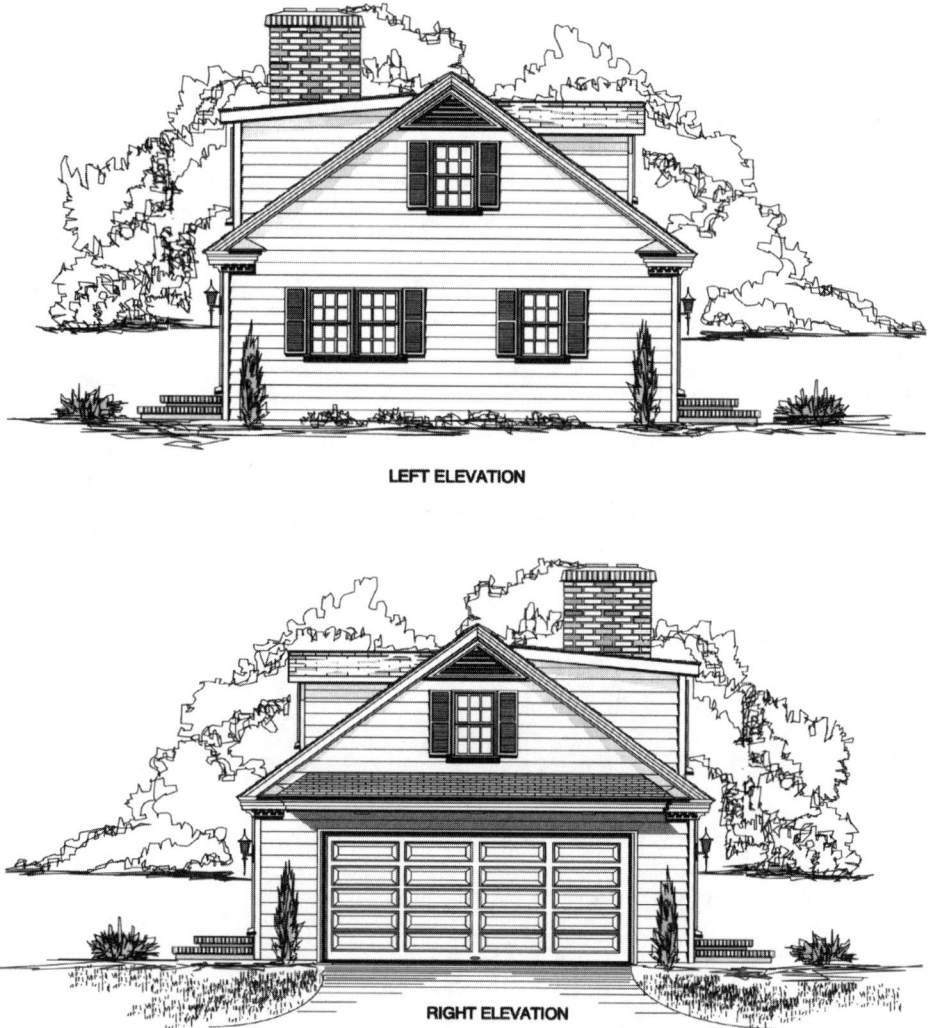

Figure 12.4   Side elevations of the building in Fig. 12.3.

In architectural drawings, it is often necessary to show more than one section on the same drawing. The different section lines must be distinguished by letters, numbers, or other designations placed at the ends of the lines (Fig. 12.5), in which the sections are lettered A-A and B-B. These section letters are generally heavy and large so as to stand out on the drawings. To further avoid confusion, the same letter is usually placed at each end of the section line. The section is named according to these letters—that is, Section A-A, Section B-B, etc.

A longitudinal section is taken length-wise while a cross section is usually taken straight across the width of an object. Sometimes, however, a section is not taken along one straight line. It is often taken along a zigzag line to show important parts of the object.

A sectional view, as applied to architectural drawings, is a drawing showing the building, or portion of a building, as though cut through, as if by a saw, on some imaginary line. This line can be either vertical (straight up and down) or horizontal. Wall sections are nearly always made vertically so that

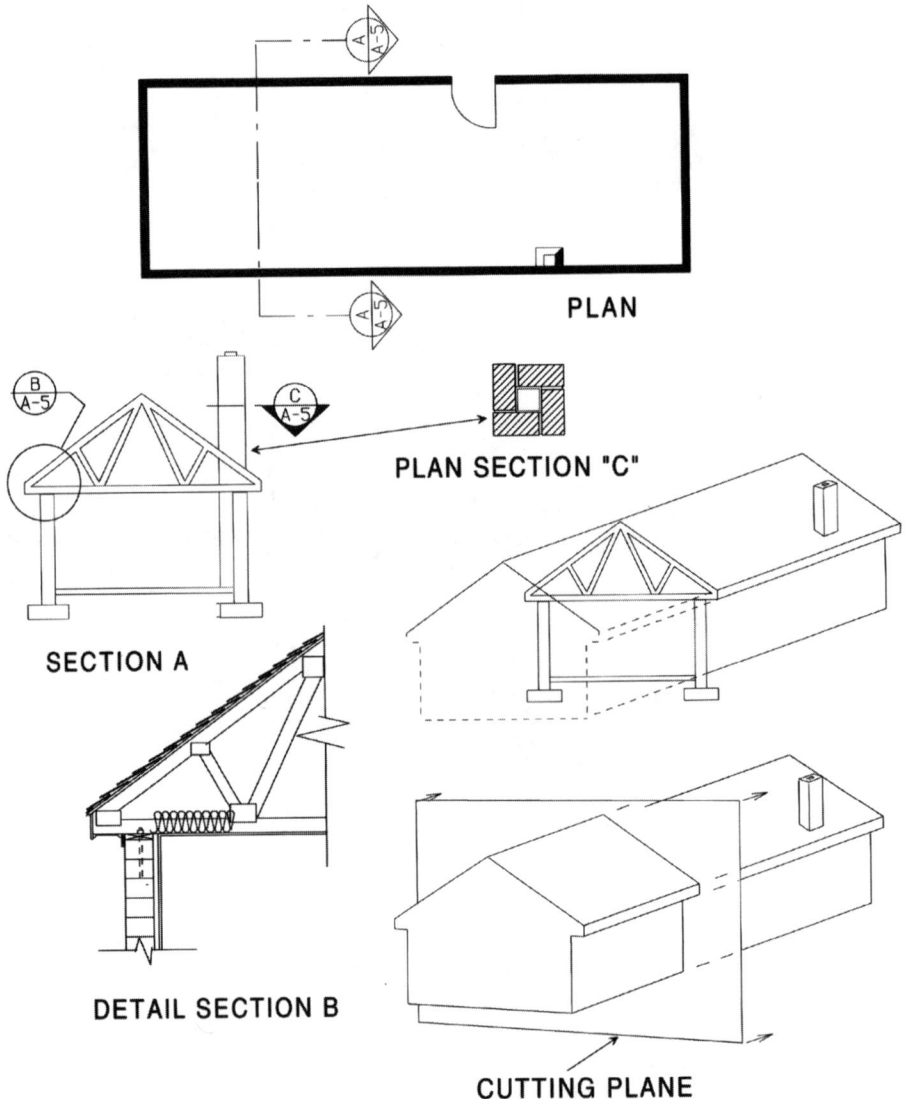

PLAN

PLAN SECTION "C"

SECTION A

DETAIL SECTION B

CUTTING PLANE

Figure 12.5 The principles of showing sections on working drawings and prints.

the cut edge is exposed from top to bottom. Wall sections are very important for construction workers because they answer the questions related to how a structure is built. The floor plans of a building show how each floor is arranged, but the wall sections tell how each part is to be constructed and usually indicate the material to be used. The security technician needs to know this information when laying out the security/fire-alarm system to determine the location and mounting techniques of the alarm devices, the routing of the conductors, estimating the quantity of materials, and choosing wiring methods that comply with the latest edition of the NEC.

Electrical/Electronics Security Drawings

The ideal electrical/electronics drawing should show in a clear, concise manner exactly what is required of the workers. The amount of data shown on such drawings should be sufficient, but not overdone. This means that a complete set of drawings could consist of only one $8^{1}/_{2}'' \times 11''$ sheet, or it could consist of several dozen $24'' \times 36''$ (or larger) sheets, depending on the size and complexity of the given project. A shop drawing, for example, might contain details of only one piece of equipment (Fig. 12.6), while a set of security-system working drawings for a large commercial installation might contain dozens of drawing sheets detailing the layout and location of all security devices in the building, along with equipment, controls, connection diagrams, schematic diagrams, equipment schedules, and a host of other pertinent data.

In general, electrical/electronics working drawings for a given project serve three distinct functions:

- They give contractors an exact description of the project so that materials and labor can be estimated in order to form a total cost of the project for bidding purposes.

- They give workers on the project instructions as to how the security system is to be installed.

- They provide a "map" of the security system once the job is completed to aid in maintenance and troubleshooting for years to come.

Security-system drawings from manufacturers and consulting engineering firms will vary in quality from sketchy, incomplete drawings to neat, very complete drawings that are easy to understand. Few, however, will cover every exact detail of the system. Therefore, a good knowledge of installation practices must go hand in hand with interpreting security-system working drawings.

Sometimes security-system contractors will have drafters prepare special supplemental drawings for use by the contractors' employees. On certain projects, these supplemental drawings can save supervision time in the field once the project has begun.

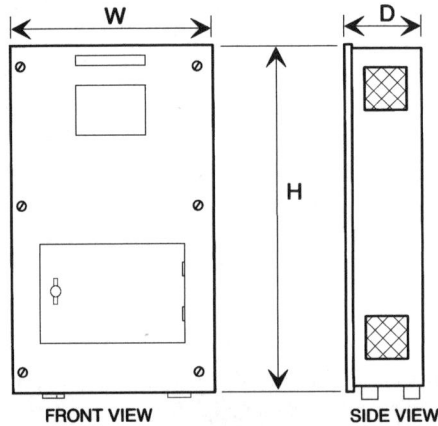

Model	Dimensions, Inches		
Number	H	W	D
1200	29	18	4
1800	29	18	4
2000	29	18	4
2500	33	22	6
3000	41	29	8

Figure 12.6 A shop drawing of a fire-alarm control panel. This gives the necessary dimensions for "roughing-in" the panel during the construction or renovation of the building.

Drawing Layout

Although a strong effort has been made to standardize drawing practices in the industry, seldom will working drawings—prepared by different architectural or engineering firms—be identical. Similarities, however, will exist between most sets of prints, and with a little experience, you should have little trouble interpreting any set of drawings.

Most drawings used for building construction projects will be drawn on drawing paper from 11″ × 17″ to 24″ × 36″ in size. Each drawing sheet will have border lines framing the overall drawing and a title block (Fig. 12.7). Notice that the type and size of title blocks vary with each firm preparing the drawings. In addition, some drawing sheets will also contain a revision block near the title block, and perhaps an approval block. This information is normally found on each drawing sheet, regardless of the type of project or the information contained on the sheet.

Title block

The title block for a print or technical drawing is usually boxed in the lower right-hand corner of the drawing sheet; the size of the block varies with the size of the drawing and also with the information required (Fig. 12.8).

In general, the title block of a security-system drawing should contain the following:

■ Name of the project

■ Address of the project

■ Name of the owner or client

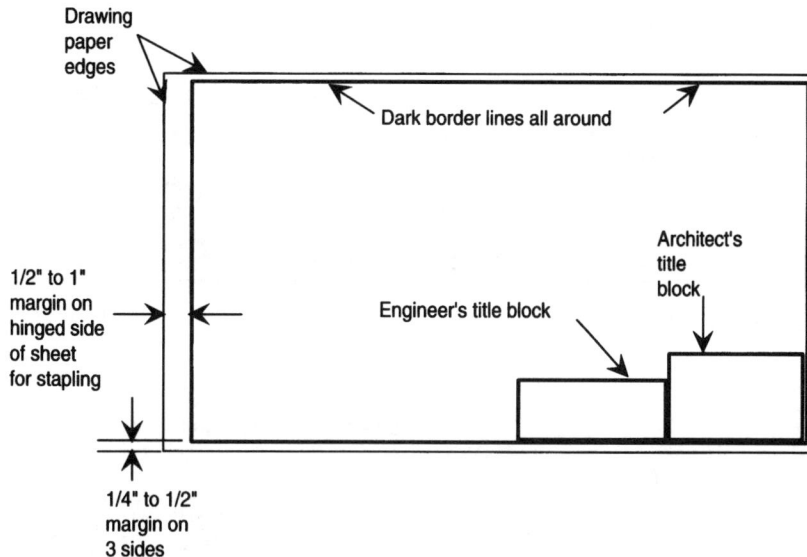

Figure 12.7 Typical layout of a drawing sheet.

- Name of the architectural and/or engineering firm
- Date of completion
- Scale(s)
- Initial of the drafter, checker, and designer, with dates under each
- Job number
- Sheet number
- General description of the drawing

Every architectural and engineering firm has its own standard for drawing titles, and they are often preprinted directly on the tracing paper or else printed on "stick-on" paper, which is placed on the drawing.

Often the consulting engineering firm will also be listed, which means that an additional title block will be applied to the drawing—usually next to the architect's title block. In some cases, the security equipment supplier will also appear. Figure 12.9 shows completed architectural and engineering title blocks as they appeared on an actual drawing.

Approval block

The "approval block," in most cases, will appear on the drawing sheet as shown in Fig. 12.10. The various types of approval blocks (drawn, checked, etc.) will be initialed by the appropriate personnel. This type of approval block is usually part of the title block and appears on each drawing sheet.

Figure 12.8 Typical architect's title block.

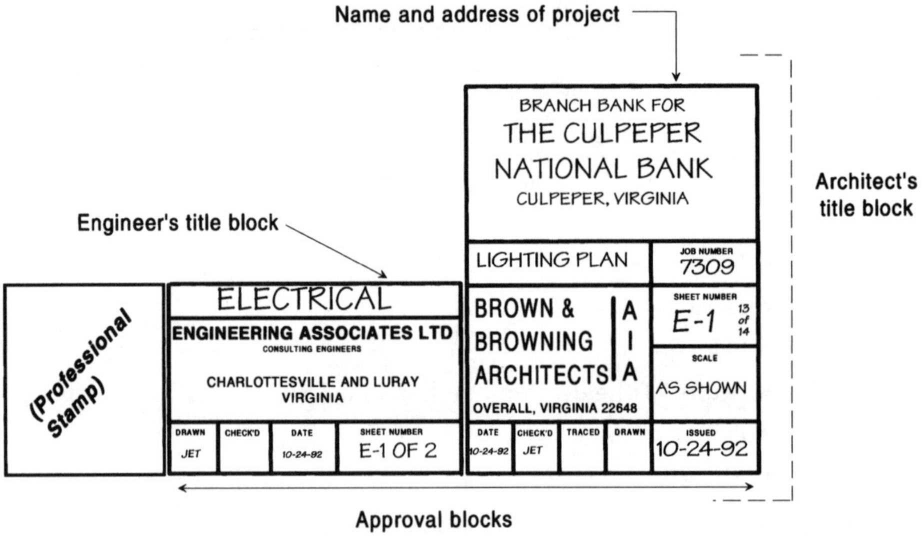

Figure 12.9 Completed engineering and architectural title blocks.

On some projects, authorized signatures are required before certain systems may be installed, or even before the project begins. An approval block (such as the one shown in Fig. 12.11) indicates that all required personnel has checked the drawings for accuracy, and that the set meets with everyone's approval. Such approval blocks usually appear on the front sheet of the print set and can include:

- Professional stamp—registered seal of approval by the architect or consulting engineer.

- Design supervisor's signature—the person who is overseeing the design.

Name and address of project ⟶

BRANCH BANK FOR
THE CULPEPER
NATIONAL BANK
CULPEPER, VIRGINIA

Architect's
title block

Engineer's title block

LIGHTING PLAN | 7309

ELECTRICAL
ENGINEERING ASSOCIATES LTD

CHARLOTTESVILLE AND LURAY
VIRGINIA

BROWN &
BROWNING
ARCHITECTS

A
I
A

E-1

OVERALL, VIRGINIA 22949

AS SHOWN

(Professional Stamp)

JET | | 10-24-82 | E-1 OF 2 | | JET | | 10-24-82

Approval blocks

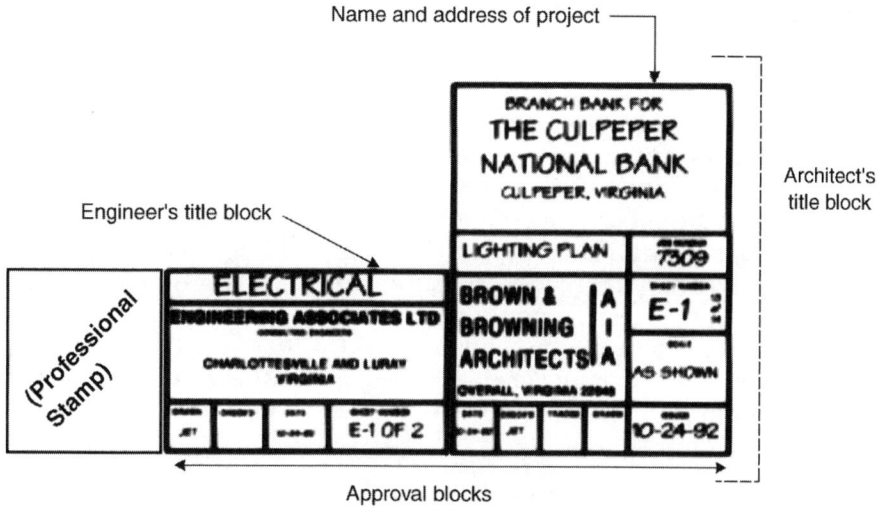

Figure 12.10 One type of approval block used on electrical drawings.

Professional Stamp	DESIGN SUPERVISOR	DATE
	DRAWN	DATE
	CHECKED	DATE
	APPROVED	DATE
	OWNER'S APPROVAL	DATE

Figure 12.11 An alternate approval block.

- Drawn (by)—signature or initials of the person who drafted the drawing and the date it was completed.
- Checked (by)—signature or initials of the person(s) who reviewed the drawing and the date of approval.
- Approved—signature or initials of the architect/engineer and the date of the approval.
- Owner's approval—signature of the project owner or the owner's representative along with the date signed.

Revision block

Sometimes security-system drawings will have to be partially redrawn or modified during the installation of a system. It is extremely important that such modifications are noted and dated on the drawings to ensure that the workers have an up-to-date set.

Busy offices should study computer-aided drafting software and the ability to e-mail updates directly to the project and print them out on site. In some situations, sufficient space is left near the title block for dates and description of revisions. In other cases, a revision block is provided (again, near the title block), as shown in Fig. 12.12. But these two samples are by no means the only types or styles of revision blocks that will be seen on electrical working drawings. Each architect/engineer/designer/drafter has his or her own method of showing revisions, so expect to find deviations from those shown.

Drafting Lines

All drafting lines have one thing in common: they are the same color. However, good easy-to-read contrasting lines can be made by varying the width of the lines or else "breaking" the lines in some uniform way.

Figure 12.13 on the next page shows common lines used on architectural drawings. However, these lines can vary. Architects and engineers have strived for a common "standard" for the past century, but unfortunately, their goal has yet to be reached. Therefore, you will find variations in lines and symbols from drawing to drawing, so always consult the legend or symbol list when referring to an architectural or electrical drawing. Also carefully inspect each drawing to ensure that line types are used consistently.

REVISIONS				
REV	DESCRIPTION	DR	APP'D	DATE
1	FIXTURE No. 3 IN LIGHTING-FIXTURE SCHDL	BD	GLC	10/12/92

Professional Stamp

ELECTRICAL
DISTRICT HOME
LAUNDRY BUILDING
AUGUSTA COUNTY, VIRGINIA

G. LEWIS CRAIG, ARCHITECT
WAYNESBORO, VIRGINIA

COMM. NO.	DATE	DRAWN	CHECKED	REVISED
7215	9/6/92	GK	GLC	

SHEET NO.

E-1

Figure 12.12 One method of showing revisions on working drawings.

A brief description of the drafting lines shown in Fig. 12.13 follows:

Light Full Line - This line is used for section lines, building background (outlines), and similar uses where the object to be drawn is secondary to the electrical system.

Medium Full Line - This type of line is frequently used for hand lettering on a drawing. It is further used for some drawing symbols, circuit lines, etc.

Heavy Full Line - This line is used for borders around title blocks, schedules and for hand lettering drawing titles. Some types of symbols are frequently drawn with the heavy full line.

Extraheavy Full Line - This line is used for border lines on architectural/engineering drawings.

Centerline - A centerline is a broken line made up of long and short dashes alternately spaced. It indicates the centers of objects (such as holes, pillars, or fixtures). Sometimes, the centerline indicates the dimensions of a finished floor.

Hidden Line - A hidden line consists of a series of short dashes closely and evenly spaced. It shows the edges of objects that are not visible in a particular view. The object outlined by hidden lines in one drawing is often fully pictured in another drawing.

Dimension Lines - These are thinly drawn lines used to show the extent and direction of dimensions. The dimension is usually placed in a break inside of the dimension lines. Normal practice is to place the dimension lines outside the object's outline. However, sometimes it might be necessary to draw the dimensions inside the outline.

Short Break Line - This line is usually drawn freehand and is used for short breaks.

Long Break Line - This line, drawn partly with a straightedge and partly with freehand zigzags, is used for long breaks.

Match Line - This line is used to show the position of the cutting plane. Therefore, it is also called *cutting plane line*. A match or cutting plane line is an extra heavy line with long dashes alternating with two short dashes. It is used on drawings of large structures to show where one drawing stops and the next drawing starts.

Secondary Line - This line is frequently used to outline pieces of equipment or to indicate reference points of a drawing that is secondary to the drawing's purpose.

Property Line - This line consists of one long and two short dashes alternately spaced. It indicates land boundaries on the site plan.

Other uses of the lines just mentioned include the following:

Extension Lines - Extension lines are lightweight lines that start about $\frac{1}{16}$" away from an object's edge and extend out. A common use of extension lines is to create a boundary for dimension lines. Dimension lines meet extension lines with arrowheads, slashes, or dots. Extension lines that point from a note or other reference to a particular feature on a drawing are called *leaders*. They usually end in either an arrowhead, a slash or a dot, and might include an explanatory note at the end.

Light full line	————————————
Medium full line	————————————
Heavy full line	————————————
Extraheavy full line	————————————
Centerline	— · — · — · — · — · —
Hidden	··············
Dimension line	←———————3.00"———————→
Short break line	
Long break line	
Match line	— — — — — —
Secondary line	– – – – – – – –
Property line	— · · — · · — · · —

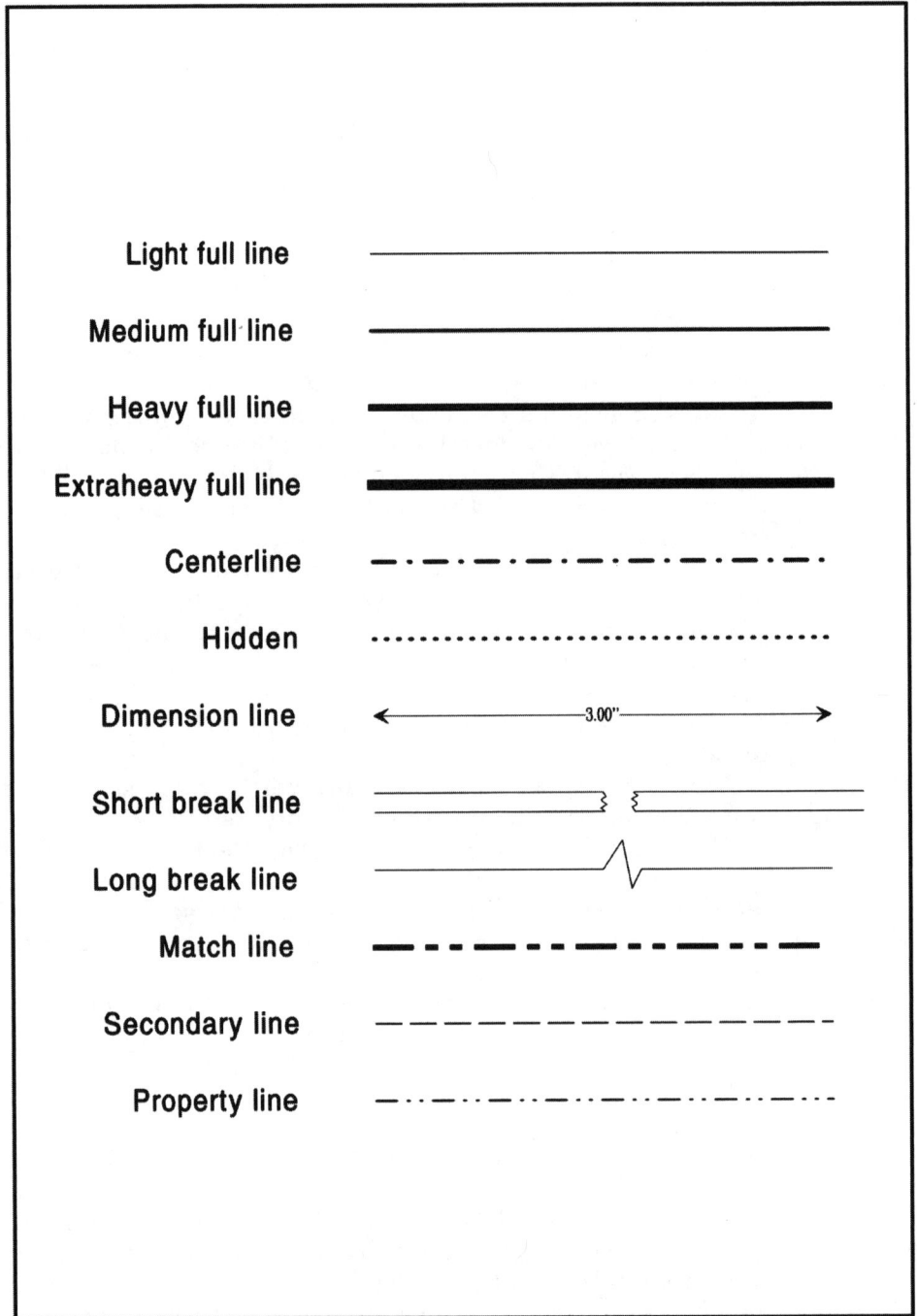

Figure 12.13 Typical drafting lines.

Section Lines - These are often referred to as *cross-hatch lines*. Drawn at a 45-degree angle, these lines show where an object has been cut away to reveal the inside.

Phanton Lines - Phantom lines are solid, light lines that show where an object will be installed. A future door opening or a future piece of equipment can be shown with phantom lines.

Electrical drafting lines

Besides the architectural lines shown in Fig. 12.13, consulting electrical engineers, designers, manufacturers of security equipment and components, and their drafters use additional lines to represent circuits and their related components. Again, these lines can vary from drawing to drawing, so check the symbol list or legend for the exact meaning of lines on the drawing with which you are working. Figure 12.14 shows lines used on some electrical drawings.

Electrical Symbols

The electrician must be able to read and understand electrical working drawings, which includes a thorough knowledge of electrical symbols and their application.

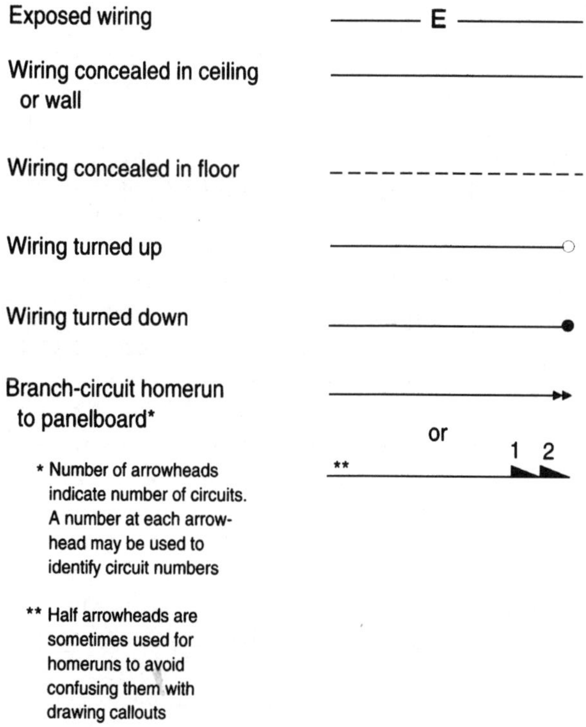

Exposed wiring

Wiring concealed in ceiling
   or wall

Wiring concealed in floor

Wiring turned up

Wiring turned down

Branch-circuit homerun
   to panelboard*

   * Number of arrowheads
   indicate number of circuits.
   A number at each arrowhead may be used to
   identify circuit numbers

** Half arrowheads are
   sometimes used for
   homeruns to avoid
   confusing them with
   drawing callouts

Figure 12.14 Circuit lines used on electrical drawings.

An electrical symbol is a figure or mark that stands for a component used in the electrical system. For example, Fig. 12.15 shows a list of electrical symbols that are currently recommended by the American National Standards Institute (ANSI). It is evident from this list of symbols that many have the same basic form, but, because of some slight difference, their meaning changes. A good procedure to follow in learning symbols is to first learn the basic form and then apply the variations for obtaining different meanings.

It would be much simpler if all architects, engineers, security equipment manufacturers, electrical designers, and drafters used the same symbols. However, this is not the case. Although standardization is getting closer to a reality, existing symbols are still modified and new symbols are created for almost every new project. Therefore, it is very important to review the symbols on working drawings with care.

The electrical symbols described in the following paragraphs represent those found on actual electrical working drawings throughout the United States and Canada. Many are similar to those recommended by ANSI and the Consulting Engineers Council/US; others are not. Understanding how these symbols were devised will help you to interpret unknown electrical symbols in the future.

Some of the symbols used on security-system drawings are abbreviations, such as *WP* for *weatherproof* and *AFF* for *above finished floor*. Others are simplified pictographs. In some cases, the symbols are combinations of abbreviations and pictographs. Several symbols used on security-system drawings are shown in Fig. 12.16.

Circuit and feeder wiring symbols are getting closer to being standardized. Most circuits concealed in the ceiling or wall are indicated by a solid line; a broken line is used for circuits concealed in the floor or ceiling below; and exposed raceways are indicated by short dashes or else the letter E placed in the same plane with the circuit line at various intervals.

The number of conductors in a cable, conduit, or raceway system might be indicated in the panelboard schedule under the appropriate column, or the information might be shown on the floor plan.

Symbols for communication and signal systems, as well as symbols for light and power, are drawn to an appropriate scale and accurately located with respect to the building; this reduces the number of references made to the architectural drawings. Where extreme accuracy is required in locating outlets and equipment, exact dimensions are given on larger-scale drawings and shown on the plans.

Each different category in an electrical system is usually represented by a distinguishing basic symbol. To further identify items of equipment or outlets in the category, a numeral or other identifying mark is placed within the open basic symbol. In addition, all such individual symbols used on the drawings should be included in the symbol list or legend.

SWITCH OUTLETS

Single-Pole Switch — S

Double-Pole Switch — S₂ → S_2

Three-Way Switch — S_3

Four-Way Switch — S_4

Key-Operated Switch — S_K

Switch w/ Pilot — S_P

Low-Voltage Switch — S_L

Switch & Single Receptacle — S

Switch & Duplex Receptacle — S

Door Switch — S_D

Momentary Contact Switch — S_{MC}

RECEPTACLE OUTLETS

Single Receptacle

Duplex Receptacle

Triplex Receptacle

Split-Wired Duplex Recep.

Single Special Purpose Recep.

Duplex Special Purpose Recep.

Range Receptacle — R

Special Purpose Connection or Provision for Connection. Subscript letters indicate Function (DW - Dishwasher; CD - Clothes Dryer, etc.) — DW

Clock Receptacle w/Hanger — C

Fan Receptacle w/Hanger — F

Single Floor Receptacle

*Numeral or Letter within symbol or as a subscript keyed to List of Symbols indicates type of receptacle or usage.

LIGHTING OUTLETS

Ceiling / Wall

Surface Fixture

Surface Fixt. w/ Pull Switch — PS / PS

Recessed Fixture — R / R

Surface or Pendant Fluorescent Fixture

Recessed Fluor. Fixture — R

Surface or Pendant Continuous Row Fluor. Fixtures

Recessed Continuous Row Fluorescent Fixtures — R

Surface Exit Light — X / X

Recesses Exit Light — X R / X R

Blanked Outlet — B / B

Junction Box — J / J

CIRCUITING

Wiring Concealed in Ceiling or Wall ————

Wiring Concealed in Floor — — — —

Wiring Exposed ....

Branch Circuit Homerun to Panelboard. Number of arrows indicates number of circuits in run. Note: Any circuit without further identification is 2-wire. A greater number of wires is indicated by cross lines as shown below. Wire size is sometimes shown with numerals placed above or below cross lines.

/// 3-Wire

//// 4-Wire

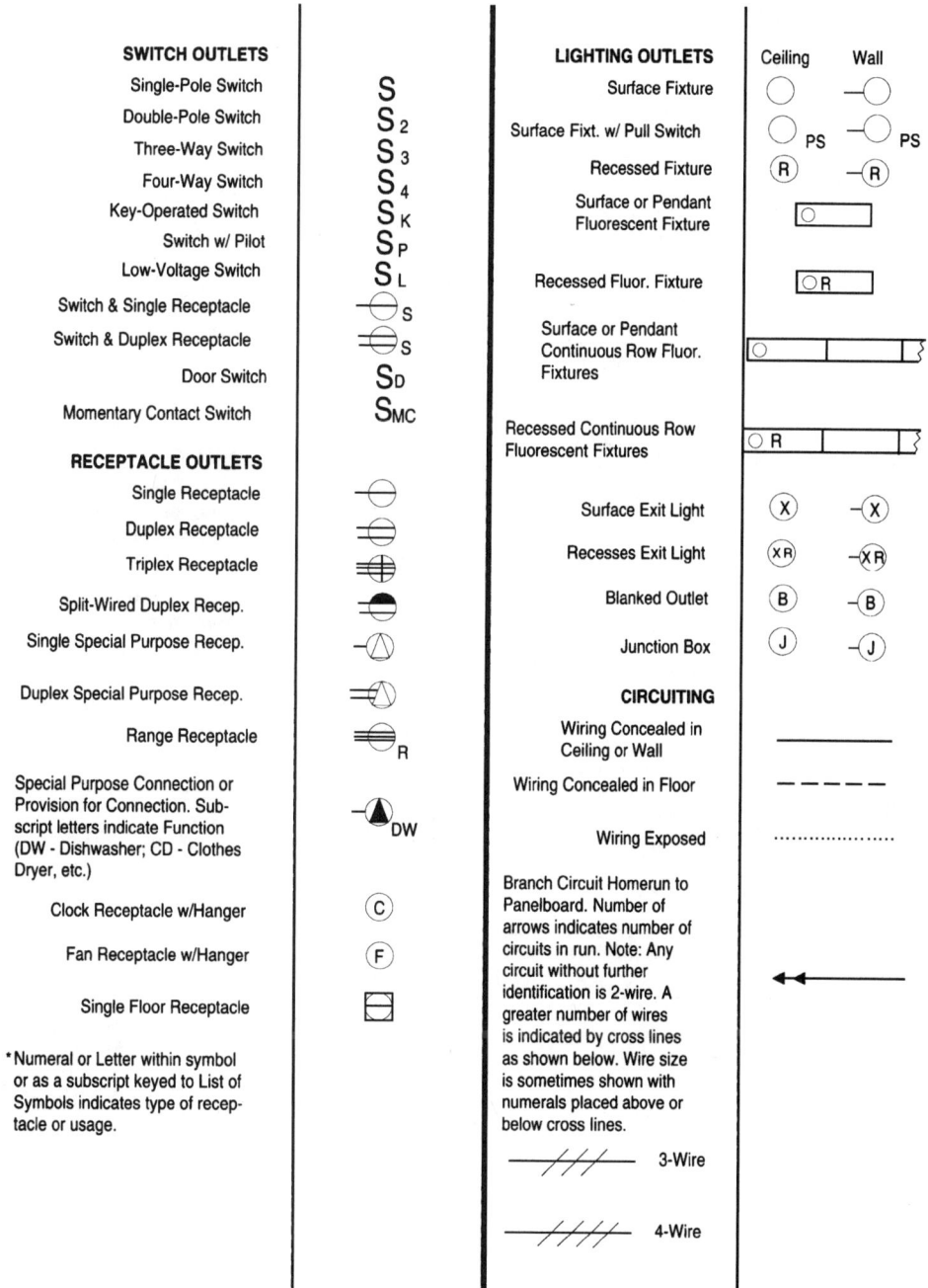

Figure 12.15 Electrical symbols recommended by ANSI.

SIGNALLING SYSTEM OUTLETS

INSTITUTIONAL, COMMERCIAL, AND INDUSTRIAL OCCUPANCIES

I   NURSE CALL SYSTEM DEVICES
(Any Type)

Basic Symbol

(Examples of Individual Item Identification
Not a Part of Standard)

Nurses' Annunciator
(Add a number after it as
+⊖24 to indicate number
of lamps)

Call Station, Single Cord,
Pilot Light

Call Station, Double Cord,
Microphone Speaker

Corridor Dome Light
1 Lamp

Transformer

Any Other Item On Same
System Use Number As
Required

II   PAGING SYSTEM DEVICES

Basic Symbol

(Examples of Individual Item Identification
Not a Part of Standard)

Keyboard

Flush Annunciator

2-Face Annunciator

Any Other Item On Same
System Use Numbers As
Required

III   FIRE ALARM SYSTEM DEVICES
(Any Type) Including Smoke
and Sprinkler Alarm Devices

Basic Symbol

(Examples of Individual Item
Identification. Not a Part of
Standard)

Control Panel

Station

10" Gong

Pre-Signal Chime

Any Other Item On Same System
Use Numbers As Required

IV   STAFF REGISTER SYSTEM DEVICES
(Any Type)

Basic Symbol

(Examples of Individual Item
Identification. Not a Part of
Standard)

Phone Operators' Register

Entrance Register - Flush

Staff Room Register

Transformer

Any Other Item On Same System
Use Numbers As Required

V   ELECTRIC CLOCK SYSTEM DEVICES
(Any Type)

Basic Symbol

(Examples of Individual Item
Identification. Not a Part of
Standard)

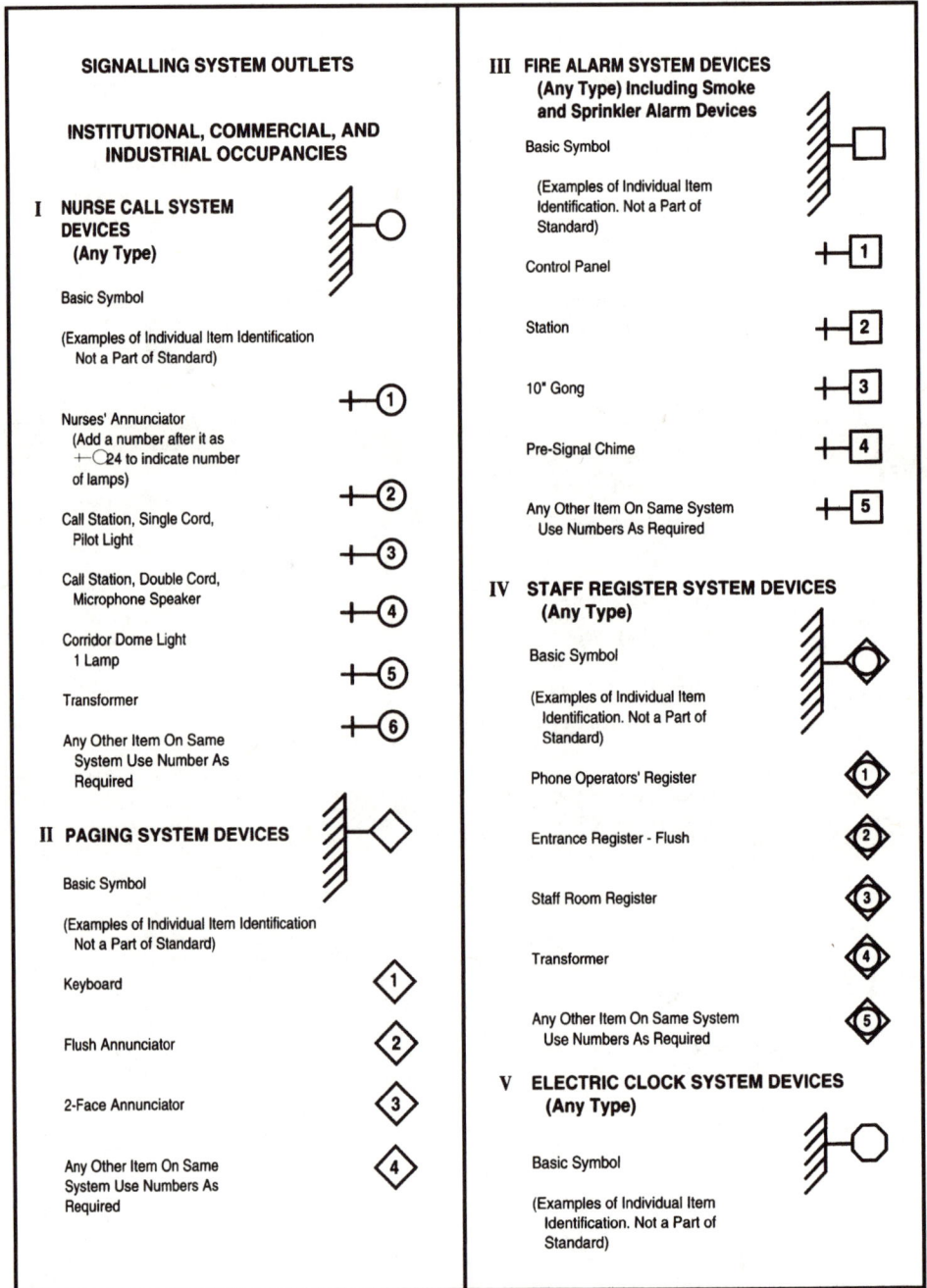

Figure 12.16   Symbols used on security-system drawings.

Master Clock

12" Secondary - Flush

12" Double Dial - Wall Mounted

18" Skeleton Dial

Any Other Item On Same System
Use Numbers As Required

VI PUBLIC TELEPHONE SYSTEM DEVICES

Basic Symbol

(Examples of Individual Item
Identification. Not a Part of
Standard)

Switchboard

Desk Phone

Any Other Item On Same
System Use Numbers As
Required

VII PRIVATE TELEPHONE SYSTEM DEVICES
(Any Type)

Basic Symbol

(Examples of Individual Item
Identification. Not a Part of
Standard)

Switchboard

Wall Phone

Any Other Item On Same System
Use Numbers As Required

VIII WATCHMAN SYSTEM DEVICES
(Any Type)

Basic Symbol

(Examples of Individual Item
Identification. Not a Part of
Standard)

Central Station

Key Station

Any Other Item On Same System
Use Numbers As Required

IX SOUND SYSTEM

Basic Symbol

(Examples of Individual Item
Identification. Not a Part of
Standard)

Amplifier

Microphone

Interior Speaker

Exterior Speaker

Any Other Item On Same System
Use Numbers As Required

X OTHER SIGNAL SYSTEM DEVICES

Basic Symbol

(Examples of Individual Item
Identification. Not a Part of
Standard)

Buzzer

Bell

Pushbutton

Annunciator

Any Other Item On Same System
Use Numbers As Required

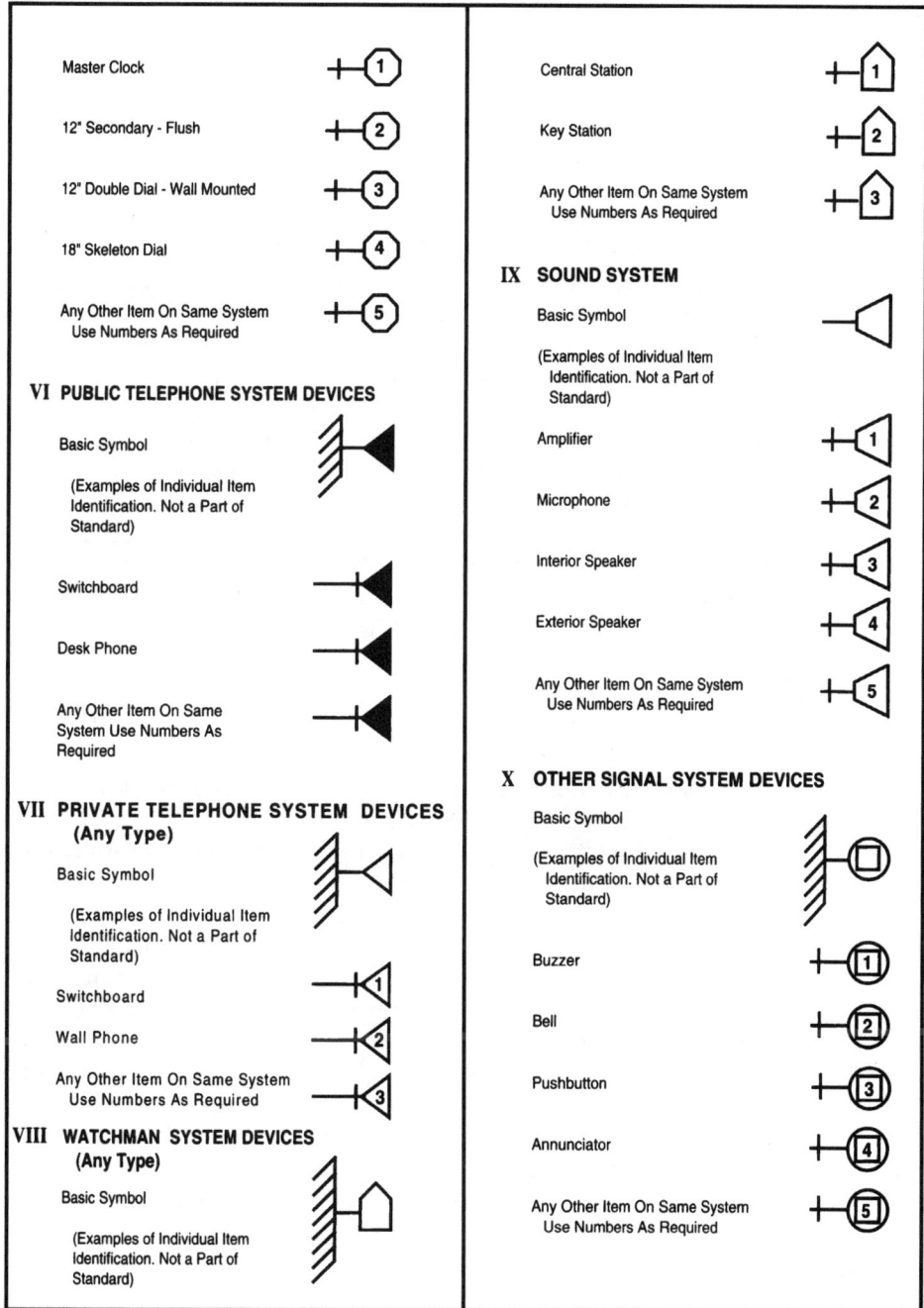

Figure 12.16 (*Continued*)

Scale Drawings

In most architectural and electrical construction drawings, the components are so large that it would be impossible to draw them in their actual size on a usable drawing sheet. Consequently, drawings are made to some reduced scale—that is, all the distances are drawn smaller than the actual dimension of the object itself, all dimensions being reduced in the same proportion. For example, if a floor plan of a building is to be drawn to a scale of $\frac{1}{4}'' = 1'\text{-}0''$, each $\frac{1}{4}''$ on the drawing would equal 1 foot on the building itself; if the scale is $\frac{1}{8}'' = 1'\text{-}0''$, each $\frac{1}{8}''$ on the drawing equals 1 foot on the building, and so forth.

When architectural and engineering drawings are produced, the scale decided upon is very important. Where dimensions must be held to extreme accuracy, the scale drawings should be made as large as practical with dimension lines added. Where dimensions require only reasonable accuracy, the object can be drawn to a smaller scale (with dimension lines possibly omitted) because the object can be scaled with the appropriate scale.

In dimensioning drawings, the dimension written on the drawing is the actual dimension of the building, not the distance that is measured on the drawing. To further illustrate this point, look at the floor plan in Fig. 12.17; it is drawn to a scale of $\frac{1}{2}'' = 1'\text{-}0''$. One of the walls is drawn to an actual length of $3\frac{1}{2}''$ on the drawing paper, but because the scale is $\frac{1}{2}'' = 1'\text{-}0''$ and since $3\frac{1}{2}''$ contains 7 halves of an inch ($7 \times 0.5 = 3\frac{1}{2}$), the dimension shown on the drawing will therefore be $7'\text{-}0''$ on the actual building.

From the previous example, you can see that the most common method of reducing all the dimensions (in feet and inches) in the same proportion is to choose a certain distance and let that distance represent one foot. This distance can then be divided into 12 parts, each of which represents an inch. If half inches are required, these twelfths are further subdivided into halves, etc. We now have a scale that represents the common foot rule with its subdivisions into inches and fractions, except that the scaled foot is smaller than the distance known as a foot and, likewise its subdivisions are proportionately smaller.

When a minor measurement is made on the drawing, it is often made with the reduced foot rule or scale. For continual scaling, computer software that scales automatically is often used. The most common reduced foot rules or scales used in security-system drawings are the architect's scale and the engineer's scale. Sometimes drawings may be encountered that use a metric scale, but the principle of using this scale is similar to the architect's or engineer's scales. Using CAD scales is simple and learning to scale with computers is covered in the instructions that come with the software.

Architect's scale

Figure 12.18 shows two configurations of architect's scales: the one on the left is designed so that $1'' = 1'\text{-}0''$; the one on the right has graduations spaced to represent $\frac{1}{8}'' = 1'\text{-}0''$. Now let's zoom in for a closer look.

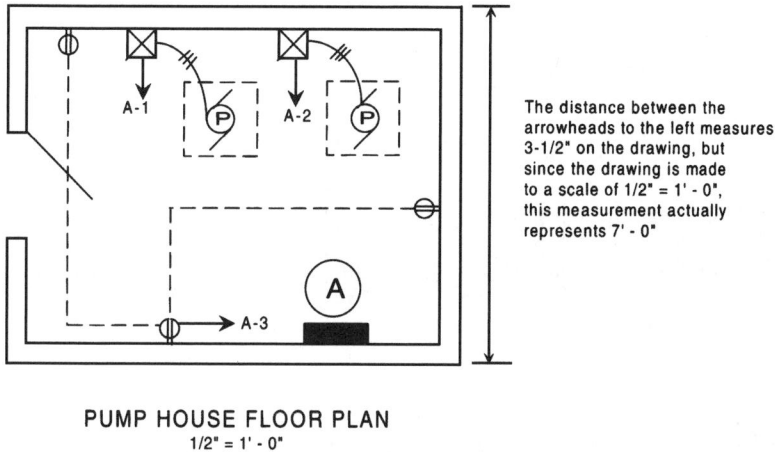

The distance between the arrowheads to the left measures 3-1/2" on the drawing, but since the drawing is made to a scale of 1/2" = 1' - 0", this measurement actually represents 7' - 0"

PUMP HOUSE FLOOR PLAN
1/2" = 1' - 0"

Figure 12.17 Floor plan showing drawing title and scale.

Notice on the one-inch scale in Fig. 12.18 that the longer marks to the right of the zero (with a numeral beneath) represent feet. Therefore, the distance between the zero and the numeral 1 equals 1 foot. The shorter mark between the zero and 1 represents $\frac{1}{2}$ of a foot (six inches).

Referring again to Fig. 12.18, look at the marks to the left of the zero. There are four different lengths of marks in this group. The longest marks are spaced three scaled inches apart and have the numerals 0, 3, 6, and 9 for use as reference points. The next longest group of lines each represent scaled inches, but are not marked with numerals. In use, you can count the number of marks to the left of the zero to find the number of inches, but after some practice, you will be able to tell the exact measurement at a glance. For example, the measurement "A" represents 5″ because it is the fifth "inch" mark to the left of the zero; it is also one "inch" mark short of the 6-in line on the scale.

The next size line that is shorter than the "inch" line is the half-inch line, and the shortest lines in the group represent $\frac{1}{4}$″. On smaller scales, however, the basic unit is not divided into as many divisions. For example, the smallest subdivision on the $\frac{1}{8}$″ = 1'-0″ scale represents 2 inches.

Types of architect's scales

Architect's scales are available in several types, but the most common include the triangular scale and the "flat" scale. The quality of architect's scales also vary from cheap plastic scales (costing a dollar or two) to high-quality wooden-laminated tools, such as produced by Keffel&Esser, Dietzgen, and others. Even if your firm draws with CAD software and does takeoffs with a digitizer, a scale is handy for checking drawings on site.

The triangular scale (Fig. 12.19) is frequently found in drafting and estimating departments or engineering and security-system contracting firms,

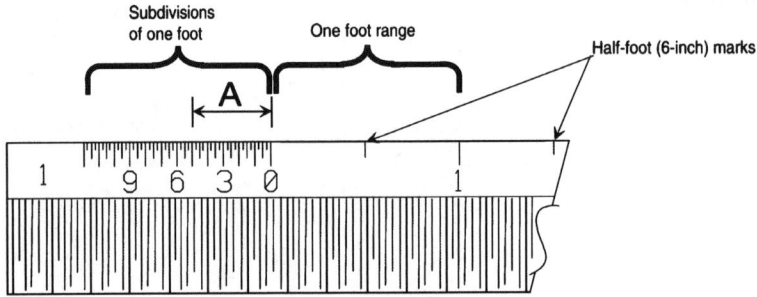

Figure 12.18   A close-up view of the 1-inch architect's scale.

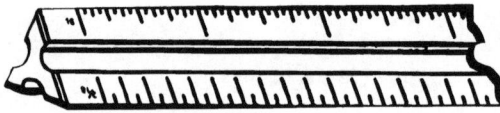

Figure 12.19   Typical triangular architect's scale.

while the flat scales are more convenient to carry on the job site by workers.

Triangular-shaped architect's scales (Fig. 12.20) have 12 different scales (two on each edge) as follows:

- Common foot rule (12 inches)
- $\frac{1}{16}$" = 1'-0"
- $\frac{3}{32}$" = 1'-0"
- $\frac{3}{16}$" = 1'-0"
- $\frac{1}{8}$" = 1'-0"
- $\frac{1}{4}$" = 1'-0"
- $\frac{3}{8}$" = 1'-0"
- $\frac{3}{4}$" = 1'-0"
- 1" = 1'-0"
- $\frac{1}{2}$" = 1'-0"
- $1\frac{1}{2}$" = 1'-0"
- 3" = 1'-0"

Two separate scales on one face might seem confusing at first, but after some experience, reading these scales becomes "second nature."

In all but one of the scales on the triangular architect's scale, each face has one of the scales spaced exactly one-half of the other. For example, on the 1″ face, the 1″ scale is read from left to right, starting from the zero mark. The half-inch scale is read from right to left, again starting from the zero mark.

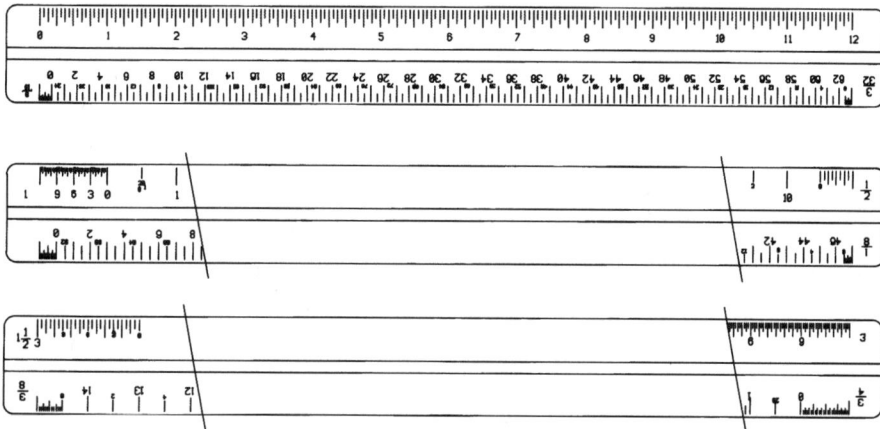

Figure 12.20 The various scales on a triangular architect's scale.

On the remaining foot-rule scale ($^1/_{16}$" = 1'0"), each $^1/_{16}$" mark on the scale represents one foot.

The "flat" architect's scale shown in Fig. 12.21 is ideal for workers on most projects. It is easily and conveniently carried in the shirt pocket, and the four scales ($^1/_8$, $^1/_4$, $^1/_2$, and 1") are adequate for the majority of projects that will be encountered.

The partial floor plan shown in Fig. 12.21 is drawn to a scale of $^1/_8$" = 1'-0". The dimension in question is found by placing the $^1/_8$" architect's scale on the drawing and reading the figures. It can be seen that the dimension reads 24'6".

Every drawing should have the scale to which it is drawn plainly marked on it as part of the drawing title. However, it is not uncommon to have several different drawings on one print sheet—all with different scales. Therefore, always check the scale of each different view found on a drawing sheet.

Engineer's scale

The civil engineer's scale is used fundamentally in the same manner as the architect's scale, the principal difference being that the graduations on the engineer's scale are decimal units, rather than feet, as on the architect's scale.

The engineer's scale is used by placing it on the drawing with the working edge away from the user. The scale is then aligned in the direction of the required measurement. Then, by looking down over the scale, the dimension is read.

Civil engineer's scales are common in the following graduations:

- 1" = 10 units
- 1" = 20 units
- 1" = 30 units

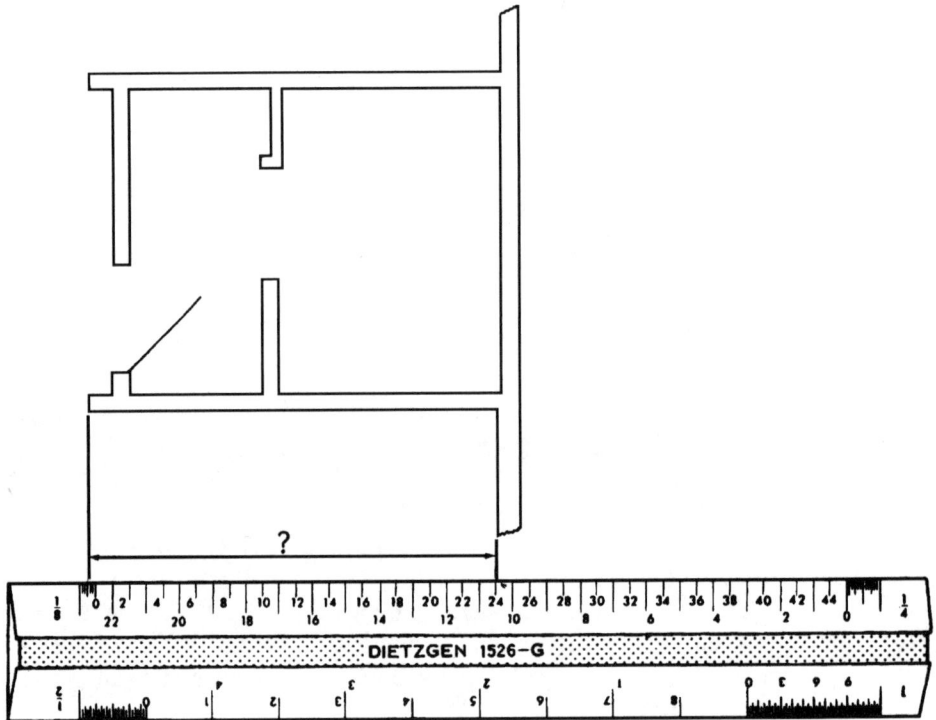

Figure 12.21 Using the 1/8″ architect's scale to determine dimensions on a drawing.

- 1″ = 40 units
- 1″ = 60 units
- 1″ = 80 units
- 1″ = 100 units

The purpose of this scale is to transfer the relative dimensions of an object to the drawing or vice versa. It is used mainly on on-site plans to determine distances between property lines, manholes, duct runs, direct-burial cable runs, etc.

Site plans are drawn to scale using the engineer's scale rather than the architect's scale. On small lots, a scale of (for example, 1″ = 10′ or 1″ = 20′) is used. This means that 1″ (actual measurement on the drawing) is equal to 10 feet, 20 feet, etc., on the land itself.

On larger drawings, where a large area must be covered, the scale could be 1 inch = 100 feet or 1 inch = 1000 feet, or any other integral power of 10. On drawings with the scale in multiples of 10, the engineering scale marked *10* is used. If the scale is 1 inch = 200 feet, the engineer's scale marked *20* is used, etc.

Although site plans are reduced in scale, depending on the size of the object and the size of the drawing sheet to be used, the actual true-length dimensions must be shown on the drawings at all times. When you are reading the draw-

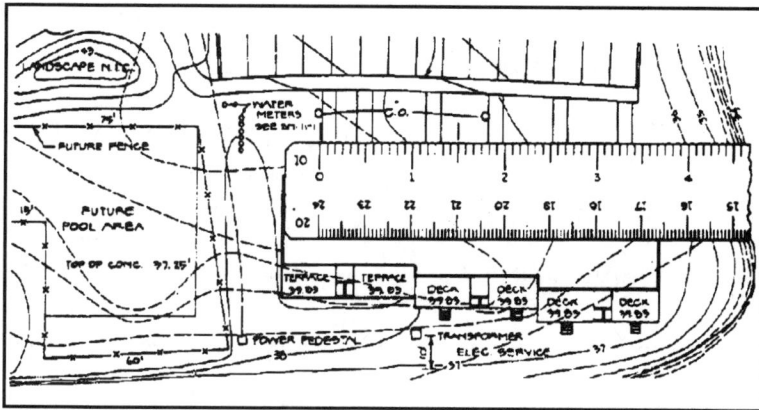

Figure 12.22   Engineer's scale.

ing plans to scale, think of each dimension in its full size and not in its reduced scale on the drawing (Fig. 12.22).

Security technicians frequently encounter plot plans when installing security systems in industrial establishments or when a network system is installed in shopping malls or tract development houses.

The metric scale

Metric scales (Fig. 12.23), are divided into centimeters (cm), with the centimeters divided into 10-divisioned millimeters (mm), or into 20-divisioned half-millimeters. Scales are available with metric divisions on one edge, but inch divisions are inscribed on the opposite edge. Many contracting firms that deal in international trade have adopted a dual-dimensioning system expressed in both metric and English symbols. Furthermore, drawings prepared for government projects frequently require metric dimensions.

Schematic Diagrams

In previous chapters we reviewed various types of electrical circuits, including series, parallel, and series-parallel. Consequently, you should already have a basic understanding of schematic diagrams. However, let's review schematic drawings briefly to refresh your memory.

In general, a schematic diagram is a picture of an electrical/electronic circuit that uses symbols to represent the various circuit components and lines to connect these components. Basic symbols used in schematic diagrams are shown in Fig. 12.24. However, as you have already learned in an earlier chapter, this list is by no means complete. In fact, you will encounter dozens of other symbols when using schematic diagrams—a different symbol for each of the components used in the circuit. If any variations are necessary, an exact description of each is usually listed in schedules or else noted on the drawings.

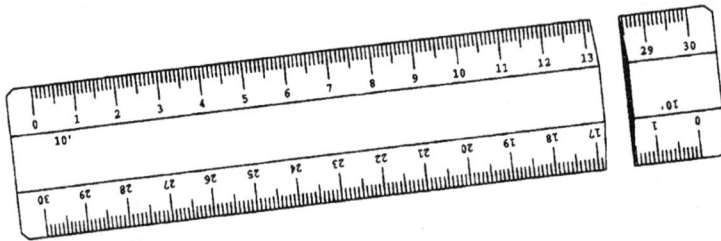

Figure 12.23 Metric scale.

———— Wire	Lamp incandescent
Conductors Connected	Fuse
Connected	Resistors Fixed
Not connected	Variable (potentiometer)
Ground	Rheostat
Cell	Switch
Battery	Voltmeter
or	Ammeter

Figure 12.24 Symbols commonly used in electrical schematic diagrams.

Schematic diagrams indicate the scheme or plan according to which components are connected for a specific purpose. They are seldom, if ever, drawn to scale as an architectural drawing would be. They appear in diagrammatic form, and the symbols rarely look exactly like the component. In the better drawings, however, the components are arranged in a neat and logical sequence so that they can be easily traced and clearly understood.

To serve all its intended purposes, the schematic diagram must be accurate. Also, it must be understood by all qualified personnel, and it must provide definite information without ambiguity.

The schematics for a security system should indicate all circuits in the system. In nearly all cases, the conductors connecting the electronic symbols will be drawn either horizontally or vertically. Rarely are they ever slanted.

A dot at the junction of two crossing wires means a connection between the two wires. An absence of a dot, in most cases, indicates that the wires cross without connecting.

Schematic diagrams are, in effect, shorthand explanations of the manner in which an electrical/electronic circuit or group of circuits operates. They make extensive use of symbols and abbreviations.

Typically, each component on a complete schematic diagram has a number to identify it. Supplementary data about each area is supplied on the diagram or on an accompanying list in the form of a schedule, which describes the component in detail or refers to a common catalog number familiar to the trade.

To interpret schematic diagrams, remember that each circuit must be closed in itself. Each component should be in a closed loop connected by conductors to a source of electric current, such as a battery, transformer, or other power supply. There will always be a conducting path leading from the source to the component and a return path leading from the component to the source. The path can consist of one or more conductors. Other components can also be in the same loop or in additional loops branching off to other devices. Again, for each component, it must be possible to trace a completed conducting loop to the source.

The common flashlight is an example of a basic electric circuit. It contains a source of electrical energy (the dry cells in the flashlight), a load (the bulb), which changes the electrical energy into light energy, and a switch to control the energy delivered to the load; that is, to turn the light off and on.

Schedules

A schedule, as applied to print reading, is a systematic method of presenting notes or lists of materials, equipment, components, etc. on a drawing in tabular form. When properly organized and thoroughly understood by both the drafter and those reading the drawings, schedules are great time-saving devices.

Refer again to the common flashlight circuit. In most cases, there are three major components in the circuit:

- Switch
- Battery
- Bulb

In a circuit as simple as this, the components could be easily indicated by note on the schematic. However, on more complex circuits, such notes would tend to make the drawing "muddy" and if there were many components of the same type, it would be easier to list them in a schedule such as the one in Fig. 12.25.

A checklist schedule for an electronic security device is shown in Fig. 12.26. This schedule allows the person installing the system to perform tests to ensure that correct connections have been made. The schedule is to be used with a voltmeter. Notice that the person testing the circuit reads the top line of the schedule; he places the negative test lead to P102, pin 2 and the positive test lead to P102, pin 1. The voltmeter reading should show 70 V dc. If not, something in the circuit is at fault.

If the same information in this schedule was included in the written specifications, more time would be required to explain how the test is made. The person building the project would have to comb through pages of written specifications to find the explanation.

From this explanation, it should also be obvious that schematic diagrams and schedules can be a great help in installing new security systems, as well as troubleshooting existing systems.

Riser Diagrams

Single-line block diagrams are used extensively to show the arrangement of security equipment and components on many security/fire-alarm installations. The riser diagram in Fig. 12.27, for example, shows a riser diagram for a fire-alarm system used in a small commercial building. The heart of this system is the main or master control panel indicated by the large square in the drawing. To this panel are connected various detector and alarm circuits.

The primary power for this fire-alarm system is taken from an unswitched three-wire distribution line of 120/240 V ac. The initiating and alarm circuits

Component Identification	Name of Item	GE Catalog Number
S1	Switch	1087-BA
DS1	3-V lamp	9875-3 V
Battery (2)	1.5 V battery	4756-1.5 V

Figure 12.25   Component schedule for a common flashlight circuit.

Negative Lead To:	Positive Lead To:	Meter Reading
P102, pin 2	P102, pin 1	70V dc
P102, pin 4	P102, pin 3	13 V dc
P102, pin 7	P102, pin 6	+20 V dc
P102, pin 7	P102, pin 8	-20 V dc
Heat Sink	S516, pin 3	5 V dc
Heat Sink	S516, pin 1	+20 V dc

Figure 12.26   Checklist schedule for a security-system component.

are connected to the neutral ground and to one leg of the main circuit. Trouble-indicator circuits are connected to a neutral ground and also to the opposite leg of the circuit.

When an automatic detector or manual station is activated, the contacts close to complete a circuit path and apply 120 V of power to the alarm control circuits in the main control panel. This includes a synchronous motor on some systems, which immediately operates various cam assemblies that cause the alarm circuit switch contacts to make and break in a coded sequence. Additional cam-controlled switches stop the motor and alarm signals after, for example, four complete rounds and actuate the alarm buzzer on the main panel.

Most fire-alarm panels contain a supplementary relay control for connection to an external auxiliary circuit to operate other devices or circuits related to the system. The relay usually has a single-pole, double-throw contact that operates in step with the master code signal. The circuit can be used to activate other auxiliary alarms or controls, such as a city fire-department connection, fan shutdown, or door release.

All of the electrical/electronic details are not shown on the riser diagram in Fig. 12.27. The riser diagram merely shows the number and relation of the components to the main control panel, along with the number of circuits and conductors. Other details are usually furnished in schedules, written specifications, or wiring diagrams.

A schematic wiring diagram like the one shown in Fig. 12.28 should always be provided during the rough-in stage of construction, as well as during actual connection of the system and the various components. Details of station mounting should also be included.

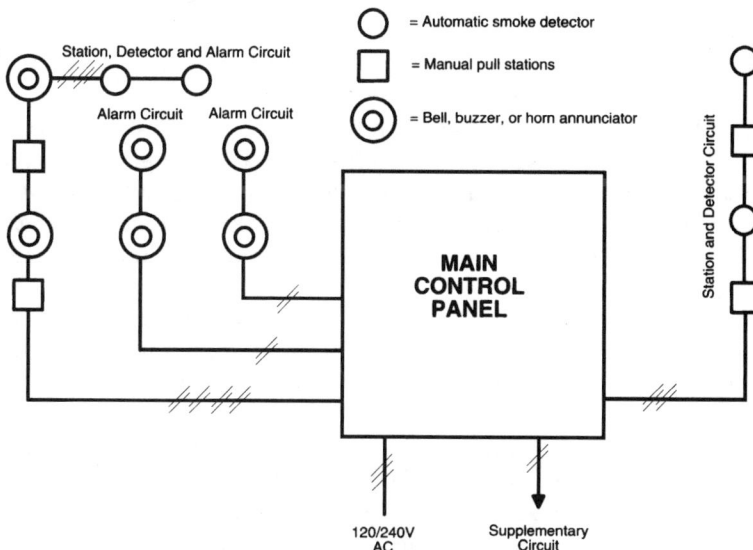

Figure 12.27 A riser diagram for a typical fire-alarm system.

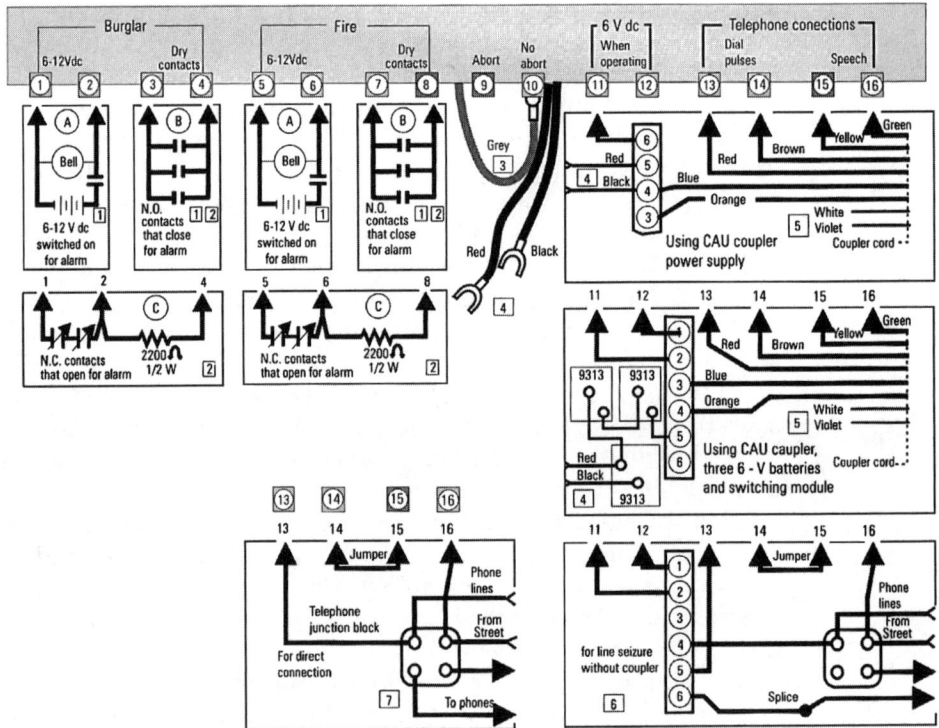

Figure 12.28 Schematic wiring diagrams should always be provided during the rough-in stage of any security system.

Drawing Details

A detail drawing is a drawing of a separate item or portion of a security system, giving a complete and exact description of its use and all the details needed to show the workers exactly what is required for the installation. The riser diagram in Fig. 12.27 indicates the installation of four smoke detectors. This is a good example of where an extra, detailed drawing is desirable. Because several types of smoke detectors are on the market, workers need to know the rough-in dimensions, mounting requirements, and circuit connections.

A detail drawing of the smoke detector such as shown in Fig. 12.29 leaves little doubt about the detector's dimensions and mounting requirements. The additional connection detail in Fig. 12.30 further facilitates the installation of the smoke detectors.

Security equipment manufacturers will sometimes provide pictorial drawings in the form of isometric or perspective views of the installation requirements of their equipment (Fig. 12.31). Even with these highly realistic drawings, schematic wiring diagrams must also be provided to detail the connections of each component in the system.

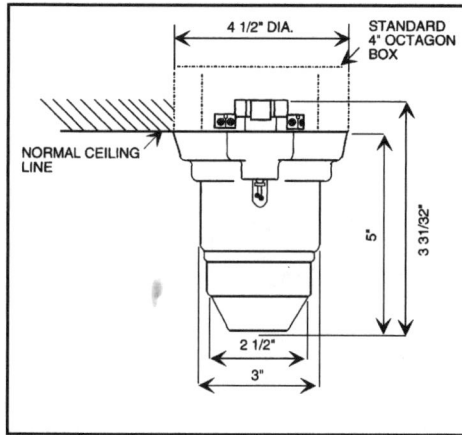

Figure 12.29 Details of smoke-detector mounting requirements.

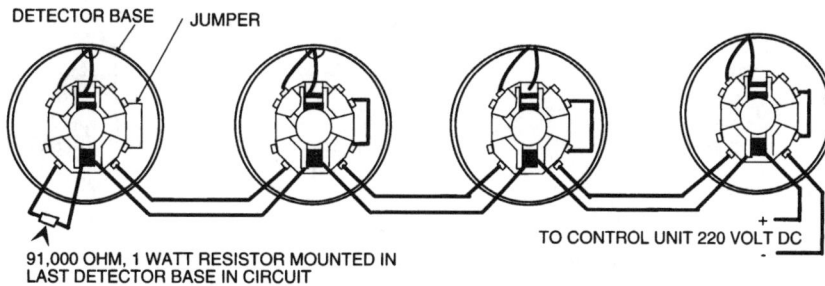

Figure 12.30 Wiring or connection details usually are provided during the rough-in stage as well as during the time of mounting the equipment.

Many older fire-alarm systems are being updated with modern ionization, infrared, and photoelectric detection capabilities. A special control panel is added to the old system to accommodate the new detectors. In such situations, it is best to have complete schematic diagrams of the existing system as well as the new connections. Sometimes these drawings will be provided by the manufacturer or engineering firm; other times the security technician will have to make a survey of the building, and prepare these drawings from scratch. These drawings (with all necessary dimensions) are usually sketched on graph paper at the job site and then put into finished form in the drafting room. Using a CAD package in a laptop can speed this process and make it much easier. Once the drawings have been prepared, the new system may be integrated with the old quite easily by experienced workers.

Once the extent of the existing system has been determined, fire-alarm equipment manufacturers will usually provide security technicians with complete instructions for the installation or modification of the existing system—often at no charge if the equipment is purchased from the manufacturer doing the design drawings.

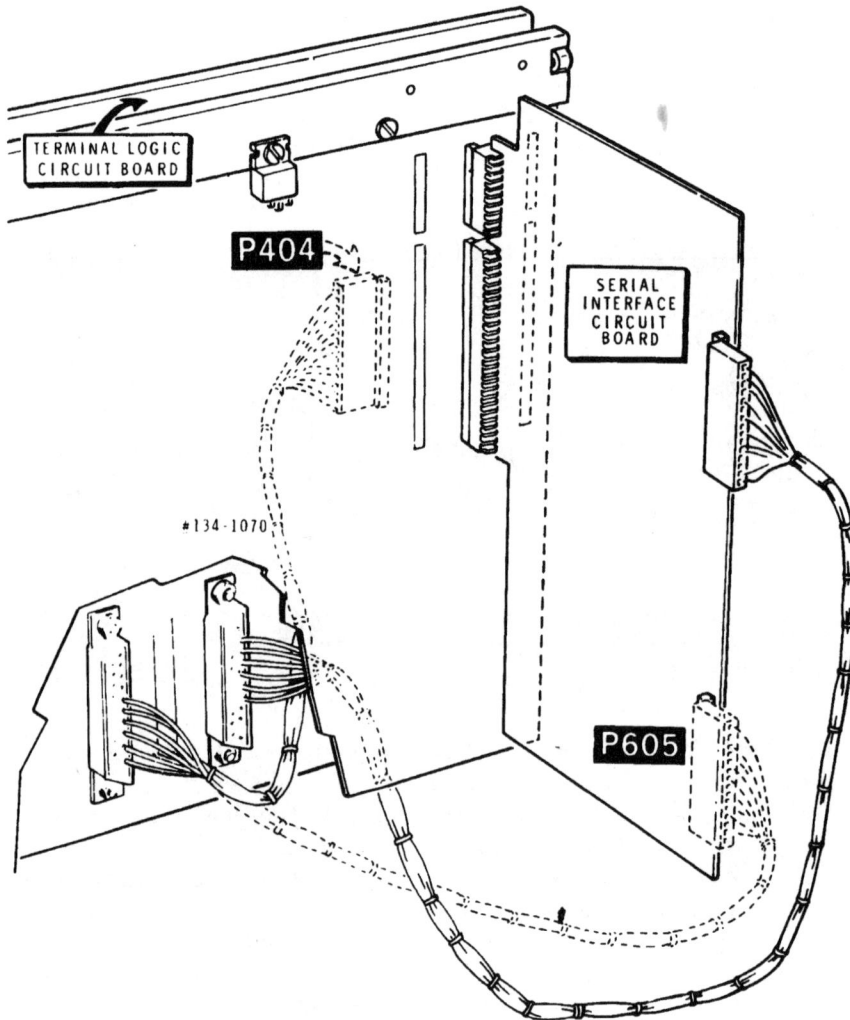

Figure 12.31   Even if highly detailed pictorial drawings are furnished with the security equipment, technicians still need a schematic wiring diagram to ensure correct connections.

Written Specifications

The specifications for a building or project are the written descriptions of work and duties required by the owner, the architect, and the consulting engineer. Together with the working drawings, these specifications form the basic construction for contract requirements for the project, including the mechanical and electrical systems associated with the building. Those who use such construction drawings and specifications must always be alert to discrepancies between the working drawings and the written specifications. Such discrepancies occur particularly when:

- Architects or engineers use standard or prototype specifications and attempt to apply them without any modification to specific working drawings.
- Previously prepared standard drawings are changed or amended by reference in the specifications only and the drawings themselves are not changed.
- Items are duplicated in both the drawings and specifications, but an item is subsequently amended in one and overlooked on the other contract document.

In such instances, the person in charge of the project has the responsibility to ascertain whether the drawings or the specifications take precedence. Such questions must be resolved, preferably before the work is started, to avoid added cost to either the owner, the architect/engineer, the contractor, or the security/electronics technician.

How specifications are written

Writing accurate and complete specifications for building construction and the building's related systems is a serious responsibility for those who design the buildings because the specifications, combined with the working drawings, govern practically all important decisions made during the construction span of every project. Compiling and writing these specifications is not a simple task—even for those who have had considerable experience in preparing such documents. A set of written specifications for a single project will typically contain thousands of products, parts and components, and methods of installing them, all of which must be covered either on the drawings and/or in the specifications document. No one can memorize all of the necessary items required to accurately describe the various areas of construction. One must rely upon reference materials: manufacturer's data, catalogs, checklists, and, best of all, a high-quality master specification.

Specification format

For convenience in writing, speed in estimating work, and ease in reference, the most suitable organization of the specification is a series of sections dealing successively with the different trades, and in each area grouping all the work of the particular trade to which the section is devoted. All the work of each trade should be incorporated into the section devoted to that trade. Those people who use the specifications must be able to find all information needed without taking too much time in looking for it. Computers, with the use of databases and their ability to search rapidly, can facilitate specification work tremendously.

The CSI format

The Construction Specification Institute (CSI) developed the Uniform Construction Index some years ago and now specifications, product information, and cost data are typically referenced with the CSI's uniform system. This format is now followed on most large construction projects in North America. All construction is divided into 16 divisions, and each division has several sections and subsections. The following outline describes the various divisions normally included in a set of specifications for building construction.

Division 1—General Requirements. This division summarizes the work, alternatives, project meetings, submissions, quality control, temporary facilities and controls, products, and the project closeout. Every responsible person involved with the project should become familiar with this division.

Division 2—Site Work. This division outlines work involving such items as paving, sidewalks, outside utility lines (electrical, plumbing, gas, telephone, etc.), landscaping, grading, and other items pertaining to the outside of the building.

Division 3—Concrete. This division covers work involving footings, concrete formwork, expansion and contraction joints, cast-in-place concrete, specially finished concrete, precast concrete, concrete slabs, etc.

Division 4—Masonry. This division covers concrete, mortar, stone, masonry accessories, etc.

Division 5—Metals. Metal roofs, structural metal framing, metal joists, metal decking, ornamental metal, and expansion control normally fall under this division.

Division 6—Carpentry. Items falling under this division include: rough carpentry, heavy timber construction, trestles, prefabricated structural wood, finish carpentry, wood treatment, architectural woodwork, and the like. Plastic fabrications can also be included in this division of the specifications.

Division 7—Thermal and Moisture Protection. Waterproofing is the main topic discussed under this division. Other related items (such as dampproofing, building insulation, shingles and roofing tiles, preformed roofing and siding, membrane roofing, sheet metal work, wall flashing, roof accessories, and sealants) are also included.

Division 8—Doors and Windows. All types of doors and frames are included under this division: metal, plastic, wood, etc. Windows and framing are also included along with hardware and other window and door accessories.

Division 9—Finishes. Included in this division are the types, quality, and workmanship of lath and plaster, gypsum wallboard, tile, terrazzo, acoustical treatment, ceiling suspension systems, wood flooring, floor treatment, special coatings, painting, and wallcovering.

Division 10—Specialties. Specialty items (such as chalkboards and tackboards; compartments and cubicles, louvers and vents that are not connected with the heating, ventilating, and air conditioning system; wall and corner guards; access flooring; specialty modules; pest control; fireplaces; flagpoles; identifying devices; lockers; protective covers; postal specialties; partitions; scales; storage shelving; wardrobe specialties; etc.) are covered in this division of the specifications.

Division 11—Equipment. The equipment included in this division could include central vacuum cleaning systems, bank vaults, darkrooms, food service, vending machines, laundry equipment, and many similar items.

Division 12—Furnishing. Such items as cabinets and storage, fabrics, furniture, rugs and mats, seating, and other similar furnishing accessories are included under this division.

Division 13—Special Construction. Such items as air-supported structures, incinerators, and other special items will fall under this division.

Division 14—Conveying Systems. This division covers conveying such apparatus as dumbwaiters, elevators, hoists and cranes, lifts, material-handling systems, turntables, moving stairs and walks, pneumatic tube systems, and also powered scaffolding.

Division 15—Mechanical. This division includes plumbing, heating, ventilating, and air conditioning and related work. Electric heat is sometimes covered under Division 16—especially if individual baseboard heating units are used in each room or area of the building.

Division 16—Electrical. This division covers all electrical requirements for the building including lighting, power, alarm and communication systems, special electrical systems, and related electrical equipment. This is the division that security technicians will use the most. Division 16 contains the following main sections:

DIVISION 16—ELECTRICAL
16050   Electrical Contractors
16200   Power Generation
16300   Power Transmission
16400   Service and Distribution
16500   Lighting
16600   Special Systems
16700   Communications
16850   Heating and Cooling
16900   Controls and Instrumentation

These sections are further subdivided into many subsections. For example, items covered under Section 16400, Service and Distribution, will usually include the project's service entrance, metering, grounding, service-entrance conductors, and similar details.

Print and specification readings, like having a thorough understanding of the code, is very important for the professional. And in the new century, with low-voltage systems playing a bigger role in buildings everyday, it is important to sharpen your skills at plan reading on a continual basis.

Using the CD

On the inside of the rear cover of the book is a sleeve with your compact disc in it. Open the packet and place the CD in the CD drive of your computer.

Load Microsoft Word or word-processing software that imports MS Word files and translates them automatically to work fully in your software. If you have any doubts about these instructions, ask your consultant to do this with you the first time.

Now you are ready to use any of the files on the CD. Go to FILE OPEN in your software and click on D: (or however your CD drive is labeled). The files will be listed, and the names are the same as they are titled in the book.

The magic of computers is at your finger tips. You can open any of the checklist files found in the book. They become word-processing files automatically. You can then save them on your hard drive. You can make copies of them, do test lists, or print them and hand them out to your staff for issuing at job sites and throughout the streamlining of your firm.

The REFERENCE file opens in exactly the same way as the checklist files. The REFERENCE LIST appears exactly as it does in the back of your book. After you have opened the file, save it to your ordinary word-processing directory. If this is confusing to you, have your consultant visit and do it with the person in order to adapt it to your system.

Block any of the names of groups or companies at the top of a listing and copy it to the search line on your Internet provider's screen. Then click your provider's Search button, go into the listing, and explore it at your leisure. You can then save it as a favorite address in your e-mail software. For example, if America Online is your Internet provider, you will save the address in your AOL favorites. Again, if you are a new user, you must get your consultant to help you wade through this process. Never waste your time playing with computers when you are a low-voltage professional. That would be like trying to manufacture alarms for one of your jobs—it would affect your cash flow drastically.

Welcome to the Net. You will never regret the money you have spent and you will never turn back.

Glossary

AA (Aluminum Association) A manufacturers' association that promotes the use of aluminum.

AAC All-aluminum conductor.

AEC The Architecture, Engineering, Contracting Industry.

AIA (American Institute of Architects) The largest organization of architects in the United States.

ASC Aluminum alloy stranded conductors.

Abrasion The process of rubbing, grinding, or wearing away by friction.

Abrasion resistance Ability to resist surface wear.

Abrasive paper Paper or cloth on which flint, garnet, emery, aluminum oxide, or corundum has been fastened with glue or some other adhesive. One use in the electrical field is to clean conductors, contacts, or terminals.

ac (alternating current) 1. A periodic current, the average of which is zero over a period; normally the current reverses after given time intervals and has alternately positive and negative values. 2. The type of electrical current actually produced in a rotating generator (alternator).

Accelerated life tests Subjecting a product to operating conditions more severe than normal to expedite deterioration, affording some measure of probable life at normal conditions.

Accelerator 1. A substance that increases the speed of a chemical reaction. 2. Something to increase velocity.

Accent lighting Lighting used to emphasize an object or specific area.

Acceptable (nuclear power) Demonstrated to be adequate by the safety analysis of the station.

Acceptance test Made to demonstrate the degree of compliance with specified requirements.

Accepted Approval for a specific installation or arrangement of equipment or materials.

Accessible Capable of being removed or exposed without damaging the building structure or finish, or not permanently closed in by the structure or finish of the building. *See* readily accessible.

ACSR (aluminum, conductor, steel reinforced) A bare composite of aluminum and steel wires, usually aluminum around steel.

Actuated equipment (nuclear power)   Component(s) that perform a protective function.

Administrative authority   An organization exercising jurisdiction over the National Electrical Safety Code.

AEIC   Association of Edison Illuminating Companies.

Aggregate   Material mixed with cement and water to produce concrete.

Aging   The irreversible change of material properties after exposure to an environment for an interval of time.

AIA   1. American Institute of Architects. 2. Aircraft Industries Association.

Air cleaner   Device used to remove airborne impurities.

Air diffuser   Air-distribution outlet designed to direct airflow into desired patterns.

Air-entrained concrete   Concrete in which a small amount of air is trapped by addition of a special material to produce greater durability.

Air flow   The distribution or movement of air.

Air gap   A small gap that is deliberately left in the magnetic core of an electromagnetic device. When the coil becomes deenergized, some magnetic flux (residual magnetism) always remains. If it were not for the air gap in the magnetic core, the residual magnetism might be sufficient to hold the armature in the sealed-in position.

Air horsepower (AHP)   Work done in moving a given volume or weight of air at a given speed.

Air oven   A lab oven used to heat by convection of hot air.

Al   Aluminum.

Alarm-indicating appliance   An appliance that produces an audible or visible alarm signal, or both, in response to a fire emergency.

Alarm initiating device   A device that, when actuated, initiates an alarm. Such devices, depending on their type, can be operated manually or actuated automatically in response to smoke, flame, heat, or waterflow.

Alarm signal   A signal indicating an emergency requiring immediate action, such as an alarm for fire from a manual station, a waterflow alarm, or an alarm from an automatic fire-alarm system.

Alarm system   A combination of compatible initiating devices, control panels, and indicating appliances designed and installed to produce an alarm signal in the event of fire, intrusion, or other similar action.

Al-Cu   An abbreviation for aluminum and copper, commonly marked on terminals, lugs, and other electrical connectors to indicate that the device is suitable for use with either aluminum conductors or copper conductors.

Alive   Energized; having voltage applied.

Alligator wrench   A wrench with toothed V-shaped jaws fixed in position.

Alloy   A substance having metallic properties and being composed of elemental metal and one or more chemical elements.

Alternator   A device to produce alternating current. Alternators range in size from small automotive types to huge types used in power plants to produce electricity for crosscountry distribution.

Alumoweld®   An aluminum clad steel wire by Copperweld Steel Corp.

Ambient   Surrounding on all sides, such as air or noise.

Ambient temperature   Temperature of fluid (usually air) that surrounds an object on all sides.

Ambient temperature compensated   A device, such as an overload relay, which is not affected by the temperature surrounding it.

American bond   Brickwork pattern consisting of five courses of stretchers followed by one bonding course of headers.

Ammeter   An electric meter used to measure current, calibrated in amperes.

Ampacity   The current-carrying capacity of conductors or equipment, expressed in amperes.

Ampere (A)   The basic Sl unit measuring the quantity of electricity.

Ampere-hour   The quantity of electricity that passes any point in a circuit in one hour.

Ampere rating   The current-carrying capacity of an overcurrent-protective device. When a fuse or circuit breaker is subjected to a current above its ampere rating, it will open the circuit after a predetermined period of time.

Ampere squared seconds (I^2t)   The measure of heat energy developed within a circuit during the fuse's clearing. It can be expressed as "melting I^2t," or "racing I^2t" or the sum of them as "clearing I^2t." I stands for effective let-through current (RMS), which is squared, and the t stands for time of opening in seconds.

Ampere-turn   The product of amperes times the number of turns in a coil.

Amplification   Procedure of expanding the strength of a signal.

Amplifier   1. A device that enables an input signal to directly control a larger energy flow. 2. The process of increasing the strength of an input.

Amplitude   The maximum value of a wave.

Analog   Pertaining to data from continuously varying physical quantities.

Analog device   A device that can be adjusted to different states. A simple example is the dimmer control for a conventional lighting fixture. This type of control can turn the light on or off, or adjusted to any position between. Thus, a dimmer control is an analog device, as opposed to a digital device, such as a common on/off snap switch.

Ancillary cashflow   A business that is a separate profit center within the main business. For example, if a security systems contractor also provides a systems maintenance business.

Angle bracket   A form of support having two faces generally at right angles to each other. A web is often added to increase strength.

Angle, roll over (overhead)   The sum of the vertical angles between the conductor and the horizontal on both sides of the traveler, excessive roll-over angles can cause premature splice failures.

Angular velocity   The average time rate of change of angular position; in electrical circuits = $2f$, and f equals frequency.

ANI (American Nuclear Insurers)   A voluntary unincorporated association of companies providing property and liability insurance for U.S. nuclear power plants; formerly NELPIA.

Annealing   The process of preventing or removing objectional stresses in materials by controlled cooling from a heated state, measured by tensile strength.

Annealing, bright   Annealing in a protective environment to prevent discoloration of the surface.

Annunciator   An electrically or mechanically operated visible signaling device having one or more target or lamp indicators to display the status of a security/fire-alarm system.

Anode   1. Positive electrode through which current enters a nonmetallic conductor, such as an electrolytic cell. 2. The negative pole of a storage battery.

ANSI (American National Standards Institute)   An organization that publishes nationally recognized standards.

Antenna   A device for transmission or reception of electromagnetic waves.

Anti-dogleg methods   Any method used to secure or strap conduit in place while bending is occurring.

Antioxidant   Retards or prevents degradation of materials exposed to oxygen (air) or peroxides.

Antisiphon trap   Trap in a drainage system designed to preserve a water seal by defeating siphonage.

Aperture seal (nuclear)   A seal between containment aperture and the electrical penetration assembly.

Appendix, NEC   The National Electrical Code (NEC) Appendices A and B are located at the end of the code book. Appendix A includes material extracted from other NFPA documents. Appendix B is not part of the requirements of the NEC and contains additional material for informational purposes only.

Appliance   Equipment designed for a particular purpose, using electricity to produce heat, light, mechanical motion, etc.; usually complete in itself, generally other than industrial use, normally in standard sizes or types.

Approved   1. Acceptable to the authority having legal enforcement. 2. Per Occupational Safety and Health Act. A product that has been tested to standards and found suitable for general application, subject to limitations outlined in the nationally recognized testing lab's listing.

Apron   Piece of horizontal wood trim under the sill of the interior casing of a window.

Arc   A flow of current across an insulating medium.

Arc furnace   Heats by heavy current flow through the material to be heated.

Arcing time   The amount of time from the instant the fuse link has melted until the overcurrent is interrupted or cleared.

Arc resistance   The time required for an arc to establish a conductive path in or across a material.

Area (A)   The square measurement of any plane surface or cross section of a conduit, air duct, etc., or the circular plane of a propeller.

Areaway Open space below the ground level immediately outside a building. It is enclosed by substantial walls.

Armature 1. Rotating machine, the member in which alternating voltage is generated. 2. Electromagnet, the member that is moved by magnetic force.

Armor Mechanical protector for cables; usually a helical winding of metal tape, formed so that each convolution locks mechanically upon the previous one (interlocked armor); can be a formed metal tube or a helical wrap of wires.

Arrester 1. A circuit device designed to check and limit voltage surges caused by lightning. 2. Wire screen secured to the top of an incinerator to confine sparks and other products of burning.

Ashlar Squared and dressed stones used to face a masonry wall; short upright wood pieces extending from the attic floor to the rafters forming a dwarf wall.

Article, NEC The National Electrical Code (NEC) is organized into nine chapters. The main body of the NEC text begins with Article 90 "Introduction," and ends with Article 820 "Community Antenna Television and Radio Distribution Systems."

Askarel A synthetic insulating oil that is nonflammable, but very toxic. It is being replaced by silicone oils.

ASME American Society of Mechanical Engineers.

Assembly method Estimating technique in which the parts of construction components such as a switch box are all combined to give a total of components within the item for rapid takeoffs and extensions of materials and labor costs.

Associated circuits (nuclear power) Nonclass IE circuits that share power supplies or are not physically separated from Class IE circuits.

ASTM (American Society for Testing and Materials) A group writing standards for testing materials and specifications for materials.

Asymmetrical Not identical on both sides of a central line; unsymmetrical.

Atom The smallest particle to which an element can be divided and still retain the properties of the element. According to our present understanding, the atom is believed to consist of a central nucleus composed of protons and neutrons, surrounded by orbiting electrons.

Attachment plug or cap The male connector for electrical cords.

Attenuation A decrease in energy magnitude during transmission.

Audible Capable of being heard by humans.

Auditable data Technical information that is documented and organized to be readily understandable and traceable to independently verify inferences or conclusions based on these records.

Auger A wood-boring tool of large size with handle attached at right angles to the tool line. Several types are made for different purposes.

Autoclave A heated pressure vessel used to bond, cure, seal, or used for environmental testing.

Automatic Operating by own mechanism when actuated by some impersonal influence; nonmanual; self-acting.

Automatic transfer equipment A device to transfer a load from one power source to another, usually from normal to emergency source and back.

Autotransformer Any transformer where primary and secondary connections are made to a single cell. The application of an autotransformer is a good choice for some users where a 480Y/277- or 208Y/120-volt, three-phase, four-wire distribution system is utilized.

Auxiliary A device or equipment that aids the main device or equipment.

AWG (American Wire Gauge) The standard for measuring wires in America.

Awl A small pointed tool for making holes for nails or screws. When used to mark metal objects, it is sometimes called *scratch awl.*

Axial Of or forming an axis; on or around an axis.

Axis The line about which a rotating body turns.

Ballast A device designed to stabilize current flow.

Bar Association The main organization of lawyers in the United States.

Battery A device that changes chemical to electrical energy, used to store electricity.

Bimetal strip Temperature regulating or indicating device that works on the principle that two dissimilar metals with unequal expansion rates, welded together, will bend as temperature changes.

Bonding bushing A special conduit bushing equipped with a conductor terminal to take a bonding jumper; also a screw or other sharp device used to bite into the enclosure wall to bond the conduit to the enclosure without a jumper when no concentric knockouts are left in the wall of the enclosure.

Bonding jumper A bare or insulated conductor used to ensure the required electrical conductivity between metal parts required to be electrically connected. Frequently used from a bonding bushing to the service equipment enclosure to provide a path around concentric knockouts in an enclosure wall; also used to bond one raceway to another.

Bonding locknut A threaded locknut for use on the end of a conduit terminal, but a locknut equipped with a screw through its lip. When the locknut is installed, the screw is tightened so that its end bites into the wall of the enclosure close to the edge of the knockout.

Braid An interwoven cylindrical covering of fiber or wire.

Branch circuit That portion of a wiring system extending beyond the final overcurrent device protecting a circuit.

Bridge A circuit that measures by balancing four impedances through which the same current flows:
- Wheatstone: resistance
- Kelvin: low resistance
- Schering: capacitance, dissipation factor, dielectric constant
- Wien: capacitance, dissipation factor

Bus The conductor(s) serving as a common connection for two or more circuits.

Bus bars The conductive bars used as the main current supplying elements of panel boards or switchboards; also, the conductive bars duct; an assembly of bus bars within an enclosure that is designed for ease of installation, has no fixed electrical characteristics, and allows power to be taken off conveniently, usually without circuit interruption.

Business plan Study of an existing or proposed business typically used to obtain capital.

BX A generic nickname for armored cable (wires with a spiral-wound, flexible steel outer jacketing). BX is a registered tradename of the General Electric Company.

Bypass Passage at one side of or around a regular passage.

Cable An assembly of two or more wires that can be insulated or bare.

Cable, aerial An assembly of one or more conductors and a supporting messenger.

Cable, armored A cable having armor. *See* Armor.

Cable, belted A multiconductor cable having a layer of insulation over the assembled insulated conductors.

Cable clamp A device used to clamp around a cable to transmit mechanical strain to all elements of the cable.

Cable, coaxial A cable used for high frequency, consisting of two cylindrical conductors with a common axis separated by a dielectric. Normally, the outer conductor is operated at ground potential for shielding.

Cable, control Used to supply voltage (usually on or off).

Cable, duplex A twisted pair of cables.

Cable, power Used to supply current (power).

Cable, pressure A cable having a pressurized fluid (gas or oil) as part of the insulation. Paper and oil are the most common insulators.

Cable, ribbon A flat multiconductor cable.

Cable, service drop The cable from the utility line to the customer's property.

Cable, signal A cable used to transmit data.

Cable, spacer An aerial distribution cable made of covered conductors held by insulated spacers, designed for wooded areas.

Cable tray A rigid structure to support cables. A type of raceway that normally has the appearance of a ladder and is open at the top to facilitate changes.

Cable, tray A multiconductor with a nonmetallic jacket, designed for use in cable trays (not to be confused with type TC cable, for which the jacket must also be flame retardant).

Cable, triplexed d Helical assembly of three insulated conductors and sometimes a bare grounding conductor.

Cable, unit A cable having pairs of cables stranded into groups (units) of a given quantity; these groups then form the core.

Cable, vertical riser Cables utilized in circuits of considerable elevation change; usually incorporate additional components for tensile strength.

Cabling   Helically wrapping together of two or more insulated wires.

Capacitance   The storage of electricity in a capacitor; the opposition to voltage change. The unit of measurement is the farad.

Capacitor   An apparatus consisting of two conducting surfaces separated by an insulating material. It stores energy, blocks the flow of direct current, and permits the flow of alternating current to a degree, depending on the capacitance and frequency.

Capillary action   The traveling of liquid along a small vessel because of surface tension.

Capstan   A rotating drum used to pull cables or ropes by friction. The cables are wrapped around the drum.

Cashflow   The money that moves through the accounts of a company. Sometimes used in reference to the funds that remain after all payables have been factored into the complete assets of the firm.

Cathode   1. The negative electrode through which current leaves a nonmetallic conductor, such as an electrolytic cell. 2. The positive pole of a storage battery. 3. In a vacuum tube, the electrode that emits electrons.

Cathode-ray tube   An electronic tube that has a screen upon which a beam of electrons from the cathode can be made to create images. For example, the television picture tube.

Cathodic protection   Reduction or prevention of corrosion by making the metal to be protected the cathode in a direct current circuit.

Cavity wall   Wall built of solid masonry units arranged to provide air space within the wall.

CB   Pronounced "see bee." An expression used to refer to "circuit breaker."

C-C   Center to center.

CCA   *See* Customer Complaint Analysis.

CD litigation   Construction defect lawsuits.

CEE   *See* International Commission on Rules for the Approval of Electrical Equipment.

Centigrade scale temperature   Scale used in metric system. Freezing point of water is 0°C; boiling point is 100°C.

CFR   *See* Code of Federal Regulations.

Change order   A request to alter a task from the description given in the contract documents.

Choke coil   A coil used to limit the flow of alternating current while permitting direct current to pass.

Circuit   A closed path through which current flows from a generator, through various components, and back to the generator.

Circuit breaker   A resettable fuse-like device designed to protect a circuit against overloading.

Circuit foot   One foot of circuit. If one has a three-conductor circuit, each lineal foot of circuit would have three circuit feet.

Circular mil   The non-SI unit for measuring the cross-sectional area of a conductor.

CL   Center line.

Clearance   The vertical space between a cable and its conduit.

Coaxial cable   A cable consisting of two conductors concentric with and insulated from each other.

Code   Short for National Electrical Code.

Code installation   An installation that conforms to the local code and/or the national code for safe and efficient installation.

Code of Federal Regulations (CFR)   The general and permanent rules published in the Federal Register by the executive departments and agencies of the federal government. The code is divided into 50 titles, which represent broad areas. Titles are divided into chapters, which usually bear the name of the issuing agency. For example, Title 30—Mineral Resources, Chapter I—MESA; Title 29—Labor, Chapter XVII—OSHA; Title 10—Energy, Chapter I—NRC.

Color code   Identifying conductors by the use of color.

Come along   A cable grip (usually of tubular basket-weave construction that tightens its grip on the cable as it is pulled) with a pulling "eye" on one end for attaching to a pull-rope for pulling conductors into a conduit or other raceway.

Completion date   The date for finishing a project that is specified in a contract.

Computer   An electronic apparatus: 1. for solving complex and involved problems, usually mathematical or logical, rapidly. 2. For storing large amounts of data.

Computer codes   Symbols given to various electrical materials.

Concealed   Rendered inaccessible by the structure or finish of the building. Wires in concealed raceways are considered concealed—even though they could become accessible by being withdrawn.

Concentricity   The measurement of the center of the conductor with respect to the center of the insulation.

Conductance   The ability of material to carry an electric current.

Conductor   Any substance that allows energy flow through it, with the transfer being made by physical contact, but excluding net mass flow.

Conductor, bare   Having no covering or insulation whatsoever.

Conductor, covered   A conductor having one or more layers of nonconducting materials that are not recognized as insulation under the National Electrical Code.

Conductor, insulated   A conductor covered with material recognized as insulation.

Conductor load   The mechanical load on an aerial conductor: wind, weight, ice, etc.

Conductor, plain   A conductor that consists of only one metal.

Conductor, segmental   A conductor that has sections isolated one from the other and connected in parallel; used to reduce ac resistance.

Conductor, solid   A single wire.

Conductor, stranded   Assembly of several wires, usually twisted or braided.

Conductor stress control   The conducting layer applied to make the conductor a smooth surface in intimate contact with the insulation; formerly called *extruded strand shield (ESS)*.

Conduit   A tubular raceway.

Conduit rill   Amount of cross-sectional area used in a raceway.

Conduit, rigid metal   Conduit made of Schedule 40 pipe, normally in 10-foot lengths.

Configuration, cradled   The geometric pattern which cables will take in a conduit when the cables are pulled in parallel and the ratio of the conduit ID to the IIC cable OD is greater than 3.0.

Configuration, triangular   The geometric pattern that cables will take in a conduit when the cables are triplexed or are pulled in parallel with the ratio of the conduit ID to the IIC cable OD less than 2.5.

Connection   1. The part of a circuit that has negligible impedance and joins components or devices. 2. A cable terminal, splice, or seal at the interface of the cable and equipment.

Connection, delta   Interconnection of three electrical equipment windings in delta (triangular) fashion.

Connection, star   Interconnection of three electrical equipment windings in star (wye) fashion.

Connector   A device used to physically and electrically connect two or more conductors.

Connector, pressure   A connector applied by using pressure to form a cold weld between the conductor and the connector.

Connector, reducing   A connector used to join two different sizes of conductors.

Consequential damage   Harm to a building or site that has been caused by construction defects.

Constant current   A type of power system in which the same amount of current flows through each utilization equipment, used for simplicity in street-lighting circuits.

Constant voltage   The common type of power in which all loads are connected in parallel, but different amounts of current flow through each load.

Construction defects   Errors in workmanship or materials on a building.

Contact   A device designed for repetitive connections.

Contactor   A type of relay.

Continuity   The state of being whole, unbroken.

Continuous load   1. As stipulated by NEC: in operation three hours or more. 2. For nuclear power: 8760 hours/year (scheduled maintenance outages permitted).

Continuous vulcanization (CV)   A system utilizing heat and pressure to vulcanize insulation after extrusion onto wire or cable; the curing tube can be in a horizontal or a vertical pole.

Contract review   The process of studying a contract document, typically, before signing.

Control   Automatic or manual device used to stop, start, and/or regulate the flow of gas, liquid, and/or electricity.

Copper   A word used by itself to refer to copper conductors. Examples: "A circuit of 500 kcmil copper" or "the copper cost of the circuit." It is a good conductor of electricity, easily formed, and easily connected to itself and other metals.

Cord   A small flexible conductor assembly, usually jacketed.

Cord set   A cord having a wiring connector on one or more ends.

Core   The portion of a foundry mold that shapes the interior of a hollow casting.

Core (cable)   The portion of an insulated cable under a protective covering.

Counter emf   The voltage opposing the applied voltage and the current in a coil; caused by a flow of current in the coil. Also known as *back emf.*

Coupling   The means by which signals are transferred from one circuit to another.

Coupon   A piece of metal for testing, of specified size; a piece of metal from which a test specimen can be prepared.

Coverage   The commitment stated in an insurance policy to take care of financial responsibility for the insured by the insurance company.

CPU (Central Processing Unit)   The main chip for a computer system.

Critical path   A method for scheduling construction projects that emphasizes the completion of key elements of the project like framing, which are referred to as milestones.

CSI (Construction Specifications Institute)   Organization that is devoted entirely to developing details related to the building of structures.

CT   Pronounced "see tee," refers to current transformer.

CU   Copper.

Current (1)   The time rate of flow of electric charges; measured in amperes.

Current, charging   The current needed to bring the cable up to voltage; determined by capacitance of the cable. After withdrawal of voltage, the charging current returns to the circuit; the charging current will be 90° out of phase with the voltage.

Current density   The current per unit cross-sectional area.

Current-induced   Current in a conductor because of the application of a time-varying electromagnetic field.

Current, leakage   The small amount of current that flows through insulation whenever a voltage is present and heats the insulation because of the insulation's resistance. The leakage current is in phase with the voltage and is a power loss.

Current limiting   A characteristic of short-circuit protective devices, such as fuses, by which the device operates so fast on high short-circuit currents that less than a quarter wave of the alternating cycle is permitted to flow before the circuit is opened, thereby limiting the thermal and magnetic energy to a certain maximum value, regardless of the current available.

Customer Complaint Analysis (CCA)   A formal investigation of a cable defect or failure.

Cut in   The connection of electrical service to a building, from the power company line to the service equipment. For example, "the building was cut in" or "the power company cut in the service."

Cycle   1. An interval of space or time in which one set of events or phenomena is completed. 2. A set of operations that are repeated regularly in the same sequence. 3. A number of different processes a system in a given state goes through before it finally returns to its initial state.

Dead   1. Not having electrical charge. 2. Not having voltage applied.

Dead-end   A mechanical terminating device on a building or pole to provide support at the end of an overhead electric circuit. A *dead-end* is also the term used to refer to the last pole in the pole line. The pole at which the electric circuiting is brought down the pole to go underground or to the building served.

Dead-front   A switchboard or panel or other electrical apparatus without "live" energized terminals or parts exposed on the front, where personnel might make contact.

Demand   1. The measure of the maximum load of a utility's customer over a short period of time. 2. The load integrated over a specified time interval.

Demand factor   For an electrical system or feeder circuit, a ratio of the amount of connected load (in kVA or amperes) that will be operating at the same time to the total amount of connected load on the circuit. An 80% demand factor, for instance, indicates that only 80% of the connected load on a circuit will ever be operating at the same time. Conductor capacity can be based on that amount of load.

Details   Diagrams that explain how a particular part of a construction project is to be built.

Detection   The process of separating the modulation component from the received signal.

Device   An item intended to carry, or help carry, but not utilize electrical energy.

Dew point   The temperature at which vapor starts to condense (liquify) from a gas vapor mixture at constant pressure.

Dielectric strength   The maximum voltage that an insulation can withstand without breaking down; usually expressed as a gradient, in volts per mil (vpm).

Digitizer   Electronic device used to count items and record areas and volumes depicted on construction drawings.

Diode   A device having two electrodes, the cathode and the plate or anode, and which is used as a rectifier and detector.

Direct current (dc)   1. Electricity that flows in only one direction. 2. The type of electricity produced by a battery.

Direct job expense   The costs of items related specifically to a job.

Disconnect   A switch to disconnect an electrical circuit or load (motor, transformer, panel) from the conductors that supply power to it. For example, "He pulled the motor disconnect" means he opened the disconnect switch to the motor.

Disconnecting means   A device, a group of devices, or other means whereby the conductors of a circuit can be disconnected from their supply source.

Distribution, statistical analysis   A statistical method used to analyze data by correlating data to a theoretical curve in order to (a) test validity of data; (b) predict performance at conditions different from those used to produce the data. The normal distribution curve is most common.

Drawing, block diagram   A simplified drawing of a system showing major items as blocks; normally used to show how the system works and the relationship between major items.

Drawing, line schematic (diagram)   Shows how a circuit works.

Drawing, plot or layout   Shows the "floor plan."

Drawing, wiring diagram   Shows how the devices are interconnected.

Drill   A circular tool used for machining a hole.

Drywall   Interior wall construction consisting of plasterboard, wood paneling, or plywood nailed directly to the studs without application of plaster.

Duty, continuous   A service requirement that demands operation at a substantially constant load for an indefinitely long time.

Duty, intermittent   A service requirement that demands operation for alternate intervals of load and no load, load and rest, or load, no load, and rest.

Duty, periodic   A type of intermittent duty in which the load conditions regularly reoccur.

Duty, short-time   A requirement of service that demands operations at loads and for intervals of time that might be subject to wide variation.

Edison base   The standard screw base used for ordinary lamps.

EEL   Edison Electric Institute.

Efficiency   The ratio of the output to the input.

Elasticity   The property of recovery to original size and shape after deformation.

Electrolyte   A liquid or solid that conducts electricity by the flow of ions.

Electrolytic condenser-capacitor   Plate or surface capable of storing small electrical charges. Common electrolytic condensers are formed by rolling thin sheets of foil between insulating materials. Condenser capacity is expressed in microfarads.

Electromagnet   A device consisting of a ferromagnetic core and a coil that produces appreciable magnetic effects only when an electric current exists in the coil.

Electromotive force (emf) voltage   Electrical force that causes current (free electrons) to flow or move in an electrical circuit. The unit of measurement is the volt.

Electron   The subatomic particle that carries the unit negative charge of electricity.

Electron emission   The release of electrons from the surface of a material into surrounding space because of heat, light, high voltage, or other causes.

Electronics   The science dealing with the development and application of devices and systems involving the flow of electrons in vacuum, gaseous media, and semiconductors.

Emitter   The part of a transistor that emits electrons.

Engine   An apparatus that converts heat to mechanical energy.

Environment   1. The universe within which a system must operate. 2. All the elements over which the designer has no control and that affect a system or its inputs and outputs.

Equipment   A general term including material, fittings, devices, appliances, fixtures, apparatus, etc. used as part of, or in connection with, an electrical installation.

Extras   Work that is not in the original contract.

Farad   The basic unit of capacitance: one farad equals one coulomb per volt.

Fatigue   The weakening or breakdown of a material because of cyclic stress.

Fault   An abnormal connection in a circuit.

Fault, arcing   A fault with high impedance causing arcing.

Fault, bolting   A fault of very low impedance.

Fault, ground   A fault to ground.

Feedback   The process of transferring energy from the output circuit of a device back to its input.

Feeder   A circuit, such as conductors in conduit or a busway run, which carries a large block of power from the service equipment to a subfeeder panel, or a branch circuit panel, or to some point at which the block or power is broken down into smaller circuits.

FICA   Federal Government's Social Security and Medicare tax.

Fish tape   A flexible metal tape to fish through conduits or other raceways to pull in wires or cables. It is also made in nonmetallic form of "rigid rope" for hand fishing of raceways.

Fitting   An accessory (such as a locknut, bushing, or other part of a wiring system) that is intended primarily to perform a mechanical, rather than an electrical, function.

Flex   Common term used to refer to flexible metallic conduit.

Flexural strength   The strength of a material in bending, expressed as the tensile stress of the outermost fibers of a bent test sample at the instant of failure.

Forecasting   Business term that refers to predicting trends in the ups-and-downs of a company's cash flow.

Frequency   The number of complete cycles that an alternating electric current, sound wave, or vibrating object undergoes per second.

Friction tape   An insulating tape made of asphalt-impregnated cloth; used on 600-V cables.

Fuse   A protecting device that opens a circuit when the fusible element is severed by heating because of overcurrent passing through. Rating: voltage, normal current, maximum let-through current, time delay of interruption.

Fuse, dual element   A fuse that having two fuse characteristics; the usual combination is having an overcurrent limit and a time delay before activation.

Fuse, nonrenewable or one-time   A fuse that must be replaced after it interrupts a circuit.

Fuse, renewable link A fuse that can be reused by replacing the meltable link after current interruption.

Fusible plug A plug or fitting made with a metal of a known low melting temperature; used as a safety device to release pressures in case of fire.

FUTA Federal government's unemployment tax.

Galvanometer An instrument to indicate or measure a small electrical current by means of a mechanical motion derived from electromagnetic or dynamic forces.

Gauge 1. Dimension expressed in terms of a system of arbitrary reference numbers; dimensions expressed in decimals are preferred. 2. To measure.

Generator 1. A rotating machine to convert from mechanical to electrical energy. 2. A machine to convert automotive-mechanical to direct current. 3. General apparatus, equipment, etc. to convert or change energy from one form to another.

GFI *See* Ground Fault Interrupter.

Grade Ground level of a site and building.

Greenfield Another name for flexible metal conduit.

Grommet A plastic, metal, or rubber doughnut-shaped protector for wires or tubing as they pass through a hole in an object.

Ground A large conducting body (as the earth) used as a common return for an electric circuit and as an arbitrary zero of potential.

Ground check A pilot wire in portable cables to monitor the grounding circuit.

Ground coil A heat exchanger buried in the ground that can be used either as an evaporator or a condenser.

Grounded Connected to earth.

Grounded conductor A system or circuit conductor that is intentionally grounded.

Ground Fault Interrupter (GFI) A protective device that detects abnormal current flowing to ground and then interrupts the circuit.

Grounding The device or conductor connected to ground and designed to conduct only under abnormal conditions.

Grounding conductor A conductor used to connect metal equipment enclosures and/or the system grounded conductor to a grounding electrode, such as the ground wire run to the water pipe at a service. It also can be a bare or insulated conductor used to ground motor frames, panel boxes, and other metal equipment enclosures used throughout an electrical system. In most conduit systems, the conduit is used as the ground conductor.

Grounds Narrow strips of wood nailed to walls as guides to plastering and as a nailing base for interior trim.

Guard 1. A conductor situated so as to conduct interference to its source and prevent the interference from having an influence on the desired signal. 2. A mechanical barrier against physical contact.

Half effect The changing of current density in a conductor due to a magnetic field extraneous to the conductor.

Half wave Rectifying only half of a sinusoidal ac supply.

Handy box   The single-gang outlet box that is used for surface mounting to enclose wall switches or receptacles, on concrete or cinder block construction of industrial and commercial buildings; also made for recessed mounting; also known as a *utility box*.

Hard drawn   A relative measure of temper; drawn to obtain maximum strength.

Hardness   Resistance to plastic deformation; resistance to scratching, abrasion, or cutting.

Harmonic   An oscillation whose frequency is an integral multiple of the fundamental frequency.

Harness   A group of conductors laced or bundled in a given configuration, usually with many breakouts.

Heat dissipation   The flow of heat from a hot body to a cooler body by convection, radiation, or conduction.

Helix   The path followed when winding a wire or strip around a tube at a constant angle.

Henry   The derived SI unit for inductance: one henry equals one weber per ampere.

Home run   The part of a branch circuit that goes from the panelboard housing the branch circuit fuse or CB and the first junction box at which the branch circuit is spliced to lighting or receptacle devices or to conductors that continue the branch circuit to the next outlet or junction box. The term *home run* is usually reserved for multioutlet lighting and appliance circuits.

Horsepower   The non-SI unit for power: 1 hp = 746 W (electric) = 9800 W (boiler).

Hot   Energized with electricity.

Hot junction   The part of the thermoelectric circuit that releases heat.

Hot leg   A circuit conductor that normally operates at a voltage above ground; the phase wires or energized circuit wires other than a grounded neutral wire or grounded phase leg.

IBEW   International Brotherhood of Electrical Workers.

IC   Pronounced "eye see." Refers to interrupting capacity of any device required to break current (switch, circuit breaker, fuse, etc.); it is the amount of current the device can interrupt without damage to itself.

ID   Inside diameter.

Identified   Marked to be recognized as grounded.

IEC   International Electrochemical Commission.

IEEE   Institute of Electrical and Electronics Engineers.

Ignition transformer   A transformer designed to provide a high-voltage current.

Impedance (A)   The opposition to current flow in an ac circuit; impedance includes resistance (R), capacitive reactance (X_c), and inductive reactance (X_L); it is measured in ohms.

Impedance matching   Matching source and load impedance for optimum energy transfer with minimum distortion.

Impulse   A surge of unidirectional polarity.

Inductance The creation of a voltage from a time-varying current; the opposition to current change, causing current changes to lag behind voltage changes; the unit of measurement is the henry.

Infrared lamp An electrical device that emits infrared rays, which are invisible rays just beyond red in the visible spectrum.

Infrared radiation Radiant energy given off by heated bodies that transmit heat and will pass through glass.

In phase The condition existing when waves pass through their maximum and minimum values of like polarity at the same instant.

Instrument A device to measure the value of the quantity under observation.

Insulated Separated from other conducting surfaces by a substance permanently offering a high resistance to the passage of energy through the substance.

Insulated Power Cable Engineers Association (IPCEA) The association of cable manufacturing engineers who make nationally recognized specifications and tests for cables.

Insulation, class rating A temperature rating descriptive of classes of insulations for which various tests are made to distinguish the materials; not necessarily related to operating temperatures.

Insulation de resistance constant (IRK) A system to classify materials according to their resistance on a 1000-foot basis at 15.5°C (60°F).

Insulation, electrical A medium in which it is possible to maintain an electrical field with little supply of energy from additional sources; the energy required to produce the electric field is fully recoverable only in a complete vacuum (the ideal dielectric) when the field or applied voltage is removed: used to save space, enhance safety, and improve appearance.

Insulation fall-in The filling of strand interstices, especially the inner interstices, which can contribute to connection failures.

Insulation level (cable) The thickness of insulation for circuits having ground fault detectors which interrupt fault currents within one minute = 100% level, one hour = 133% level, more than one hour = 173% level.

Insulation resistance (111) The measurement of the dc resistance of insulating material; can be either volume or surface resistivity. Extremely temperature sensitive.

Insulation, thermal Substance used to retard or slow the flow of heat through a wall or partition.

Integrated circuit A circuit in which different types of devices (such as resistors, capacitors, and transistors) are made from a single piece of material and then connected to form a circuit.

Integrator Any device producing an output proportionate to the integral of one variable with respect to a second variable; the second variable is usually time.

Intercalated tapes Two or more tapes of different materials helically wound and overlapping on a cable to separate the materials.

Interconnected system Operating with two or more power systems connected through the lines.

Interference   Undesired, extraneous signals.

Interlock   A safety device to ensure that a piece of apparatus will not operate until certain conditions have been satisfied.

Inverter   An item that changes dc to ac.

Ion   An electrically charged atom or radical.

Ionization   1. The process or the result of any process by which a neutral atom or molecule acquires a charge. 2. A breakdown that occurs in gaseous parts of an insulation when the dielectric stress exceeds a critical value without initiating a complete break down of the insulation system. Ionization is harmful to living tissue, and is detectable and measurable; might be evidenced by corona.

Ionization factor   The difference between the percentages of dissipation factors at two specified values of electrical stress; the lower of the two stresses is usually so selected that the effect of the ionization on the dissipation factor at this stress is negligible.

IPCEA   *See* Insulated Power Cable Engineers Association.

IR   *See* Insulation resistance.

IR drop   The voltage drop across a resistance because of the flow of current through the resistor.

IRK   *See* Insulation dc resistance constant.

Isolated   Not readily accessible to persons unless special means of access are used. *Isolating with switches* means that the switch is not a loadbreak type and must be opened only when no current is flowing in the circuit. This term also refers to transformers (an isolating transformer) used to provide magnetic isolation of one circuit from another, thereby breaking a metallic conductive path between the circuits.

Jacket   A nonmetallic polymeric close-fitting protective covering over cable insulation; the cable could have one or more conductors.

Jacket, conducting   An electrically conducting polymeric covering over an insulation.

Job cost accounting   The accounting method used to record the financial record of a business that performs projects.

Jumper   A short length of conductor, usually a temporary connection.

Junction   A connection of two or more conductors.

Junction box   Group of electrical terminals housed in a protective box or container.

Kilowatt   Unit of electrical power equal to 1000 W.

Kilowatt-foot   The product of load in kilowatts and the circuit's distance over which a load is carried in feet; used to compute voltage drop.

Kinetic energy   Energy by virtue of motion.

Kirchhoff's Laws   1. The algebraic sum of the currents at any point in a circuit is zero. 2. The algebraic sum of the product of the current and the impedance in each conductor in a circuit is equal to the electromotive force in the circuit.

Knockout   A portion of an enclosure designed to be readily removed for installation of a raceway.

KO   Pronounced "kay oh," a knockout, the partially cut opening in boxes, panel cabinets, and other enclosures that can easily be knocked out with a screwdriver and hammer to provide a clean hole for connecting conduit, cable, or some fittings.

KVA   Kilovolts times ampere.

LA   Lightning arrestor.

Labeled items   Carrying the trademark of a nationally recognized testing laboratory.

Labor unit   An amount of labor required to perform a specific task in a construction project.

Leakage   Undesirable conduction of current.

Leakage distance   The shortest distance along an insulation surface between conductors.

Leg   A portion of a circuit.

Legal action   A lawsuit that has been filed with the courts.

Leveling   A business expression that refers to removing peaks and valleys from the cashflow of a company.

Lighting outlet   An outlet intended for the direct connection of a lamp holder, lighting fixture, or pendant cord terminating in a lamp holder.

Lightning arrestor   A device designed to protect circuits and apparatus from high transient voltage by diverting the overvoltage to ground.

Limit control   Control used to open or close electrical circuits as temperature or pressure limits are reached.

Limiter   A device in which some characteristic of the output is automatically prevented from exceeding a predetermined value.

Line   1. A circuit between two points. 2. Ropes used during overhead construction.

Live-front   Any panel or other switching and protection assembly, such as a switch board or motor control center, which has exposed electrically energized parts on its front, presenting the possibility of contact by personnel.

Live load   Any load on a structure other than a dead load. Includes the weight of persons occupying the building and freestanding material.

Load   1. A device that receives power. 2. The power delivered to such a device.

Load center   An assembly of circuit breakers or switches.

Load factor   The ratio of the average to the peak load over a period.

Load losses   Those losses incidental to providing power.

Lug   A device for terminating a conductor to facilitate the mechanical connection.

Magnet   A body that produces a magnetic field external to itself; magnets attract iron particles.

Magnetic field   1. A magnetic field is said to exist at the point at which a force over and above any electrostatic force is exerted on a moving charge at that point. 2. The force field established by ac through a conductor—especially a coiled conductor.

Magnetic pole   Those portions of the magnet toward which the external magnetic induction appears to converge (south) or diverge (north).

Manufacturer's conditions Criteria set forth by a materials vendor within which they will honor warranties.

Marketing The attempts of a company to spread knowledge of its existence to its potential customers.

MCM An expression referring to conductors of sizes from 250 MCM, which stands for thousand circular mils, up to 2000 MCM. The most recent term for expressing thousand circular mils is *kcmil*.

Medium hard A relative measure of conductor temper.

Megger A term sometimes used to identify a test instrument for measuring the insulation resistance of conductors and other electrical equipment, specifically, a megohm (million ohms) meter. *Megger* is a registered trade name of the James Biddle Co.

Megohmmeter An instrument for measuring extremely high resistance.

Metal clad (MC) The cable core is enclosed in a flexible metal covering.

Mica A silicate that separates into layers and has high insulation resistance, dielectric strength, and heat resistance.

MI cable Mineral-insulated, metal-sheathed cable.

Microwave Radio waves of frequencies above one gigahertz.

Mil A unit used to measure the diameter of wire, equal to 0.001″ (25.4 cm).

MIL Military specification.

Millimeter (mm) One-thousandth of a meter.

Mil scale The heavy oxide layer formed during hot fabrication or heat treatment of metals.

Modem Equipment that connects data transmitting/receiving equipment to telephone lines: a word contraction of modulator-demodulator.

Modulation The varying of a "carrier" wave characteristic by a characteristic of a second "modulating" wave.

Moisture-resistance So constructed or treated that moisture will not readily injure.

Molded case breaker A circuit breaker enclosed in an insulating housing.

Motor An apparatus to convert from electrical to mechanical energy.

Motor, capacitor A single-phase induction motor with an auxiliary starting winding connected in series with a condenser for better starting characteristics.

Motor control Device to start and/or stop a motor at certain temperature or pressure conditions.

Mutual inductance The condition of voltage in a second conductor because of a change in current in another, adjacent conductor.

National Electrical Code (NEC) A national consensus standard for the installation of electrical systems.

National Fire Protection Association (NFPA) An organization to promote the science and improve the methods of fire protection; it sponsors various codes, including the National Code.

Natural convection   Movement of a fluid or air caused by temperature change.

Negative   Connected to the negative terminal of a power supply.

NEMA   National Electrical Manufacturers Association.

Neoprene   An oil-resistant synthetic rubber used for jackets. Originally a DuPont trade name, neoprene is now a generic term for polychloroprene.

Neutral   The element of a circuit from which other voltages are referenced, with respect to magnitude and time displacement in steady-state conditions.

Neutral block   The neutral terminal block in a panelboard, meter enclosure, gutter, or other enclosure in which circuit conductors are terminated or subdivided.

Neutral wire   A circuit conductor that is common to the other conductors of the circuit, having the same voltage between it and each of the other circuit wires and usually operating grounded. An example is the neutral of three-wire, single-phase, or three-phase, four-wire wye systems.

NFPA   *See* National Fire Protection Association.

Nineteen hundred box   A commonly used term to refer to any two-gang 4″ square outlet box used for two wiring devices or for one wiring device with a single-gang cover where the number of wires requires this box capacity.

Nipple   A threaded pipe or conduit less than 2 feet long.

Occupational Safety and Health Act (OSHA)   Federal Law #91-59G of 1970 charging all employers engaged in business affecting interstate commerce to be responsible for providing a safe working place. It is administered by the Department of Labor. The OSHA regulations are published in Title 29, Chapter XVII, Part 1910 of the CFR and the *Federal Register.*

Ohmmeter   An instrument for measuring resistance in ohms.

Ohm's law   Mathematical relationship between voltage, current, and resistance in an electric circuit.

Oscillation   The variation, usually with time, of the magnitude of a quantity that is alternately greater and smaller than a reference.

Oscillator   A device that produces an alternating or pulsating current or voltage electronically.

Oscillograph   An instrument primarily for producing a graph of rapidly varying electrical quantities.

Oscilloscope   An instrument primarily for making visible rapidly varying electrical quantities. Oscilloscopes function somewhat like TV sets.

OSHA   *See* Occupational Safety and Health Act.

Outlet   A point on the wiring system at which current is taken to supply utilization equipment.

Outline lighting   An arrangement of incandescent lamps or gaseous tubes to outline and call attention to certain features, such as the shape of a building or the decoration of a window.

Output   1. The energy delivered by a circuit or device. 2. The terminals for such delivery.

Overhead   The cost of operating a business that is not one of the direct costs required to perform a construction project.

Overlapping of trades   Portions of a construction project which require the work of more than one trade.

Overload   Load greater than the load for which the system or mechanism was intended.

Overvoltage (cable)   Voltage above normal operating voltage, usually caused by switching loads on/off, lighting, or single phasing.

Pad-mounted   A shortened expression for "pad-mount transformer," which is a completely enclosed transformer mounted outdoors on a concrete pad, without need for a surrounding chain-link fence around the metal, box-like transformer enclosure.

Panelboard   A single panel or group of panel units designed for assembly in the form of a single panel. It includes buses and can come with or without switches and/or automatic overcurrent protective devices for the control of light, heat, or power circuits of individual as well as aggregate capacity. It is designed to be placed in a cabinet or cutout box that is in or against a wall or partition and is accessible only from the front.

Payment release schedule   The method for funding the completion of work in a contract document.

Phase conductor   Any conductor other than the neutral one.

Phase leg   One of the phase conductors (an ungrounded or "hot" conductor) of a polyphase electrical system.

Phase out   A procedure by which the individual phases of a polyphase circuit or system are identified. Someone might "phase out" a three-phase circuit for a motor in order to identify phase A, phase B, and phase C. That person would then know how to connect them to the motor to get the correct phase rotation, causing the motor to rotate in the desired direction.

Phase sequence   The order in which the successive members of a periodic wave set reach their positive maximum values: (a) zero phase sequence—no phase shift; (b) plus/minus phase sequence—normal phase shift.

Phase shift   The absolute magnitude of the difference between two phase angles.

Photocell   A device in which the current-voltage characteristic is a function of incident radiation (light).

Photoelectric control   A control sensitive to incident light.

Photoelectricity   A physical action wherein an electrical flow is generated by light waves.

Photon   An elementary quantity (quantum) of radiant energy.

Photovoltaics   Products that use light to generate an electric current.

Pilot lamp   A lamp that indicates the condition of an associated circuit.

Pilot wire   An auxiliary insulated conductor in a power cable used for control or data.

Plan specifications   information provided to clarify construction drawings.

Plating   Forming an adherent layer of metal on an object.

Plug   A male connector that inserts into an outlet or jack.

Polarity   1. Distinguishing one conductor or terminal from another. 2. Identifying how devices are to be connected, such as plus (+) or minus (−) signs.

Polarization index   Ratio of insulation resistance measured after 10 minutes to the measure at one minute with voltage continuously applied.

Pole   1. That portion of a device associated exclusively with one electrically separated conducting path of the main circuit or device. 2. A supporting circular column.

Polyphase circuits   Circuits running on ac and having two or more interrelated voltages, usually of equal amplitudes, phase differences, and periods, etc. If a neutral conductor exists, the voltages referenced to the neutral conductor are equal in amplitude and phase. The most common version is that of three-phase, equal in amplitude, with phases 120′ apart.

Portable   Designed to be movable from one place to another, not necessarily while in operation.

Positive   Connected to the positive terminal of a power supply.

Potential   The difference in voltage between two points of a circuit. Frequently, one is assumed to be ground (zero potential).

Potential energy   Energy of a body or system with respect to the position of the body or the arrangement of the particles of the system.

Potentiometer   An instrument for measuring an unknown voltage or potential difference by balancing it, wholly or in part, by a known potential difference produced by the flow of known currents in a network of circuits of known electrical constants.

Power   1. Work per unit of time. 2. The time rate of transferring energy. As an adjective, the word *power* is descriptive of the energy used to perform useful work: pound feet per second, watts.

Power, active   In a three-phase symmetrical circuit, $p = 3\ VI \cos 0$; in a one-phase, two-wire circuit, $p = VI \cos 0$.

Power, apparent   The product of rms volts times rms amperes.

Power element   Sensitive element of a temperature-operated control.

Power factor   Correction coefficient for ac power necessary because of changing current and voltage values.

Power loss (cable)   Loss caused by internal cable impedance, mainly I^2R. The loss causes heating.

Precontract meeting   Meetings for discussions between the various parties before a contract document is signed.

Preconstruction conference   Meeting after contracts are signed and before construction starts for review of how the various parties will work together during a building project.

Preliminary defect list   The first listing of problems found at a building and site when a Construction Defects Lawsuit begins.

Preliminary estimate   The rough, first estimates produced during the bidding process.

Pressure motor control   A device that opens and closes an electrical circuit as pressures change.

Primary   Normally referring to the part of a device or equipment connected to the power supply circuit.

Primary control   Device that directly controls operation of a heating system.

Prime cost   The total of direct expenses required to perform a construction project.

Printed circuit   A board having interconnecting wiring printed on its surface and designed for mounting electronic components.

Process   Path of succession of states through which a system passes.

Product cutsheet   Document that describes a product to be used in a project.

Profit center   An enterprise within the company that is designed to produce a cash-flow.

Program, computer   The ordered listing of a sequence of events designed to direct the computer to accomplish a task.

Protector, circuit   An electrical device that will open an electrical circuit if excessive electrical conditions occur.

Proton   The hydrogen atom nucleus; it is electrically positive.

Prototype   The first full-size working model.

Proximity effect   The distortion of current density caused by magnetic fields; increased by conductor diameter, close spacing, frequency, and magnetic materials, such as steel conduit or beams.

Pull box   A sheet-metal boxlike enclosure used in conduit runs, either single conduits or multiple conduits, to facilitate pulling in of cables from point to point in long runs or to provide installation of conduit support bushings needed to support the weight of long riser cables or to provide for turns in multiple-conduit runs.

PV   Photovoltaic.

Pyrometer   Thermometer that measures the radiation from a heated body.

Raceway   Any channel designed expressly to hold wire, cables, or bars and used solely for those purposes.

Rack (cable)   A device to support cables.

Radar   A radio detecting and ranging system.

Radiant energy   Energy traveling in the form of electromagnetic waves.

Radiant heating   Heating system in which warm or hot surfaces are used to radiate heat into the space to be conditioned.

Radiation   The process of emitting radiant energy in the form of waves or particles.

Radiation, blackbody   Energy given off by an ideal radiating surface at any temperature.

Radiation, nuclear   The release of particles and rays during disintegration or decay of an atom's nucleus. These rays (alpha particles, beta particles, and gamma rays) cause ionization.

Radius, bending   The radii around which cables are pulled.

Rated   Indicating the limits of operating characteristics for application under specified conditions.

Reactance   1. The imaginary part of impedance. 2. The opposition to ac due to capacitance (X_c) and inductance (X_l).

Reactor   A device to introduce capacitive or inductive reactance into a circuit.

Receptacle   A contact device installed at an outlet for the connection of an attachment plug and flexible cord to supply portable equipment.

Recorder   A device that makes a permanent record, usually visual, of varying signals.

Rectifiers   Devices used to change alternating current to unidirectional current.

Rectify   To change from ac to dc.

Red-leg   1. The phase conductor of a three-phase, four-wire, delta-connected system that is not connected to the single-phase power supply. 2. The conductor with the highest voltage above ground, which must be identified (as per NEC) and is commonly painted red to provide such identification.

Relay   A device designed to change a circuit abruptly because of a specified control input.

Relay, overcurrent   A relay designed to open a circuit when current in excess of a particular setting flows through the sensor.

Remote-control circuits   The control of a circuit through relays and other means.

Resistance   The opposition in a conductor to current; the real part of impedance.

Resistor   A device whose primary purpose is to introduce resistance.

Resonance   In a circuit containing both inductance and capacitance, a condition in which the inductive reactance is equal to and cancels out the capacitance reactance.

Rheostat   A variable resistor that can be varied while energized; normally one used in a power circuit.

ROM   Read-only memory.

Romex   General Cable's trade name for type NM cable; but used generically by some electrical workers to refer to any nonmetallic sheathed cable.

Roughing in   The first stage of an electrical installation, when the raceway, cable, wires, boxes, and other equipment are installed: electrical work that must be done before any finishing or cover-up phases of building construction can be undertaken.

SBA (Small Business Administration)   National branch of the Federal Government that assists small business in many ways, from management advice to finding funds.

Self-inductance   Magnetic field induced in the conductor carrying the current.

Semiconductor   A material that has electrical properties of current flow between a conductor and an insulator.

Sensor   A material or device that goes through a physical change or an electronic characteristic change as conditions change.

Separable insulated connector   An insulated device to facilitate power cable connections and separations.

Service cable   The service conductors made up in the form of a cable.

Service conductors   The supply conductors that extend from the street main or transformers to the service equipment of the premises being supplied.

Service drop   Run of cables from the power company's aerial power lines to the point of connection on a customer's premises.

Service entrance   The point at which power is supplied to a building, including the equipment used for this purpose (service main switch, panel, or switchboard; metering devices; overcurrent protective devices; conductors for connecting to the power company's conductors; and raceways for such conductors).

Service equipment   The necessary equipment, usually consisting of a circuit breaker or switch and fuses and their accessories, located near the point of entrance of supply conductors to a building and intended to constitute the main control and cut out means for the supply to the building.

Service lateral   The underground service conductors between the street main, including any risers at a pole or other structure or from transformers, and the first point of connection to the service-entrance conductors in a terminal box, meter, or other enclosure with adequate space, inside or outside the building wall. Where there is no terminal box, meter, or other enclosure with adequate space, the point of connection is the entrance point of the service conductors into the building.

Service raceway   The rigid metal conduit, electrical metallic tubing, or other raceway that encloses the service-entrance conductors.

Sheath   A metallic close-fitting protective covering.

Shield   The conducting barrier against electromagnetic fields.

Shield, braid   A shield of interwoven small wires.

Shield, insulation   An electrically conducting layer to provide a smooth surface in intimate contact with the insulation outer surface; used to eliminate electrostatic charges external to the shield and to provide a fixed, known path to ground.

Shield, tape   The insulation shielding system whose current-carrying component is thin metallic tapes, now normally used in conjunction with a conducting layer of tapes or extruded polymer.

Short-circuit   An often-unintended low-resistance path through which current flows around, rather than through, a component or circuit.

Shunt   A device having appreciable resistance or impedance connected in parallel across other devices or another apparatus to divert some of the current. Appreciable voltage exists across the shunt and appreciable current might exist in it.

Signal   A detectable physical quantity or impulse (such as a voltage, current, or magnetic field strength) by which messages or information can be transmitted.

Signal circuit   Any electrical circuit supplying energy to an appliance that provides a recognizable signal.

Sign off   The signing of a construction document.

Single-phase motor   An electric motor that operates on single-phase alternating current.

Single-phasing   The abnormal operation of a three-phase machine when its supply is changed by the accidental opening of one conductor.

Solenoid   An electric conductor wound as a helix with a small pitch; coil.

Solidly grounded   No intentional impedance in the grounding circuit.

Solid state   A device, circuit, or system that does not depend on the physical movement of solids, liquids, gases, or plasma.

SP   Single pole.

Specs   Abbreviation for the word "specifications," which is the written description of the scope and details of an electrical installation and the equipment to be used in the system.

Starter   1. An electric controller for accelerating a motor from rest to normal speed and for stopping the motor. 2. A device used to start an electric discharge lamp.

Starting relay   An electrical device that connects and/or disconnects the starting winding of an electric motor.

Starting winding   Winding in an electric motor used only during the brief period when the motor is starting.

Static   Interference caused by electrical disturbances in the atmosphere.

Stator   The portion of a rotating machine that includes and supports the stationary active parts.

Steady state   When a characteristic exhibits only negligible change over a long period of time.

Strand   A group of wires, usually twisted or braided.

Supervised circuit   A closed circuit having a current-responsive device to indicate a break or ground

Surge   1. A sudden increase in voltage and current. 2. Transient condition.

Switch   A device for opening and closing or for changing the connection of a circuit.

Switch, ac general-use snap   A general-use snap switch suitable only for use on alternating-current circuits and for controlling resistive and inductive loads (including electric discharge lamps) not exceeding the ampere rating at the voltage involved.

Switchboard   A large single panel, frame, or assembly of panels having switches, protective devices, buses, and usually instruments mounted on the face or back or both. Switchboards are generally accessible from the rear and from the front and are not intended to be installed in cabinets.

Switch, general-use   A switch intended for use in general distribution and branch circuits. It is rated in amperes and is capable of interrupting its rated voltage.

Switch, general-use snap   A type of general-use switch so constructed that it can be installed in flush device boxes or on outlet covers or otherwise used in conjunction with wiring systems recognized by the National Electrical Code.

Switch, isolating   A switch intended for isolating an electrical circuit from the source of power. It has no interrupting rating and is intended to be operated only after the circuit has been opened by some other means.

Switch, knife   A switch in which the circuit is closed by a moving blade engaging contact clips.

Switch-leg   The part of a circuit that runs from a lighting outlet box where a luminaire or lampholder is installed down to an outlet box which contains the wall switch that turns the light or other load on or off; it is a control leg of the branch circuit.

Switch, motor-circuit   A switch, rated in horsepower, capable of interrupting the maximum operating overload current of a motor having the same horsepower rating as the switch at the rated voltage.

Synchronous machine   A machine in which the average speed of normal operation is exactly proportional to the frequency of the system to which it is connected.

Synchronous speed   The speed of rotation of the magnetic flux produced by linking the primary winding.

Synchrotron   A device for accelerating charged particles to high energies in a vacuum. The particles are guided by a changing magnetic field while they are accelerated in a closed path.

System   A region of space or quantity of matter undergoing study.

Tachometer   An instrument for measuring revolutions per minute.

Takeoff   Refers to the process of counting and the quantities of materials and labor found in the plans and specifications for a building project.

Tap   1. A splice connection of a wire to another wire (such as a feeder conductor in an auxiliary gutter) where the smaller conductor runs a short distance (usually only a few feet, but can be as much as 25 feet) to supply a panelboard or motor controller or switch. Also called a *tap-off,* indicating that energy is being taken from one circuit or piece of equipment to supply another circuit or load. 2. A tool that cuts or machines threads in the side of a round hole.

Telegraphy   Telecommunication by the use of a signal code.

Telemetering   Measurement with the aid of intermediate means that permits interpretation at a distance from the primary detector.

Telephone   The transmission and reception of sound by electronics.

Thermal cut out   An overcurrent protective device containing a heater element in addition to and affecting a renewable fusible member that opens the circuit. It is not designed to interrupt short-circuit currents.

Thermally protected (as applied to motors)   Refers to the words "thermally protected" appearing on the nameplate of a motor or motor-compressor. It means that the motor is provided with a thermal protector.

Thermal protector (as applied to motors)   A protective device that is assembled as an integral part of a motor or motor compressor and that, when properly applied, protects the motor against dangerous overheating from overloading and failing to start.

Three-phase system   A three-phase, alternating-current system containing three individual circuits or phases. Each phase is timed so that the current alternations of the first phase are one-third of a cycle (120′) ahead of the second and two-thirds of a cycle (240′) ahead of the third.

Tickler   A flagging device used to notify personnel that it is time to perform an activity.

Transformer   A device used to transfer energy from one circuit to another. It is composed of two or more coils linked by magnetic lines of force.

Trusses   Framed structural pieces consisting of triangles in a single plane for supporting loads over spans.

Utilization equipment Equipment that utilizes electric energy for mechanical, chemical, heating, lighting, or other similar useful purposes.

Ventilated Provided with a means to permit enough circulation of air to remove an excess of heat fumes or vapors.

Volt The practical unit of voltage of electromotive force. One volt sends a current of one ampere through a resistance of one ohm.

Voltage Voltage is the force, pressure, or electromotive force (emf) that causes electric current to flow in an electric circuit. Its unit of measurement is the volt, which represents the amount of electrical pressure that causes current to flow at the rate of one ampere through a resistance of one ohm. Voltage in an electric circuit is often considered similar to water pressure in a pipe or water system.

Voltage drop The voltage drop in an electric circuit is the difference between the voltage at the power source and the voltage at the point at which electricity is to be used. The voltage drop (loss) is created by the resistance of the connecting conductors.

Voltage-to-ground In grounded circuits, the voltage between the given conductor and that point or conductor of the circuit which is grounded. In ungrounded circuits, the greatest voltage between the given conductor and any other conductor of the circuit.

Watertight So constructed that moisture will not enter the enclosing case or housing.

Watt The unit of measurement of electrical power or rate of work; 756 W is equivalent to 1 hp. The watt represents the rate at which power is expended when a pressure of 1 V causes current to flow at the rate of 1 A. In a dc circuit or in an ac circuit at unity (100%) power factor, the number of watts equals the pressure (in volts) multiplied by the current (in amperes).

Weatherproof So constructed or protected that exposure to the weather will not interfere with successful operation.

Web Central portion of an I beam.

Worker hours The time spent by a work person performing a task.

Resource Directory

This resource directory is a true twenty-first-century tool for the security systems professional. Even if you are not online currently, we suggest that you go ahead and have your computer consultant hook you up right away.

Go ahead and buy a fast, beefy, up-to-date computer and use the Net to search for the software, the other systems, and all of the other products that you need. The timesavings in estimating alone will recoup the capital outlay very quickly. The rest is just gravy for your firm: research on the Net, scheduling, shopping materials, job cost accounting, change order tracking, etc. If you shop carefully for your consultant, take your time and do a careful job of covering your firm's needs. You will never regret automating your firm.

Once you are online, open the CD and boot up the RESOURCE DIRECTORY file and save it to your hard drive. Then, all you have to do is highlight a URL at the bottom of a resource listing and copy it to the search bar on the screen of your Internet provider. Hit Search and the listing will come up on your monitor. The listed resources are simply a beginning; you will undoubtedly add many personal selections to your database of Net contacts.

Associations

Alabama Alarm Association
Automatic Control Devices, Huntsville (205) 534-4215; ADCO, Gadsden (205) 546-3341; ADS of Huntsville, Huntsville (205) 721-0122; Certified Alarm Company, Sheffield (205) 383-1225; DNJ Satellite, Athens (205) 233-2006.
http://www.bham.net/alarm/dealers.html

Alarmsbc
Licensed Alarm Companies in BC, Canada. Equipment showcase with many pictures, free user manuals. Many links related to alarm industry.
http://www.alarmsbc.com/

Automatic Fire Alarm Association
The Automatic Fire Alarm Association, AFAA, is a national association providing fire alarm training and resources to the American public.
http://www.afaa.org/

Central Station Alarm Association
The Central Station Alarm Association is the trade association for the professional security monitoring industry. Our companies install and monitor burglar and fire alarms.
http://www.csaaul.org/

Fire Safety Information
The National Fire Protection Association offers fire-safety tips for families.
http://www.nfpa.org/education/Consumers_and_Families/Fire_Safety_Information/
fire_safety_information.html

Georgia Burglar & Fire Alarm Association, Inc.
Georgia Burglar & Fire Alarm Association, Inc., P.O. Box 1300, Duluth, GA 30096
Phone: 770-447-6229 or 800-783-0588; Fax: 770-447-4513; E-mail: gbfaa@gbfaa.org.
http://www.gbfaa.org/

Home Automation Association
Trade association for businesses involved with intelligent home controls and home
automation products, services, utilities, and resources.
http://www.homeautomation.org/

International Association of Home Safety and Security Professionals, Inc
Trade association provides information about locks, burglar and fire alarms; self-
defense; child, fire, and travel safety, etc.
http://www.iahssp.org/

Louisiana Burglar and Fire Alarm Association
Slide 11 of 37, Louisiana Burglar and Fire Alarm Association State Meeting, May 18,
2000. Michael C. McLean, Staff Fire Alarm Specialist Office of the State Fire
Marshal Code.
http://www.dps.state.la.us/sfm/FireAlarm/April_13_2000/sld011.htm

Los Angeles Burglar and Fire Alarm Association
Trade association of alarm and security companies serving the Southern California
area. Both consumers and security companies are welcome here.
http://www.labfaa.com/

Metropolitan Burglar and Fire Alarm Association of New York
One of the oldest associations of burglar and fire-alarm professionals in the United
States. Many of our founding members can trace back their history to the very
beginnings.
http://www.mbfaa.com/

National Burglar and Fire Alarm Association
The burglar and fire alarms, access-control, and closed-circuit TV home page.
http://www.alarm.org/

National Security Association, Links Page
National Security Association Important Links page.
http://www.nsai.org/nsalinks.htm

New Jersey Burglar and Fire Alarm Association
You can be literally saved by the bell with the information on this site. The goal of
NJBAAA is to help consumers find alarm manufacturers and distributors who are
up to code.
http://www.njbfaa.org/

Oregon Burglar and Fire Alarm Association
An association for those who ensure the safety of the general public through the use of
burglar and fire-alarm systems.
http://www.obfaa.org/

PBFAA: Pennsylvania Burglar and Fire Alarm Assn. Home
The home page of the PBFAA (Pennsylvania Burglar and Fire Alarm Association).
http://www.pbfaa.com/

Security, Building Automation, and Fire Alarm Industry Links
Browse a directory of the security, building automation, and fire-alarm industry
business listings, publications, and links.
http://www.dorsie.com/

TBFAA Homepage
Texas Burglar and Fire Alarm Association.
http://www.tbfaa.org/

Building Automation

Amerilon Home and Building Automation
Home and building automation products, specializing in LonWorks network-control
technology.
http://amerilon.com

Automation Panels
Building automation equipment, electrical control panels, lighting-control equipment,
and energy-management equipment, for every budget.
http://www.attf.com

Avio Digital, Inc.
Creators of the MediaWire home network, which carries audio, video, telephony, data,
and home control over existing telephone wiring.
http://www.aviodigital.com/

Blue Earth Research
Designs and manufactures the Micro-440e, Micro-485, and Xplor PDC series of 8051
and 8052 BASIC programmable controllers for automation and control.
http://www.BlueEarthResearch.com/

Building Automation
Using low-cost, reliable Direct Digital Controls (DDC) increases overall system
reliability, reduces wiring costs, and provides for system expansion.
http://www.aem-online.com/building_automation.htm

CABA
CABA (the Continental Automated Buildings Association) is North America's key
source for information, education, and networking relating to home and building
automation.
http://www.caba.org/

Computrols, Inc.
Provides building automation systems to control lighting, building access, HVAC
systems, and fire alarms.
http://www.computrols.com

Dorsie.com
Browse a directory of the security, building automation, and fire-alarm industry business listings, publications, and links.
http://www.dorsie.com/

Enertec-BAS
Home Site for this manufacturer and distributor of Windows-based, direct digital control building automation, and energy-management control system. Learn more about the product and its design, or sign up to be a dealer.
http://enertec-BAS.com

H I Solutions
Find out how this firm meets building management needs with hardware and software-based automation systems for industrial, commercial, office, retail, educational, and hospital facilities.
http://www.hi-s.com

Home Automation
Locate resources on the latest technologies in automation systems for the house.
http://www.looksmart.com/eus1/eus65300/eus317862/eus325859/eus304

Home Toys: Home Automation and Home Networking Resources
Comprehensive information library: home automation, home networking, security, and entertainment. Online magazine, searchable directory, shareware, and link library.
http://www.hometoys.com/

iHome
A look inside a fully automated family home. Includes software information, a virtual tour, and profiles of the family members.
http://www.ihome.com.au/

IntelliHome Home Automation
Home automation products at the lowest prices in Europe.
http://www.intellihome.be

Siemens Building Technologies
Siemens Building Technologies (with its divisions, Landis and Staefa, Cerberus, Facility Management, and Project Business) provides building productivity.
http://www.sibt.com

Smarthome.com
World's largest distributor for home automation! Free catalog for X10, home security, video surveillance, audio video, phones, automatic controls, pet care, and much more.
http://www.smarthome.com/

Smart House
Home automation capitalizing on and integrating many previously existing technologies. Video surveillance, remote climate and lighting controls, as well as local-area networks are all brought together for you.
http://www.smart-house.com/home.asp

Toshi Automation Systems
Expert solutions for complete building automation requirements.
http://www.ultraglobal.com/toshi

Wadsworth Home Automation Electronics
Wadsworth makes electronic equipment for the home automation, security, and building controls industries.
http://www.datawld.com/we/

Computers

Computer room wiring

ALCATEX
ALCATEX specializes specifically in all aspects of environmental protection, communication cabling, rectifier chargers, and the integration of an array of other services.
http://sep-a.com/alcatex/aol-data-center.htm

Beijing Tenworld Computer Room Equipment Co., Ltd.
Manufacturing access floor system with China aerospace; hi-tech products covering raised floor, net floor, free floor, anti-static floor, static control floor, construction, electronic, office environment, etc.
http://www.chinaaccessfloor.com/

Computer Motion Home Page
Development, manufacture, and sale of computer and robotic medical devices used in the surgical operating room.
http://www.computermotion.com/

Computer Room Construction
A national listing of computer room construction, as seen in the Electronic Blue Book.
http://www.thebluebook.com/cl/all687.htm

Computer Rooms
Systems for computer rooms and high-technology applications.
http://www.gamor.com/Files-Floors/Computers%20rooms.htm

Computer Room Security Systems
Computer room security systems and service in the Chicago metropolitan area from Reliable Fire Equipment Company.
http://www.reliablefire.com/businesstypes/computer_room_security.

Comp-Utility Computer Room Design
Offering a single source for telecom and computer room products and services, concentrating on customer satisfaction through technical expertise and uncompromising support.
http://www.comp-utility.com/

Cycomm International Inc.
A world-wide technology leader in the manufacture of rugged and secure computing equipment. The company's PCMOBILE is becoming the product of choice for the public.
http://www.cycomm.com

Harris Floor Tech, Inc.
Specialist in raised-access floor installations for computer rooms, equipment rooms, and general office areas. They provide materials and installation for new construction.
http://www.inc.com/users/harrisfloors.html

KBK Communications
KBK specializes in network cabling, computer room construction, LAN and WAN equipment, etc.
http://www.kbkcomm.com/references.cfm

Kwik Clamp
Earthquake bracing for computer equipment cabinets, monitors, PCs, relay racks, and related equipment located in data centers and computer rooms for protection during seismic activity.

Liebert
Distributor of support products including power supplies and air conditioners for computer rooms.
http://www.liebert.co.nz/

Mark Electronics Supply, Inc.
Computer room equipment supplier in Beltsville, MD.
http://www.thomasregional.com/cc/markes/computer_room_equipment.html

Minimizing contamination in the computer room
Several measures that you can take to minimize contamination.
http://www.accessfloor.com/micro/measure.html

Rancho Santa Fe Technology
Troubleshooting, repair, and maintenance for desktop and server systems and networks ensure that mission-critical information systems work day in and day out.
http://www.rsft.com/

Rushan Computer Room Equipment Factory
The site for a computer room equipment manufacturer.
http://www.chinarufeng.com/index.htm

Sensaphone 1104 Desktop Monitoring System
Sensaphone Model 1104, the basic programmable desktop system for smaller site-monitoring applications.
http://www.sensaphone.com/pages/1104.html

Four-Star Systems
Lanstar server racks, computer LAN work centers, Cybex KVM switches, computer furniture, printer stands, cable management, mobile carts, training tables, etc.
http://www.fourstarsystems.com/

Software for security systems contractors

Accu-Build: Construction Software
If you're ready to run your business smoothly and easily with a software package designed especially for the contracting business, take a look at Accu-Build Construction Software RFP.
http://www.accu-build.com/

American Contractor Construction Software and Service Management Software
Contractor software for accounting and job cost tracking. Full range of integrated functions for construction accounting, payroll, job cost tracking, purchase orders, and inventory.
http://www.amercon.com/

BuildingOnline
Access databases covering all aspects of the building trade, including architects, contractors, builders, and software. Get a free newsletter.
http://www.buildingonline.com/

BuyBuddy
BuyBuddy's mission is to revolutionize the way people shop. More than just a shopping comparison engine, BuyBuddy is an unbiased and intelligent consumer information service.
http://www.buybuddy.com/sleuth/8/1/1060103/479026/

CADgraphics
Offers software that allows businesses to display alarms and troubles in fire-alarm and security systems.
http://www.fire-alarm-software.com/

Calculated Industries
Calculated Industries offers specialty calculators and electronic measuring tools for professionals, including the construction master, ElectriCalc.
http://www.calculated.com/products/construction/

CDCI, Construction Data Control, Inc.
Since 1979, CDCI has provided integrated accounting and estimating software to more than 14,000 construction professionals. All CDCI programs are fully integrated.
http://www.cdci.com/

Corecon
Corecon is construction estimating, project management, and job cost accounting software for the general contractor, subcontractor, developer.
http://www.corecon.com

Estimating Solutions
Learn more about this estimating software for contractors and how to buy it. Download a sample version and also get estimating tips.
http://www.estimatingsolutions.com.au

FindAccountingSoftware
Looking for the best accounting system? With FindAccountingSoftware, you can review hundreds of systems. Do-it-yourself or complete their free online RFP.
http://www.findaccountingsoftware.com

Fire Alarm Annunciator Software
CADgraphics software connects many types of fire-alarm and security equipment to one computer. The user sees color-coded lists, instructions, and graphics.
http://www.fire-alarm-software.com

FIREHOUSE Software

FIREHOUSE Software is the fastest-selling Windows-based fire-department records management software in the industry.
http://www.firehousesoftware.com

ISS fully-integrated construction management software for homebuilders

Fully integrated construction management software. Accounting, job cost, purchasing, scheduling, payroll, and estimating. Ideal for homebuilders, remodelers, and subcontractors.
http://www.iss-software.com/

Job Bid Plus

Job Bid Plus tracks every aspect of your business—and you don't have to change the way you're currently doing business to use it.
http://www.cms.net/Job.Bid.html

Litning Software

Estimating and managing systems for contractors.
http://www.litning.com

Omware, Inc.

Offers the Master Builder estimating, project management, and accounting system for general, engineering, subcontracting, and service contractors.
http://www.themasterbuilder.com

Everdream

Let Everdream be your IT department: hardware, software, Internet, 24/7 support, and more.
http://ad.doubleclick.net/clk;2329338;5346401;e?http://www.everdrx

Profitool Construction Accounting Software

Comprehensive construction accounting and job cost software for the large contractor. Clients range from heavy/highway to large specialty and general contractors.
http://www.profitool.com

REAL Contractor Estimating for Quicken and QuickBooks

Software contractor estimating software for construction pros who work with QuickBooks and QuickBooks Pro.
http://www.litning.com/

Scalex Corporation

Scalex Corporation is the world's leading manufacturer of digital plan scalers. These innovative tools take scale measurements off prints, plans, and maps.
http://store.yahoo.com/builderscentral/planmeastool.html

Silent Knight 5670 Software Suite

The Silent Knight 5670 Software Suite introduces you to the world of simple Windows-based programming of your fire-alarm system.
http://www.silentknight.com/HTMLfiles/products/fire/5670.htm

Software Contractors' Guild

An organization of Software Contractors and Consultants offering services worldwide. Members' resumes are available online.
http://www.scguild.com/

Software for the Burglar and Fire Alarm Industry
Complete billing and management software for the alarm dealer/installer.
http://www.ajeffsvending.com/csseast/software.html

Synergetic Data Systems
SDSI is the premier developer of software tools for the BBx, ProvideX, and Unix marketplace. Since 1988, SDSI report writing, database, Web development, and laser- and pdf-forms tools have helped nearly 5,000 companies improve their productivity.
http://www.synergetic-data.com/main.cgi/news

Low-Voltage Equipment and Supplies, Miscellaneous

ABB Low Voltage Systems Ltd
Visit the link for details.
http://www.apgate.com/elec/company/ct_8940.htm

Advanced Monolithic Systems, Inc.
Advanced Monolithic Systems is a manufacturer of power-management ICs, analog semiconductors, low dropout voltage (LDO) regulators, voltage references, and dc/dc converters.
http://www.advanced-monolithic.com/html/new.html

Alstom
Low-voltage switchboards and motor-control centers for system voltages up to 950 volts and busbars rated to 3000 amperes. Outdoor distribution kiosks for urban reticulation and metering. Overhead busbar systems for power distribution.
http://www.alstom.co.za/alstomelectric/low_voltage.html

AP-477 Low-Voltage Embedded Design
This application discusses why low-voltage devices and systems are beneficial and how to design low-voltage systems. Power supplies, mixed-voltage systems, single-voltage systems, and power-management schemes are included.
http://developer.intel.com/design/intarch/applnots/272324.htm

Armor Group/Low-Voltage Systems Technology, Inc.
Blue Book ad.
http://ww2.thebluebook.com/ny/htm/0053608700003.shtml

Delta Distributing
This outdoor lighting company sells low-voltage lighting systems for residential, commercial, and marine applications.
http://www.deltadistributing.com/html/lighting.htm

Direct Low Voltage Supply
Complete product descriptions join downloadable owners' manuals on this shopping site. Pick and choose from a wide variety of monitoring and security items, including recorders and cameras.
http://www.directlvs.com/

EDN Access: Noise budgets help maintain signal integrity in low-voltage systems
Low-swing I/O circuits, with their low noise margins, present a significant challenge to system designers.
http://www.ednmag.com/ednmag/reg/1996/071896/15df3.htm

Guide to Low-Voltage and Limited-Energy Systems
Click to download the book *Guide to Low-Voltage and Limited-Energy Systems* for free (PDF 890kB).
http://mikeholt.com/low/low.htm

Johnson Systems, Inc.
Newsy site about the manufacturer of low-voltage lighting-control products and systems for upgrades of theaters, television studios, and facilities with aging dimming-control systems.
http://www.johnsonsystems.com/

Low-voltage power supplies
Low-voltage power supplies and high-voltage distribution.
http://www1.cern.ch/Atlas/SUB_DETECTORS/TILE/TDR/html/TDR-243.htm

Low-Voltage Home Pre-Wire Guide
Installing low-voltage (audio/video, phone, and LAN) wiring in a new home, for the do-it-yourselfer.
http://www.wildtracks.cihost.com/homewire/

Low-Voltage Solutions, Inc.
Design and installation of voice, data, sound, and security systems.
http://www.lvsolutions.com

Okonite single-conductor power and control cables
Premier manufacturer of high-quality insulated electric wire and cable since 1878.
http://www.okonite.com/Product_Catalog/low_voltage_cables2.html

Orion Italia
Electrical measurement and protection relays for medium- and low-voltage systems.
http://www.orionitalia.com/

Pace Technologies, Inc.
Power systems testing specialists: testing, repairing, evaluating, and calibrating for all types of switchgears, relays, breakers, transformers, and high- and low-voltage equipment.
http://www.pacepower.com/

Tdr_Stylesheet
High- and low-voltage systems.
http://uscms.fnal.gov/uscms/Subsystems/HCAL/hcal_tdr/ch12/

http://www.thebluebook.com/ny/htm/0053608700005.shtml

Technocopia
Technocopia's home automation protocols comparison reader letters.
http://www.technocopia.com/read-19991109-protocols.html

Voltage Audio Distribution
The benefits of using a voltage audio transmission system in broadcast facilities is investigated.
http://rlhess.home.mindspring.com/be/aes-80.htm

Photovoltaic Low-Voltage Systems

Advanced System Manufacturing, Inc.
ASM is a provider of equipment to produce amorphous silicon photovoltaic thin-film devices.
http://www.nbci.comimg

Atlantic Solar Products
Atlantic Solar is a manufacturer of packaged solar electric power systems and a multi-line distributor of photovoltaic panels, mounting structures, charge controllers, batteries, and much more.
http://www.atlanticsolar.com

Danish Solar Energy CE
Distributor of Photovoltaic (PV) systems and hybrid solar-wind generator systems, etc.
http://dansksolenergi.dk

Entech
Solar technology company is a maker of concentrating photovoltaic systems for both ground-based and space-based systems.
http://www.entechsolar.com/

Occidental Power
Occidental Power is installing the biggest solar electric system in San Francisco at The Rosebud Agency in the Haight/Ashbury. The installed power will be a nominal 10 kW.
http://www.oxypower.com/pvbig.html

Photovoltaic Systems Assistance Center
The Photovoltaic Systems Assistance Center (PVSAC), located at Sandia National Laboratories, is a national resource for technical information about photovoltaic systems.
http://www.sandia.gov/pv/pvsac.htm

Solar Panels and Wind Generators
Solar (PV) panels and wind generators: Siemens, Solarex, Kyocera, Unisolar, BP Solar, Southwest Windpower, etc.
http://www.windsun.com/PV_Stuff/solar_electric.htm

Types of photovoltaic systems.
http://www.alt-energy.com/catalog/pvtypes.html

UPVG Photovoltaic Data Summary and Analysis
Some very interesting real-time data on several grid-tied solar electric power systems in about 15 different states. PV systems range from less than 1 kW to more than 100 kW.
http://209.21.219.20/

Utility Photovoltaic Group
This consortium of 150 organizations is interested in tapping the enormous potential for the use of photovoltaics by electric utilities. Find data on electricity production from solar electric systems, information on photovoltaics, and a meeting schedule.
http://www.ttcorp.com/upvg/

Windsun
Solar electric (photovoltaic) power panels and deep-cycle batteries, PV panels, inverters, charge controls, and everything else you need for off grid or backup power.
http://www.windsun.com/

Security Systems—Equipment and General Information

Accurate Security Cameras and Alarm Systems
Home and business security, surveillance equipment, spy cameras, and burglar alarms.
http://www.accurate-alarm-systems.com/

ADT Home Alarm Security Systems
Home burglar alarm security systems.
http://www.freeburglaralarms.com

A Home Security Directory
Browse this directory of burglar and fire-alarm service providers by state. Explore possible home insurance cost savings of alarming your home.
http://www.geloan.com/advertisers/HomeSecurity.html

Alarmax Security Systems
You can buy components or take a complete system for free (the latter requires a monitoring contract) from the Canadian company. They will also help design your security system right over the phone.
http://www.alarmaxsecurity.com

Alarm Parts and Supply Co., Inc.
Provides burglar alarms, fire alarms, CCTV, motion detectors, glass-break detectors, sounders, medical alarms, panic buttons, and all related alarm items.
http://alarmparts.com/

Alarmsbc
Licensed Alarm Companies BC, Canada. Equipment showcase with many pictures, free user manuals, and many links related to alarm industry.
http://www.alarmsbc.com/

American Security Co.
American Security Co. is a reseller of home security equipment, such as home automation systems, smoke hoods, watchdog alarms, smoke detectors, alarm systems and controls, burglar alarms, etc. There's an online catalog and free shipping, too.
http://www.americansecurityco.com

Atlas Alarm Corporation
Services include: installation, inspections, and central station monitoring of security, burglar, fire, smoke, alarm systems, closed-circuit TV, controlled card key access systems, and emergency lighting and exit signs.
http://www.atlasalarm.com/amfgs.html

Burglar Alarm Price UK
Visit the link for details.
http://www.mamma.com/Mamma?query=Burglar+Alarm+Price+Uk&dhid=3930

Burglar Alarm System at TBO-Tech
This page features wireless home alarm systems, including infrared security systems, glass breakage alarms, entrance alarms, door jammers, and much more.
http://www.tbotech.com

Burglar Alarms
Offers online sales of burglar alarms.
http://www.burglaralarms.net/tangomerchant/alarms/wc_init.qry

CCTV surveillance camera systems and loss-prevention products
All Secure Camera Systems offer high-quality CCTV camera video surveillance, burglar alarms, access control, and hidden security cameras. Information regarding installation, pricing, and digital video are also available.
http://www.camerasite.com

Comprof Enterprises
Manufactures and supplies Capabler brand series of cylindrical locksets, secret coded locksets, alarms, and anti-burglary locksets.
http://www.capabler.com/

Conventional Fire Alarms
Features conventional fire-alarm control panels.
http://www.acmefire.com/conventi.htm

CT Businesses: Home Page for Alarm Component Sales
Full line of electronics: hard wire, wireless alarms, alarm panels, key pads, hold up buttons, panic buttons, and medical alert batteries.
http://www.alarmsales.com/

Dictograph Homepage
Homeowners have come to recognize the value of having this symbol on their homes, and so have burglars!
http://www.dictograph.com/

Edwards Systems Technology (EST)
Manufactures fire-detection devices.
http://www.est.net/

Electronic Watch Burglar Alarm
This unique alarm system uses the same advanced technology as some police radar systems. Microwave technology sees through walls to determine when someone is approaching.
http://www.saveonshopping.com/FLS/watchdogs.html

Fire Alarm Systems
Manufacturer of fire bells, electric fire alarm bells, and wall-mount fire-alarm bells.
http://www/umei.com/fire-alarm-bells-s.htm

Fire Equipment and Safety Manufacturers, Exporters, and Suppliers from India
Manufacturers, exporters, and suppliers from India.
http://www.indiamart.com/indianexporters/s_safety.html

First Alert Professional
First Alert Professional provides burglar alarms.
http://www.firstalertpro.com/door_burglar_alarms.html

Gentex Corporation
Makers of commercial fire-protection products.
http://www.gentex.com

Home Alarm System Security
Manufacturers and distributors of burglar alarm systems for home or business.
http://www.homealarms.com.au/

KK Security
Alarm system components for professional and DIY burglar alarm installers. The equipment is all professional, as used by the security alarm trade.
http://www.k-k.co.uk/alarm/

Meir Alarms
Meir Alarms offers everything in low-voltage equipment: fire, burglary, intercom, CCTV, telephone, central vac., computers, etc.
http://www.meiralarms.com/

Mills Security Alarm Systems, Inc.
Offers burglar, fire, hold up, and medical alarms; access control and closed circuit television systems; and 24-hour UL-listed central station monitoring.
http://www.mills-security.com

Mountain West Supply Co.
Mountain West Supply Company has, for the past 23 years, provided the finest equipment for self protection and protection of property at the most reasonable prices.
http://www.securitysupplies.com/

Reliable Fire Equipment
Fire-suppression systems, fire extinguishers, smoke detectors, fire-alarm systems, first-aid equipment, and fire-protection systems.
http://www.reliablefire.com

S&A Security Systems
Although this company installs burglar alarms and automatic gates for individual homeowners, the preponderance of products is aimed at industrial users. S&A manufactures card-access systems, surveillance cams, motion detectors, and other high-tech devices.
http://www.2mws.com.my/sass/

S&S Security Systems, Inc.
Offering burglar and fire-alarm equipment.
http://www.sssecurity.com

QVS Electrical Wholesale
Sells cables and flex wiring necessities, wiring accessories, computer accessories, RCD accessories, weatherproof accessories, chrome and brass accessories, door bells and chimes, and timers and controllers.
http://www.qvselectricalwholesale.co.uk/

Security
Offers general-security alarm systems and devices, security doors and windows, and vaults and safes.
http://www.nlc-bnc.ca/resource/presv/ebibl8.htm

Security Store
This site offers installation kits and supplies for the home or office.
http://www.norcoalarms.com

ServiceMagic.com
Find a pro to install or repair a burglar alarm or security system. Choose from features including video surveillance and motion detectors.
http://www.servicemagic.com/cgi-bin/servicemagic/scripts/Index.js

Sinutex Burglar Alarm Co Ltd.
Hong Kong security system and burglar alarm supplier. Licensed to supply and install burglar alarms, glassbreak detectors, CCTV, and infrared motion detectors.
http://www.sinutex.com

STARK ELECTRONICS
This major index page links to hundreds of pages with thousands of electronic products.
http://www.ultranet.com/starkel

Stealth Security
Information and an order form for this economical alarm system is provided here.
http://www.dancris.com/golfstor/

Suffolk Alarms
Ademco, ITI, and Napco burglar alarm systems.
http://www.suffolkalarms.com/mcart/index.cgi?&%3Btask=show&

System Sensor
Manufacturer of commercial fire-detection and prevention equipment.
http://www.systemsensor.com/

SystemsWebStore.com
Set up your own professional security system with these do-it-yourself alarms, CCTV, and central vacuums. Includes toll free support.
http://www.systemswebstore.com

TBO-Tech
TBO-Tech features wireless home alarm systems, including infrared security systems, glass-breakage alarms, entrance alarms, door jammers, etc.
http://www.tbotech.com

Tyco Electronic Product Group
Producer and supplier of fire, security, and safety products.
http://www.tycofire.com/tycoepg/

UK Burglar and Fire Alarm Test Equipment
The biggest resource on the Net for alarms and related test equipment.
http://www.actmeters.com/about.htm

Uniforce Burglar Alarm System
Customer-designed "station pressure-mat alarm system" for MTRC 7 Stations.
http://www.uniforce.com.hk/selected.htm

US Alarm Systems Inc.
Residential commercial security systems, burglar alarms, audio video surveillance systems (CCTV), and CAT5 wiring for homes and businesses.
http://www.usalarm.com

Wholesale Burglar
The oldest nationwide wholesale security alarm central station, consistently selected the most preferred by surveys to alarm companies.
http://emergency24.com

Zhongshan Dongfeng Wei Xin Shi Electronic Factory
Supplies quality burglar alarm systems and more.
http://home.tradeeasy.com/wxs/index.html

Index

ABOUT THE AUTHORS

JOHN E. TRAISTER (deceased) was involved in the electrical industry for more than 35 years and wrote several best-selling McGraw-Hill books, including the *Handbook of Electrical Design Details* and the *NEC Handbook*. He was the original author of *Security/Fire Alarm Systems.*

TERRY KENNEDY is an analyst for Liberty Mutual and scrutinizes the causes of defects in major construction lawsuits. He has had a hands-on involvement with the construction industry for over 30 years—carpenter, electrician, general contractor, construction manager, estimator— a thorough knowledge of the industry. Mr. Kennedy is a freelance writer, has been a featured speaker and lecturer, has recently completed the 2nd edition of the *Roofing Handbook*, and is working on the *Roofer's Instant Answer Book* for McGraw-Hill.